226
Topics in Current Chemistry

Springer

Berlin
Heidelberg
New York
Hong Kong
London
Milan
Paris
Tokyo

Colloid Chemistry I

Volume Editor: Markus Antonietti

With contributions by
L. M. Bronstein, R. A. Caruso, C. C. Co, P. Davidson,
S. Förster, J.-C. P. Gabriel, C. Göltner-Spickermann,
H.-P. Hentze, E. W. Kaler, L. Liz-Marzán, J. C. Loudet,
C. A. McKelvey, P. Mulvaney, P. Poulin

Springer

The series *Topics in Current Chemistry* presents critical reviews of the present and future trends in modern chemical research. The scope of coverage includes all areas of chemical science including the interfaces with related disciplines such as biology, medicine and materials science. The goal of each thematic volume is to give the non-specialist reader, whether at the university or in industry, a comprehensive overview of an area where new insights are emerging that are of interest to a larger scientific audience.

As a rule, contributions are specially commissioned. The editors and publishers will, however, always be pleased to receive suggestions and supplementary information. Papers are accepted for *Topics in Current Chemistry* in English.

In references *Topics in Current Chemistry* is abbreviated Top. Curr. Chem. and is cited as a journal.

Springer WWW home page: http://www.springer.de
Visit the TCC home page at http:/link.springer.de/series/tcc/
or http://link.springer-ny.com/series/tcc/

ISSN 0340-1022
ISBN 3-540-00415-7
DOI 10.1007/b10853
Springer-Verlag Berlin Heidelberg New York

Library of Congress Catalog Card Number 74-644622

Springer-Verlag Berlin Heidelberg New York
a member of BertelsmannSpringer Science+Business Media GmbH

http://www.springer.de

Cover design: KünkelLopka, Heidelberg/design & production GmbH, Heidelberg
Typesetting: Fotosatz-Service Köhler GmbH, 97084 Würzburg

02/3020 ra – 5 4 3 2 1 0 – Printed on acid-free paper

Volume Editor

Prof. Dr. Markus Antonietti
Max Planck Institute of Colloids and Interfaces
14424 Potsdam
Germany
E-mail: pape@mpigk-golm.mpg.de

Editorial Board

Topics in Current Chemistry Also Available Electronically

For all customers with a standing order for Topics in Current Chemistry we offer the electronic form via SpringerLink free of charge. Please contact your librarian who can receive a password for free access to the full articles by registration at:

http://link.springer.de/orders/index.htm

If you do not have a standing order you can nevertheless browse through the table of contents of the volumes and the abstracts of each article at:

http://link.springer.de/series/tcc
http://link.springer-ny.com/series/tcc

There you will also find information about the

- Editorial Board
- Aims and Scope
- Instructions for Authors

Preface

Over the last forty years, good old-fashioned colloid chemistry has undergone something of a revolution, transforming itself from little more than a collection of qualitative observations of the macroscopic behaviour of some complex systems into a discipline with a solid theoretical foundation and a whole toolbox of new chemical techniques. It can now boast a set of concepts which go a long way towards providing an understanding of the many strange and interesting behaviour patterns exhibited by natural and artificial systems on the mesoscale. In other words: colloid chemists have acquired a great deal of experience in the generation and control of matter with tools that are specific on a scale of some nanometers to micrometers. Modern concepts such as self-organisation, hierarchical set-up of materials, nanoparticles, functional surface engineering, interfacing and cross-talk of complex chemical objects, all of which are now in the toolbox of this 90-year old science.

It is the aim of the present issue of "Topics in Current Chemistry" to highlight some of the most attractive recent developments in colloid chemistry which are expected to have broader relevance and to be interesting to a more general readership. The contributions focus both on tools and procedures as well as on potential applications.

In this first volume, Stephan Förster opens with a quite general introduction which exemplifies the thinking in the field: how to bridge the "nanogap" between the largest structures organic chemistry can offer and the smallest structures engineering can provide by encoding the self-assembly within the structure of the primary element. Förster is an expert in employing amphiphilic block copolymers as tectonic units, and the prospects dealt with indeed span from advanced catalysis to biomedical applications.

Similar looking block-copolymers but also other self-organising templates are used by Christine Göltner-Spickermann for her "nanocasting", i.e. the replication of complex soft-matter structures into inorganic and metallic frameworks. The resulting porous materials are, at all events, inverted "hard-copies" of the primary structure with exciting potential applications for basic physical chemistry (experiments under confined conditions), catalysis and chromatographic separation.

The aspect of rationally designed catalysts on the basis of active metallic nanoparticles supported by polymeric and nanoporous inorganic systems is elaborated in the contribution by Lyudmila Bronstein. Here, colloid chemistry is used to generate local environments which make a nanoparticle under reaction more durable or the reaction more specific. The extension of nanocasting to extended membrane-like and gel-like supports via "nanocoating" is presented in the contribution by Rachel Caruso. Here, the organic template is reflected in a 3D connectivity of oxidic, crystalline nanoparticles, a technique which has the potential to revolutionize the way to construct photocatalysts, solar cells, and nanostructured catalytic supports in general.

Another way to organise inorganic particulate matter is delineated in the contribution of Jean Christophe Gabriel. Here, the classical self-organization principle of lyotropic liquid crystals is extended to inorganic colloids with a distinct shape and mutual interactions. Such organised arrays with high order are more durable and carry a potential inorganic functionality (electronic and magnetic properties) which can stimulate fresh ideas in nanotechnology.

Philippe Poulin uses liquid crystals in an unconventional fashion to generate highly organized arrays of droplets by demixing where defects control ripening and further growth. Besides providing beauty, this might give rise to a number of new matrix-assisted structuring processes, as also used indirectly in the contribution by Eric Kaler. Here, organised high-concentration surfactant phases are used as templates for organic polymerisation reactions and the Poulin-mechanism is nicely trapped by the polymer structures produced which are not copies of the original template (as in case of nanocasting), but nanostructures organised by the controlled demixing process.

This volume is closed by a contribution by Paul Mulvaney and Luis Liz-Marzán who used the nanocoating of Au-nanoparticles as a first step in the rational design of nanohybrids with very special optical and photonic properties. The ability to hybridise metals and inorganic oxides on the colloidal scale while there is no direct bond on a molecular level nicely underlines the potential of modern colloid chemistry to employ and merge very different building blocks on the mesoscale and to realise hybrid situations and high performance systems which are otherwise not accessible.

Beside introducing the diverse nano-specific instruments, it is the special elegance of the field that the structures are getting large enough to reveal their beauty. In this sense, I wish all the readers an interesting but also entertaining time reading this material.

Markus Antonietti, Potsdam February 2003

Contents

Contents of Volume 227
Colloid Chemistry II

Volume Editor: Markus Antonietti
ISBN 3-540-00418-1

Top Curr Chem (2003) 226: 1–28
DOI 10.1007/b10822

Amphiphilic Block Copolymers for Templating Applications

Stephan Förster

Institut für Physikalische Chemie, Universität Hamburg, Bundesstrasse 45, 20146 Hamburg, Germany
E-mail: forster@chemie.uni-hamburg.de

Amphiphilic block copolymers self-assemble into a variety of supramolecular structures that can be used as templates for the preparation of nanoparticles and mesoporous materials. Size and shape of nanoparticles and mesopores can be controlled via block lengths and polymer concentration. Templates can be used for the preparation of noble metals, semiconductors, and ceramic nanostructures. Concepts of consecutive and hierarchical templating allow one to prepare complex nanostructures for applications in catalysis, electro-optics, and particle separation.

Keywords. Block copolymers, Self-assembly, Templates, Nanoparticles, Mesoporous materials

List of Abbreviations and Symbols

[A]	Concentration of monomer A
[B]	Concentration of monomer B
[I]	Initiator concentration
A	Insoluble polymer block
B	Soluble polymer block
BCC	Body-centered cubic lattice
d	Long period, unit cell dimension
FCC	Face-centered cubic lattice
HEX	Hexagonally packed cylinders
LAM	Lamellae
MLAM	Modulated lamellae
N_A	Degree of polymerization of the A-block
N_B	Degree of polymerization of the B-block
P2VP	Poly-2-vinylpyridine
P4VP	Poly-4-vinylpyridine
PB	Polybutadiene
PEO	Polyethyleneoxide
PI	Polyisoprene
PLAM	Perforated lamellae
PMAc	Polymethacrylic acid
PPO	Polypropyleneoxide
PS	Polystyrene
Z	Aggregation number
χ	Flory-Huggins interaction parameter

1 Introduction

The template-directed synthesis of nano- and micromaterials has considerably evolved in recent years. Efforts were driven by the demand to produce in a controlled and reproducible way nano- and microstructures with applications in materials science, catalysis, and pharmacy. The idea to use templates is an old idea of mankind to produce reliably and reproducibly household goods and artworks needed in everyday life. The requirements for the template can easily be seen in the production of pottery. A template is filled or covered with a soft precursor material to bring the material into the desired form. Through a chemical reaction or a physical process the precursor material hardens and the template can be removed to obtain the desired product. The template has the advantage that its size and shape can be reproducibly controlled, in contrast to the desired material, where such control is more difficult.

Many of the requirements that are so obvious in pottery production are much more subtle when thinking of templating in nano- and microdimensions. Templates can be as small as single molecules, e.g., in zeolite formation and molecular imprinting, or very large such as bioceramics with micron-size pores for the templated formation of bone tissue. The templates are often produced by controlled self-assembly or controlled growth kinetics. In any case, nano- and micro templates have also to be filled or coated. This is done by making use of specific chemical and physical interactions that act on appropriate length scales to put the precursor material into the desired place. The precursor material then has to be converted to a mechanically stable form, which survives the removal of the template. For this either physical reactions like crystallization or chemical reactions that cross-link molecules into networks are used.

In recent years there has been considerable progress in the use of block copolymers as templates for the controlled synthesis of nano particles and mesoporous inorganic materials. Fresh impetus has been given by the progress made in controlling the self-assembly of block copolymers to engineer a variety of templates of different size and topology. At the same time, versatile synthetic routes have been developed to produce nanoparticles and ceramics in polymer microdomains. Today this field has advanced very rapidly. Ingenious concepts making use of consecutive and hierarchical templating to produce an astonishing variety of nano- and microstructures have been developed.

Many reviews and feature articles covering different aspects of this field have already appeared and the interested reader is encouraged for further reading to examine [1–5]. This review article will present recent developments in using block copolymers as templates for the synthesis of nanostructured materials. After an outline of the basic concepts of templating and the self-assembly of block copolymers, we will cover the synthesis of nanoparticles and mesoporous structures. At the end strategies for multiple templating and other future developments will be discussed.

2 Templating

Many concepts of templating have their direct analogy in the centuries-old process of bell-manufacturing. Bell-manufacture is a many-step templating process (Fig. 1):

1. A solid core is coated with clay to form the "wrong" bell. In this step the core acts as an *exo-template*.
2. The "wrong" bell is again used as an exo-template – it is coated with clay to form the "cope". Both "wrong" bell and "cope" are then solidified by heating. The "wrong" bell is then extracted to form the mold.
3. The mold represents an *endo-template*, i.e., it is filled with molten metal (usually bronze) which solidifies upon cooling. After removal of the mold the final bell is obtained.

Bell-manufacturing is a consecutive templating procedure where the casts are again used as templates in a subsequent templating step. It makes use of exo- and endo-templates. The concepts of exo- and endo-templates were first discussed in [1]. Endo-templates are filled, whereas exo-templates are covered to produce the desired form (see Table 1). In the case of bicontinuous templates this distinction is not possible. Between consecutive templating steps there is usually a solidifi-

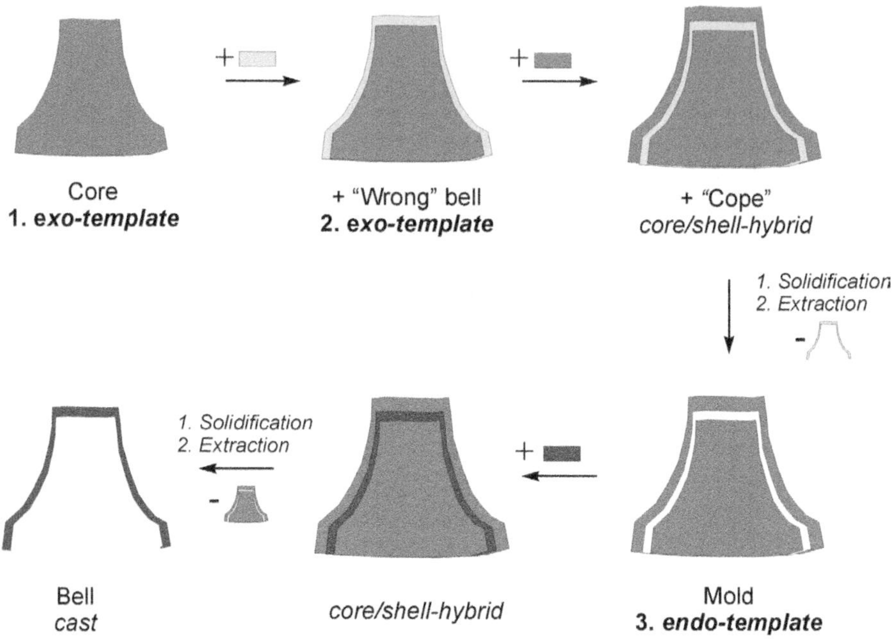

Fig. 1. Scheme of the classical bell-manufacturing process

Table 1. Different topologies that can be prepared from endo- and exo-templates

Dimensionality of the template	Cast from endo- and exo-templates	
0-D	Particle	Spherical pore
1-D	Fiber	Cylindrical pore
2-D	Sheet	Slit pore
3-D	Bicontinous structure	Bicontinous pore structure

cation/extraction process to prepare the next template. The solidification steps have to be carefully controlled in order to prevent faults and the built-up of mechanical tension.

For the preparation of nano- and microstructures there are numerous examples of endo- and exo-templates that have been used for the synthesis of particle- and pore structures from the nanometer to the micrometer scale. Templates can be inorganic, organic, or biological materials as are the materials or casts produced therefrom. The hardening process can be physical (crystallization from the melt, glass formation) or chemical (cross-linking, polycondensation). Templates usually have to be extracted in order to obtain the desired material.

Figure 2 gives an overview of commonly used templates arranged according to their size. They cover length scales from several Angströms to several hundred micrometers. Many of these templates are formed by self-assembly or by

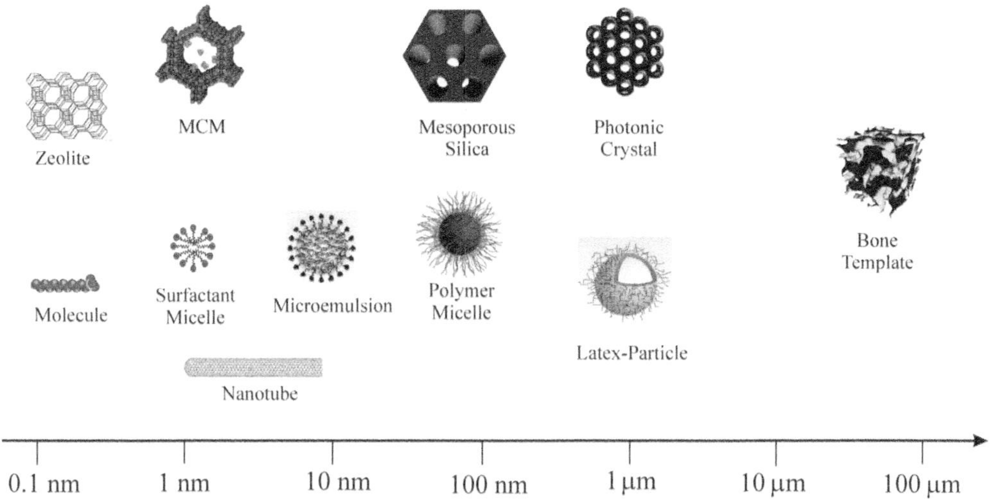

Fig. 2. Templates used for the preparation of nano- and microstructures

specific control of particle growth kinetics. This applies for nanotubes as well as for micelles, microemulsions, or latex particles. The "price" to pay for the use of self-assembled templates is a certain size-distribution. However, the relative standard deviation is in many cases only a few percent and fractionation methods allow a further decrease in this value, so that in some cases only one species is obtained, e.g., in the case of the C_{60}-fullerenes.

3 Amphiphilic Block Copolymers

Self-assembling macromolecules are particularly well suited for applications as nano- and micro-templates. Macromolecules allow control of the size and topology of the template over many decades in length scale. The simplest primary macromolecular structure which permits one to carry out these functions are diblock copolymers. The last few years have seen considerable progress in the development of methods to synthesize block polymers, some of them applicable on an industrial scale.

3.1 Synthesis

During the last decade, more and more advanced techniques of "living" or controlled polymerization to prepare block copolymers have become available. It has become possible to prepare block copolymers of various architectures, solubility, and functionality [6]. Architectures comprise diblock, triblock, and multiblock copolymers arranged linearly, as stars or grafts. The solubilities vary from polar solvents such as water to media with very low cohesion energies such as silicon oil or fluorinated solvents. Control of functionality has become important, motivated by the necessity to stabilize metallic, semiconductor, ceramic, or biological interfaces. The different synthetic routes to block copolymers comprise living anionic, living cationic, and living radical polymerization. Combining different polymerization techniques provides a large variety of block copolymers [7]. This is done by attaching functional groups which start the polymerization of the next polymer block (macroinitiators).

There has long been a large number of strategies to prepare block copolymers by living anionic polymerization [8]. Recent years have seen the development of new methods of polymerization, the living cationic [9] and the living radical polymerization [10] (e.g., RAFT technique [11–14]). New monomers have become accessible to controlled polymerization giving rise to the preparation of new classes of amphiphilic or functional polymers. The lower reactivity of the living chain ends enables polymerization to be accomplished with less technical effort (inert gas, high vacuum, rigorous exclusion of air and humidity). Therefore these polymerization techniques are of considerable interest from the technical point of view. At present the preparation of block copolymers by means of controlled radical polymerization in aqueous emulsions is being investigated. Some block copolymers such as the well-known Pluronic (BASF Wyandotte) and Kraton (Shell) block copolymers serving as emulsifiers or as thermoplastic elastomers are already being prepared on an industrial scale.

Living polymerization techniques allow one to prepare block copolymers with exactly predetermined degrees of polymerization and low polydispersities. The degree of polymerization N of the polymer blocks depends on the molar ratio of initiator $[I]$ to monomer $[A]$, $[B]$ concentration, $N_A=[A]/[I]$ and $N_B=[B]/[I]$. The polydispersities are mostly in the range $M_w/M_n<1.1$, which corresponds to a relative standard deviation in the degree of polymerization of less than 30%.

3.2 Self-Assembly

In the following we will describe the self-organization of block copolymers in dilute solution, concentrated solution, and bulk. All these structures have been or can be used as templates for the preparation of nano particles and porous structures.

3.2.1 *Micelles and Vesicles*

Surfactants as well as block copolymers form micelles [15–20] in dilute solutions. Due to the chemical structure of the blocks, block copolymers form micelles not only in polar solvents such as water, but also in very nonpolar media such as fluorinated hydrocarbons or supercritical CO_2. Block copolymer micelles have a core consisting of the insoluble A-blocks and a shell or corona of the soluble B-blocks (Fig. 3).

Block copolymers micelles have a well-defined, controllable size and shape. The size of the micelle depends on the aggregation number, which is determined by the length of the polymer blocks. Figure 4 shows the systematic dependence

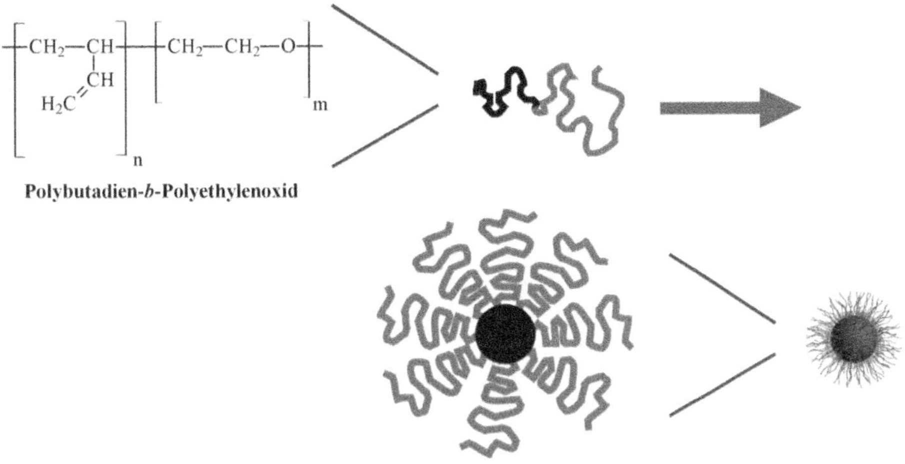

Fig. 3. Formation of a micelle (aggregation number Z=8) of AB-block copolymers

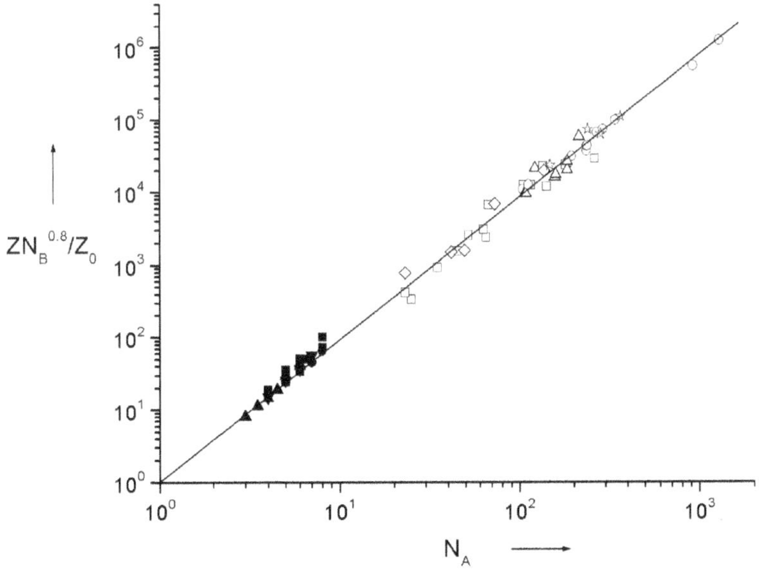

Fig. 4. Aggregation numbers Z as functions of the degree of polymerization N_A of the insoluble block. *Open circles, open squares*: diblock copolymers (PS-P4VP, PS-PMAc), *open triangles*: triblock copolymers (PMAc-PS-PMAc), *inverted open triangles*: graft copolymers (PSMSA-g-PEO), *open stars*: PS-P2VP heteroarm star polymers, *open diamonds*: PS-PAAc heteroarm star polymers *filled circles*: nonionic surfactants (C_xE_y), *filled squares*: cationic surfactants ($RNMe_3Br$), *filled triangles, filled inverted triangles*: anionic surfactants ($ROSO_3Na$, RSO_3Na) (according to [5])

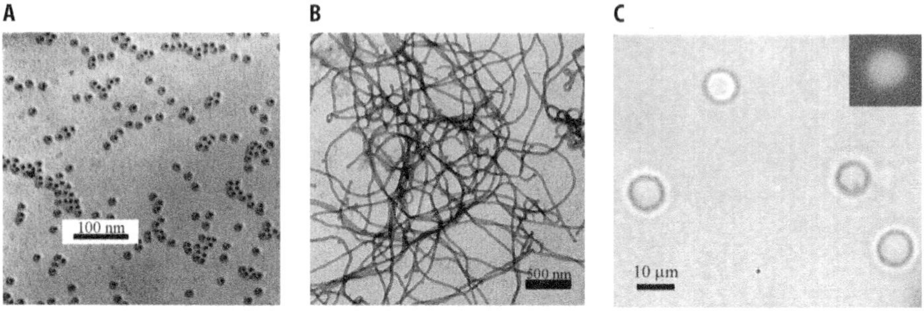

Fig. 5. **A, B** Electron micrographs. **C** optical micrographs of the three most frequent forms of self-organization of block copolymers. A – spherical micelles (PS-PI/DMF) [26], B – cylindrical micelles (PB-PEO/water) [27], C – vesicles (P2VP-PEO/water) [30]. C (above right) shows a single vesicle in which the solution of a fluorescent dye was encapsulated

between the aggregation number Z, i.e., the number of block copolymers in a micelle, and the degree of polymerization of the soluble block, N_A [21].

The experimental results can be described by a scaling law $Z = Z_0 N_A^{\alpha} N_B^{-\beta}$ where α=2 and β=0.8. Z_0 mainly depends on the enthalpy of mixing between the insoluble polymer block A and the solvent. Scaling laws with broken exponents such as 0.8 can often be found in polymer systems. The above equation describes the formation of micelles for diblock- [21, 22], triblock- [22], graft- [23], and heteroarmstar copolymers [24, 25] as well as for low molecular weight cationic, anionic, and non-ionic surfactants over a range of three orders of magnitude in block length N. Z_0 is tabulated for many block copolymers, being $Z_0 \approx 1$ for most systems which allows the aggregation numbers and diameter of the micelle to be calculated and controlled via the degrees of polymerization of the blocks. This is of relevance to the preparation of nanocolloids and porous ceramics via templating procedures.

The compartmentalization of micelles into a core/shell structure can well be observed in the electron micrographs in Fig. 5A [26]. The micelles have a total diameter of 30 nm, the core diameter being 10 nm.

Variation of block lengths further allows one to control the shape of the micelles. Block copolymers with large soluble B-blocks form spherical micelles, whereas cylindrical micelles and vesicles result from block copolymers with continuously smaller soluble blocks. Cylindrical micelles, e.g., of poly(butadiene-*b*-ethyleneoxide) (Fig. 5B) may have lengths of several micrometers [27]. Block copolymer vesicles were observed with diameters from 100 nm up to several micrometers. Compared to lipid vesicles, block copolymer vesicles are mechanically and thermodynamically much more stable [28, 29] and are well suited as templates. Figure 5C shows vesicles of poly(2-vinylpyridine-*b*-ethylene oxide) (P2VP-PEO) with diameters of more than 10 μm (giant vesicles) [30].

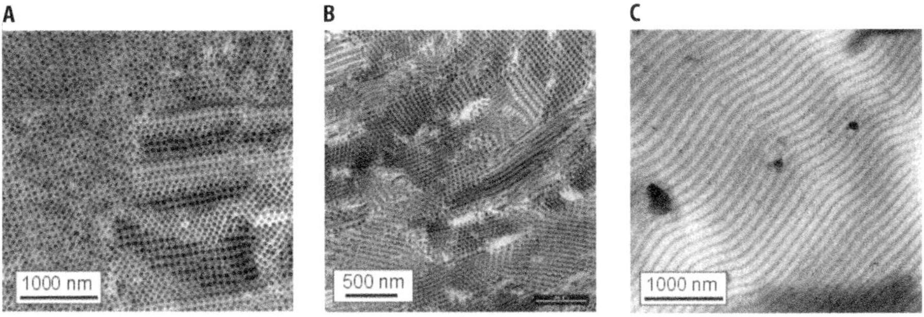

Fig. 6 A–C. Electron micrographs of a: **A** cubic; **B** hexagonal; **C** lamellar superlattice of block copolymers in lyotropic liquid-crystalline phases. The structure of the lyotropic phases was fixed by γ-irradiation [27]

3.2.2 *Lyotropic Phases*

At higher concentrations block copolymers form lyotropic liquid-crystalline phases. Their range of stability can depend strongly on temperature. In aqueous solutions polyethylene oxide (PEO) is usually the soluble block. An increase of temperature reduces the solubility of the PEO block which can result in phase transitions into different phases. Most of the present knowledge on lyotropic phase behavior of block copolymers was obtained from studies of Pluronics, i.e., poly(ethyleneoxide-*b*-propyleneoxide-*b*-ethyleneoxide) (PEO-PPO-PEO) [31]. Phase diagrams of block copolymers with shorter chains resemble those of low-molecular surfactants.

Figure 6 shows electron micrographs of lyotropic block copolymer phases of poly(butadiene-*b*-ethylene oxide) (PB-PEO) in water. The dark regions are the hydrophobic PB domains. There are cubic phases of spherical micelles on a *bcc*-lattice, hexagonally arranged cylindrical micelles, and lamellar phases. The size of the unit cell lies between 30 and 100 nm and can be controlled via the water content. Homogeneously oriented domains have dimensions of up to several micrometers. Lyotropic phases can be applied for the preparation of porous materials (Sect. 5). For templating purposes it is possible to control the morphology and the periodic length via block lengths and polymer concentration.

3.2.3 *Bulk Phases*

Also in bulk block copolymers microphase-separate into ordered liquid crystalline phases. A variety of phase morphologies such as lamellae (LAM), hexagonally ordered cylinders (HEX), arrays of spherical microdomains (BCC, FCC), modulated (MLAM) and perforated layers (PLAM), ordered bicontinuous structures such as the gyroid, as well as the related inverse structures have been documented. The morphology mainly depends on the relative block length. If, for instance, both blocks are of identical length, lamellar structures are preferred.

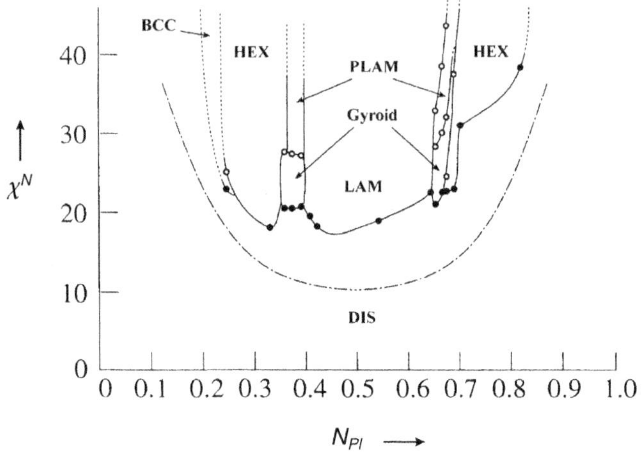

Fig. 7. Experimentally determined phase diagram of poly(styrene-*b*-isoprene) [32, 33]. By adjusting the volume ratio f_{PI} of the block copolymers, the total degree of polymerization N, and the temperature ($\chi \sim 1/T$), the individual superlattices can be specifically prepared. The structures of the individual phases are depicted in Fig. 9

The stability ranges of the different structures can be represented in a phase diagram as shown in Fig. 7 for the system poly(styrene-*b*-isoprene) (PS-PI) [32, 33].

By varying the degree of polymerization, the periodic length scale of the superlattice can be adjusted over a range from some nanometers up to several hundred nanometers. Figure 8 shows with the example of PB-PEO how the lamellar long period *d* is connected with the degree of polymerization *N*. The

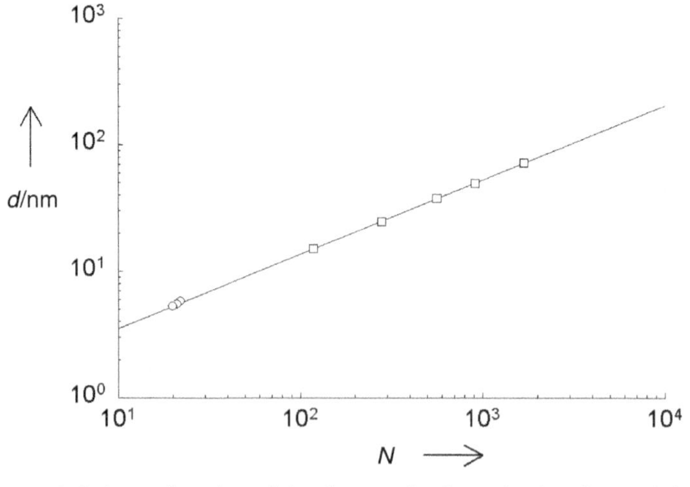

Fig. 8. Long period *d* as a function of the degree of polymerization degree (*N*) of PB-PEO (*open squares*) and nonionic surfactants (*open circles*) [27]. Via the degree of polymerization the periodic length of the nanostructure can be specifically adjusted from 1 to 100 nm

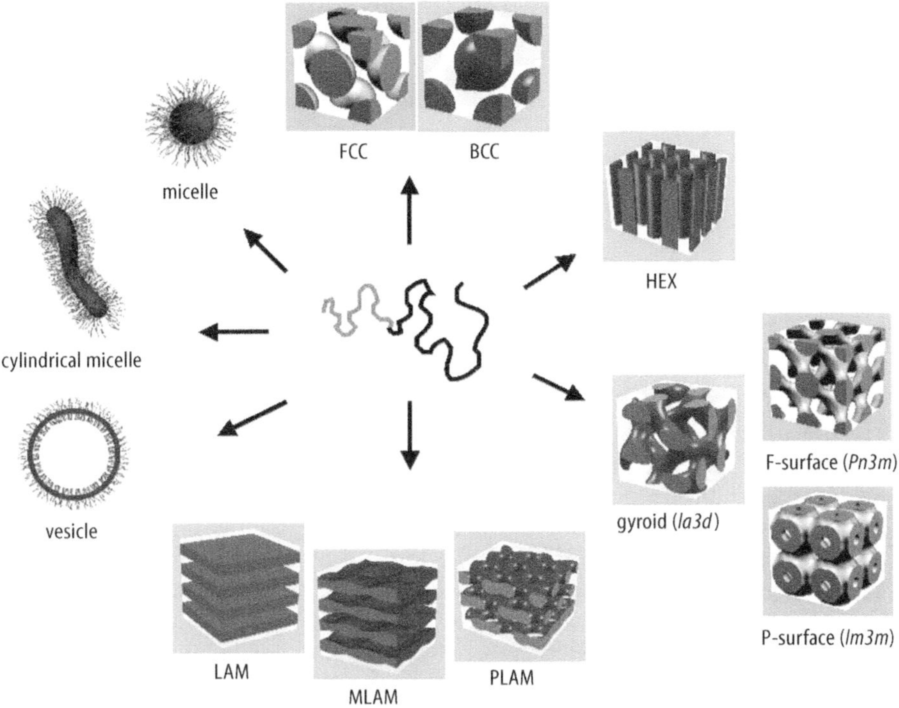

Fig. 9. Self-organization structures of block copolymers and surfactants: spherical micelles, cylindrical micelles, vesicles, fcc- and bcc-packed spheres (FCC, BCC), hexagonally packed cylinders (HEX), various minimal surfaces (gyroid, F-surface, P-surface), simple lamellae (LAM), as well as modulated and perforated lamellae (MLAM, PLAM) (with permission from [5])

data points at the lower end of the length scale result from short-chain non-ionic surfactants. This demonstrates the common universal behavior of surfactants and amphiphilic block copolymers. In accordance with theoretical predictions it can be described by a scaling law $d=d_0N^{2/3}$ (d_0=0.9 nm) over three orders of magnitude of N. This allows one to choose a superlattice of tailor-made size via the degree of polymerization of the macromolecule.

The broad variety of self-organized structures of block copolymers is shown in Fig. 9. Particulate structures such as spherical and cylindrical micelles as well as vesicles form in dilute solution. Spherical micelles with cubic packing (FCC, BCC), hexagonally packed cylindrical micelles (HEX), and lamellar phases (LAM) form in substance, in lyotropic phases as well as in ternary systems. Besides, there are modulated (MLAM) and perforated layer phases (PLAM) as well as cubic bicontinuous structures like the gyroid [34]. Using the phase diagrams one can specifically adjust the morphology of these structures via block lengths and polymer concentration. Thus it is possible to prepare tailor-made nanostructured templates of certain domain size and topology.

3.3 Filling and Coating of Microdomains

Templating involves – in the first step – filling or coating of the template. This can be achieved by solubilization of a compound into the template interior or distributing it around the template. For the desired distribution of the precursor it is advantageous to use secondary valence forces to partition the substance in the desired microdomain or at the desired interface. Commonly used forces comprise:

- Cohesion energy density, solubility, e.g., toluene/polybutadiene
- Acid-base interaction, e.g., $HAuCl_4$/polyvinylpyridine, ZnO/polymethacrylic acid
- Complex formation, e.g., Au/thiole, Pd/polybutadiene
- H-bridges, e.g., SiOH/polyethylene oxide

and thus the classical repertoire of physi- and chemisorption.

If solubilization takes place mainly via differences in the cohesion energy density, the solubilization capacity ϕ_{max} can be derived from thermodynamic arguments to be $\phi_{max}\sim\chi^b$, where χ is the Flory-Huggins interaction parameter and b a positive empiric constant. Cohesion energy densities and their relative χ-values have been tabulated in standard works. Nagarajan et al. found this relation when investigating the solubilization of aliphatic and aromatic hydrocarbons [35], alcohols, ethylene esters, ketones, and aldehydes all of which are typical components present in food flavoring [36]. Since the thermodynamic driving forces involving cohesion energy densities are relatively weak, it is advantageous to use additionally specific interaction to place the precursor into or at the desired microdomain.

Block copolymers can be made to solubilize or adhere to inorganic materials. This is of special relevance to the template synthesis of inorganic colloids or the controlled distribution of precursor materials. The incorporation of inorganic materials into polymeric domains can be achieved through ligands. These ligands are able to stabilize the contact surface between metals and inorganic materials. The ligands can be chosen according to Pearson's hard-soft acid-base (HSAB) principle. In the HSAB sense, most metals are soft and acidic and are therefore well-stabilized by soft bases such as thiols and phosphines, known as "capping agents" from the synthesis of small metal clusters. Many semiconductors and metal oxides are fairly soft and most of them are bases, therefore fairly soft acidic ligands such as alcohols, phosphates, and carboxylates provide good stabilization effects [37]. The stabilization of inorganic polyacids, which are of relevance as intermediates in sol/gel processes, can proceed via H-bridges, e.g., polyethylene oxide. The solubilization of polymers [38, 39] and biopolymers [40] can be achieved via Coulomb interactions.

The specific interface in materials with nanodomains being very large, it is indispensable to stabilize it sufficiently in order to achieve thermodynamically stable systems. The interface per volume is given by $A_v=D\phi/R$ where D is the dimensionality of the domains (D=3 for spheres, D=2 for cylinders, D=1 for lamellae), ϕ is the volume fraction of the nanodomains, and R is the radius or half diameter of the domain. Thus a material with spherical domains (D=3) of

1 nm diameter (R=0.5 nm) at a volume fraction of 50% (ϕ=0.5) has an interface of 3000 m^2/cm^3, which is nearly the size of a football field that has to be stabilized. The following example of the adhesion between natural rubber and brass-plated steel for the steel cord tire application is an illustration of successful interface stabilization. Because of their difference in hardness (in the HSAB sense), there is only poor adhesion between natural rubber and steel. Interestingly, brass-plated steel was found to bind much more effectively to natural rubber. Sulfur, the crosslinking or vulcanizing agent, actually also functions as a chemical adhesive between natural rubber and brass to form the Cu_xS/ZnS interphases [41]. Alkylsulfides are soft bases which preferably bind to soft acids like Cu. The binding of rubber to metallic cord fibers in this manner is responsible for the performance of the present generation of tires. This is a good example of the challenges one has to meet when constructing stable nanostructured materials.

3.4 Solidification

In templating procedures it is necessary to cast the template and the desired material into a mechanically stable form. Chemical and physical processes can be used for solidification. The increase of the molecular weight resulting from cross-linking can in many cases destabilize the template structure. This represents a problem mainly with templates of low molecular weight surfactants. Only in very few cases it was possible to stabilize lyotropic phases by using cross-linkable low molecular weight surfactants [42, 43]. Block copolymer superlattices are thermodynamically more stable and are easier to stabilize. An overview of the most common cross-linking reactions for block copolymers is given in Table 2 [27, 44–62].

Table 2. Various cross-linking reactions that can be used to stabilize templates and casts

Cross linking	Reaction	Example
Photochemical	[2+2]-Addition	Anthracene, cinnamic acid, diacetylene [44–47]
	Radical recombination	Photoinitiator, e.g., irgacure
Chemical	Polycondensation	Epoxy resin [48, 49], polyester [54]
	Sol/gel-process	Silicic acid
	Quaternization	1,4-Dibromobutane, *p*-dibromoxylylene [50–52]
Thermal	Radical recombination	Redox initiator, e.g., Ce^{4+} [53, 54] fragmentation, e.g. AIBN, dibenzoyl peroxide [55–59]
Radiochemical		^{60}Co [27, 60]
Secondary valence bond	H-bonds	Cytosine/guanine
	Ligands	Chelates
	Coulomb	Simplex formation [61, 62]
Physical	Supercooling, freezing	Cooling below glass transition temperature ("frozen micelles"), fast crystallization

A simple fixation of lyotropic polymer phases is achieved by means of γ-irradiation [27, 60]. Cross-linking by γ-irradiation is possible when the rate of depolymerization is lower than that of cross-linking. If this condition is met for both blocks of a block copolymer, this method is very efficient in combining high cross-linking efficiency with high penetration depth and minimum change of the thermodynamic state. It is, for example, well suited for fixation of the superlattices of lyotropic block copolymer phases so that the obtained gels can be swollen or dried maintaining their structure. With this method it was possible to carry out the electron microscopy investigation of the lyotropic phases in Fig. 6.

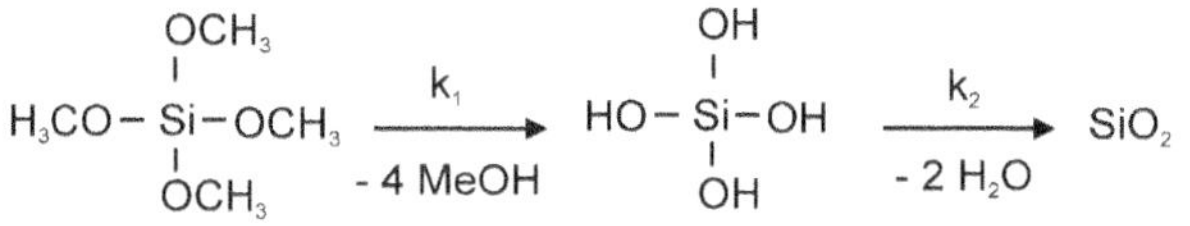

Scheme 1. Hydrolysis and condensation of a silicious acid precursor (tetramethoxysilane, TMOS)

A particular versatile method to prepare stable templates and casts is the *sol-gel process*. In this process an inorganic ester is hydrolyzed to form the free acid. Afterwards the acid condenses to form a three-dimensional network or gel (Scheme 1). Usually it is desirable to have a fast hydrolysis which has a large reaction constant k_1 followed by a slower polycondensation with a smaller constant k_2. This allows one to homogenize the solution and cast it into the desired form before it solidifies. The values of k_1 and k_2 can in many cases be adjusted by changing pH and temperature of the solution.

At the beginning of the sol-gel process single acid molecules with some partially condensed oligomers or clusters are present in the solution, which at this stage is called a *sol*. During the polycondensation reaction the size of the clusters increase until they span the whole sample to form a mechanically stable *gel*. This point is called the gel-point. Further cross-linking and cluster formation mechanically stabilizes the gel. The development from sol to gel can be described as a percolation process as schematically shown in Fig. 10. The growth of the clusters is shown as a function of the extent of the cross-linking reaction. For this example of a two-dimensional percolation on a square lattice the gel-

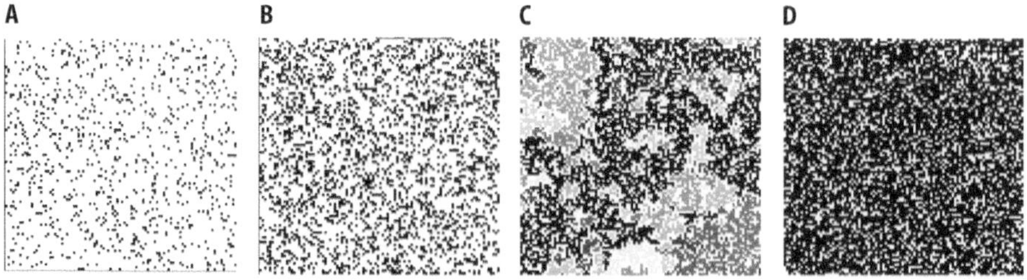

Fig. 10 A–D. Percolation model of the sol-gel process at different extent of cross-linking: **A** 10%; **B** 30%; **C** 57%; **D** 70%. The cluster spanning the whole sample at the gel point is indicated in C

point occurs at 59.27% for very large, macroscopic systems. Percolation models provide a good description of different stages during a sol-gel or solidification process. There are numerous examples of sol-gel chemistry that can be used to prepare a variety of interesting materials.

4
Nanoparticles

Nanostructures represent the transition from atom to solid. Their special optical, electronic, magnetic, and chemical properties are the result of size quantization effects as well as of the high number of surface atoms and the subsequent special surface states. These properties favor nanostructures to be applied on the sectors of signal transmission, data and energy storage, as well as catalysis.

An important goal of materials science is the controlled and specific synthesis of well-defined nanoparticles. It is essential to obtain particles with uniform diameters and shapes and – for the purpose of particular applications – to functionalize the surface and embed the particles in a superstructure. To produce inorganic nanoparticles, bulk materials are mechanically crushed or the particles are synthesized from *precursor* compounds by controlling the crystal growth kinetics. In order to stabilize the extremely large specific surface, appropriate ligands have to be added.

Nanoparticles can also be template-synthesized in the microcompartments of self-organized systems. This offers the advantage to restrict the size growth of the particles to a definite diameter and to prevent the particles from aggregating into larger sizes. If the microcompartments are arranged on a superlattice, this kind of synthesis also leads the nanoparticles to become integrated into the lattice. This gives rise to the formation of nanostructured inorganic/polymer hybrid materials. The extremely large inorganic/polymer interface can be stabilized by binding appropriate ligands to the polymer blocks.

The endo-template pathway in Fig. 11 shows schematically how to proceed when preparing nanocolloids in block copolymer microdomains. The synthesis is similar to that in usual reaction flasks. The precursor material is brought into the template by solubilization. Adequate sorption provided, this can be achieved by simple stirring, impregnation, or swelling. The next step is the chemical conversion to the desired material. For this purpose in some cases an additional reagent has to be added, a sufficient permeability through the material being necessary which is directly connected with the solubilization and the diffusivity

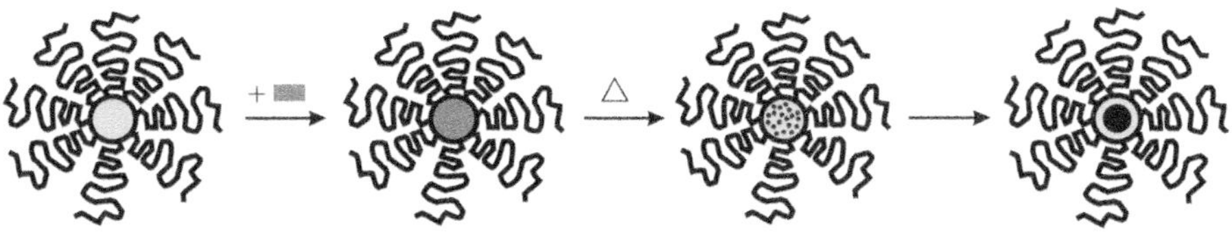

Fig. 11. Endo-templating scheme for the preparation of nanoparticles within block copolymer micelles

of the matrix. Both factors are prerequisite for an optimized synthesis. For high diffusivities gaseous reagents are often used which quickly diffuse into the compartment. Alternatively the compartment material itself can be used as reagent [63].

After a chemical reaction has taken place, small inorganic primary particles are formed which are homogeneously distributed in the microdomain. Depending on the strength of the ligands, the primary particles can be stabilized at this stage or they can further grow inside the microdomain. The size of the resulting inorganic particles depends on the conditions of nucleation and growth (concentration, size of domain, viscosity of the matrix, stabilization of the interface). Thus nano single crystals of uniform size can be prepared in the microdomains under appropriate conditions. Via the quantity of inorganic *precursor* material per microdomain the particle size can exactly be adjusted. For many applications, e.g., electro-optics and catalysis, the material can remain in the compartment or in the matrix.

4.1 Metal Nanoparticles

Small metal clusters represent the transition from metal atom to solid. The cluster size determines the optical, electronic, and chemical properties of the metal. Thus the standard potential of the silver atom $(Ag)_1$ increases via n-atomic silver cluster $(Ag)_n$ up to solids $(Ag)_\infty$ from −1.8 V to +0.8 V. Collective oscillations of the electrons lead to plasmon resonances which in the case of metal clusters belonging to the Ib-group (Cu, Ag, Au) cause strong absorptions in the optical range (400 nm$<\lambda_{max}<$600 nm). As the refractive index takes on high values near a strong absorption band, such systems are being investigated for applications in non-linear optics. Most of the atoms of small metal clusters are surface atoms. Special surface states combined with the large specific surface area lead to extraordinary catalytic properties.

In order to prepare metal clusters in polymer templates, the interface between both materials has to be stabilized sufficiently. According to the HSAB concept, metals mostly are soft acids and it should therefore be possible to stabilize them well with soft bases. Thus polymer blocks functionalized with N-, S-, or P-ligands are applied. Labile metal complexes with hard bases (chlorides, acetates) serve as *precursor* materials which, by complexing with softer ligands, can be solubilized into the polymer templates. There must not be too strong complexing between the *precursor* and the ligand as otherwise there will be no further reaction to the elementary metal.

Figure 12 shows electron microscopy images of micelles of poly(styrene-*b*-4-vinylpyridine) (PS-P4VP). For the preparation of Au nanocolloids, $HAuCl_4$ was solubilized as *precursor* into the P4VP micelle core (Fig. 12A). After reaction with a strong reducing agent ($LiAlH_4$) has taken place, small primary Au particles with a diameter of about 1 nm originate in the micelle core (Fig. 12B). Upon slow reduction with Et_3SiH as a reducing agent one single crystal with a diameter of 3 nm arises per micelle core (Fig. 12C). The micelle shells prevent further growth to colloidal aggregates [71]. In this way a number of metal colloids such

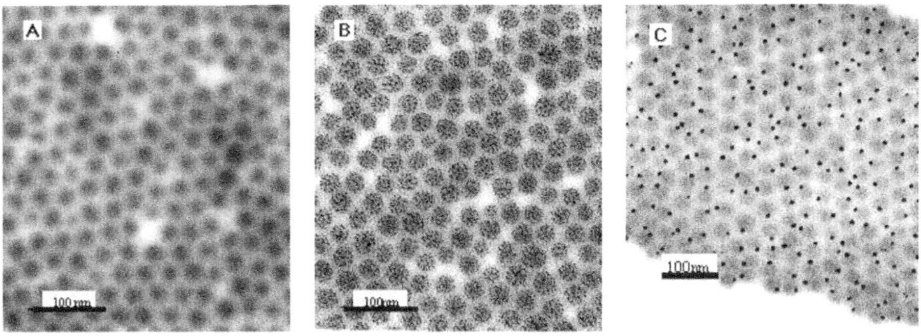

Fig. 12A–C. Preparation of gold nanocolloids in micellar block copolymer systems [4]: A micelles loaded with the *precursor* ($HAuCl_4$); **B** Au-nanocolloids after fast reduction; C Au-nanocolloids after slow reduction

as Au [64–74], Ag [64, 65, 73], Pd [65, 73, 75–78], Pt [65, 74, 75, 77, 79], Cu [65], Ni [65], Pb [65], Rh [73, 79], and Co [80] can be prepared. Co-cluster with a diameter below 20 nm consist of a magnetic single domain (Weiß region). Solutions of such clusters are called magnetic liquids or ferrofluids. They show superparamagnetic behavior. Ferrofluids are being investigated with regard to their suitability as sealing liquids and magnetic inks for printers as well as with regard to medical applications (artificial muscles, tumor diagnostics, and cancer therapy). Being fixed in a matrix, the clusters keep the direction of magnetization and store magnetic information.

Of special significance are the catalytic properties of small metal clusters. At their surface such clusters have a large number of atoms with a low coordination number to which substrates bind. Catalytic reactions are being studied in hydrogenation, hydrosilylation, hydration, and the Heck reaction. Metal clusters are also of importance with regard to redox and electron transfer processes such as the photochemical decomposition of water (fuel cells) and photocatalytic hydrogenation.

The outstanding features of metal clusters prepared in block copolymer micelles [81] are their high catalytic activity combined with high stability. Such micellar catalyst systems can be recovered after reaction by precipitation or ultrafiltration. In many cases high selectivity and stability have been observed. Cyclohexadiene, for instance, is selectively hydrogenated by Pd colloids just to cyclo-octene [69]. High activity and stability of such catalyst particles have been reported for the Heck-reaction with unusually high *turnover* numbers of 70,000 mol product per mol catalyst.

4.2
Semiconductor and Ceramic Nanoparticles

Reducing the diameter of semiconductor particles to a few nanometers results in a change of their electronic properties. Confinement to the nanometer range is possible in one, two, and three dimensions, i.e., in the form of *quantum wells*, *quantum wires*, and *quantum dots*. These quantum structures serve as control and storage elements which can be switched with one single electron or its spin which enables very high computing speeds to be achieved. Diminishing semiconductor nanoparticles to the nanometer scale results among other things in an:

- Increase of the band gap by several eV (size quantization effect) which leads to a blue shift of the absorption and of the photo- and electroluminescence
- An increased energy of photogenerated electrons which is utilized, e.g., in photovoltaics
- An increase of the optical absorption coefficient significant for applications related to UV protection

Usually semiconductor structures are prepared by means of *top-down* technologies such as photolithography or molecular beam epitaxy. Recently *bottom-up* methods have been successfully studied and have reached a high standard in the preparation of quantum dots or nanoparticles by using self-organization processes and by controlling growth kinetics. Monodisperse nanoparticles of CdS, CdSe, CdTe, InP, and InAs which are able to organize in superlattices have been prepared. It is also possible to synthesize semiconductor nanoparticles in the microdomains of block copolymers that are used as templates [71]. Semiconductor clusters are mainly generated by sulfidic precipitation with H_2S. These comprise CdS [82–84], ZnS [83, 85, 86], PbS [83, 87], ZnF [86], CuS [83], CoS [83], FeS [83], and ZnO [88].

Size control is an important goal when preparing semiconductor nanoparticles. This can be achieved by varying the size of the block copolymer domains and using them as templates. In smaller domains we obtain smaller semiconductor particles which, owing to the size quantization effect, have absorption edges with smaller wavelengths. Figure 13 exhibits this blue shift of the absorption band with decreasing particle size of CdS clusters. A CdS solid has an

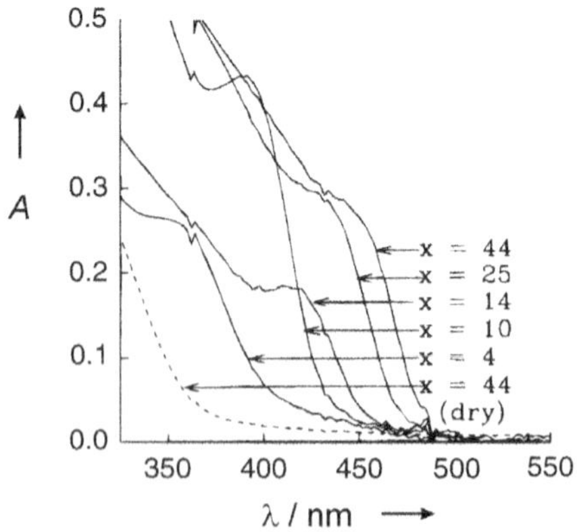

Fig. 13. Blue shift of the absorption edge of CdS with decreasing particle size ("size quantization effect") [82]. The particle size was adjusted via the degree of polymerization x of the polymethacrylic acid block

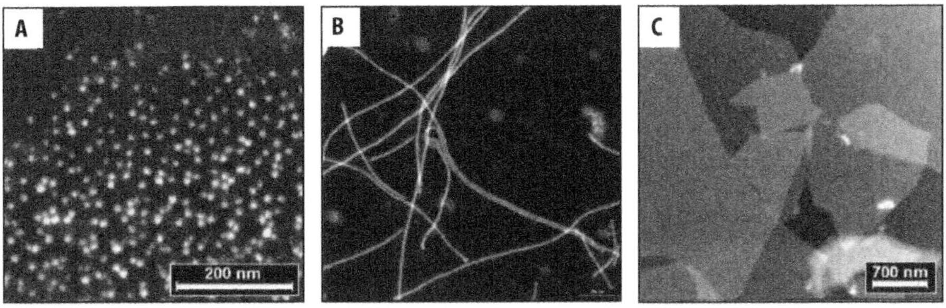

Fig. 14. Silica nanoparticles of spherical, fibrous and platelet shape (after [93])

absorption edge at λ=480 nm. Small CdS nanoparticles have an absorption edge shifted below 400 nm. Synthesis was made in polymethacrylic acid domains of poly(styrene-*b*-methacrylic acid) (PS-PMAc).

Semiconductor particles with tunable optical properties are well-suited for applications in optoelectronics and communication technology where with higher transfer frequencies and by a more efficient use of the optical spectrum larger band widths for data transfer can be achieved. Quantum dots have recently been used as fluorescent labels for biological tagging experiments in diagnostics and for visualizing molecular processes in living cells [89, 90]. TiO_2 particles with diameters of 10–30 nm are studied with regard to their application in photovoltaics (Graetzel cells) or as photocatalysts for the photooxidation of waste water contaminations [91].

When the sol-gel process is carried out in the microdomains of nanostructured polymer matrices, this reaction leads to silica/polymer hybrid materials. By calcination the polymer matrix can be removed and the pure silica nanoparticles are obtained. In this way spherical, cylindrical, and pellet-shaped silica particles can be prepared in a controlled way [92, 93] as shown in Fig. 14.

Polymer nanocomposites are being investigated with regard to their application as substitute of metals in motor industry, for the reinforcement of polymer fibers and tires, as scratch-resistant and inflammable coatings, and as biocompatible materials for prosthetic use.

5 Mesoporous Structures

According to their pore size, one distinguishes between microporous (d<2 nm), mesoporous (2≤d<50 nm), and macroporous (d≥50 nm) materials (see Fig. 15). Among microporous inorganic materials zeolites belong to the most important ones. These are crystalline aluminosilicates with a framework structure consisting of cavities and channel systems of defined size and shape. According to the kind of zeolite the pore diameter is between 0.4 and 1.5 nm. Owing to this structure, zeolites have particular properties like the ability to exchange ions and to desorb reversibly water, properties profited by when employing zeolites as ion exchangers and molecular sieves. Above all zeolites are applied as catalysts. In

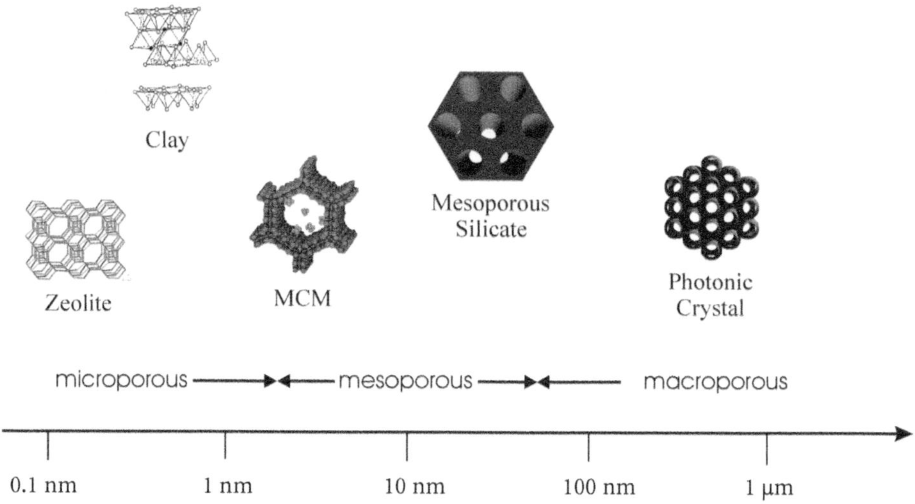

Fig. 15. Typical length scales of three-dimensional porous structures (with permission from [5])

the sector of petrochemistry (ZSM-5, zeolite Y) there is an annual turnover of more than 30 billions of US$. The catalytic properties of the zeolites result from Brönsted acid centers at the inner surface and from the electrostatic field inside the cavities and the channels.

Mesoporous materials with pore diameters of more than 2 nm were developed by Mobil Co. in 1992 and were called MCM materials [94, 95]. They are amorphous silicates or aluminosilicates with pore sizes between 1.6 and 10 nm, which represents a considerable advantage compared with zeolites because much bigger molecules can be received and made to react.

Similar to some zeolite syntheses, short-chain surfactants are used as templates in order to synthesize MCM. In aqueous solutions these surfactants form micellar phases, where the dissolved silicate species is made to condense. This way amorphous silicate walls, about 1 nm thick, originate between the micelles. Depending on the concentration of the surfactant, lyotropic, lamellar, hexagonal, and cubic phases develop which leads to varying silicate structures. After thermal removal of the surfactant by calcination, materials with free pore systems are obtained.

Lyotropic phases of block copolymers are well-suited as templates for the preparation of mesoporous inorganic materials with still larger pore diameters. Size and shape of the pores are predetermined by the hydrophobic templating domains in the lyotropic phase. Figure 16A shows an electron micrograph of a hexagonal H_I-phase of PB-PEO in water. The morphology has been stabilized by γ-irradiation. The diameter of the hydrophobic, hexagonally packed PB-cylinders is 28 nm. The sol-gel process carried out in the aqueous phase followed by calcination leads to the mesoporous silica structure shown in Fig. 16B [60]. The cylindrical, hexagonally arranged pores have diameters of 29 nm which demonstrates the good control of the templating procedure.

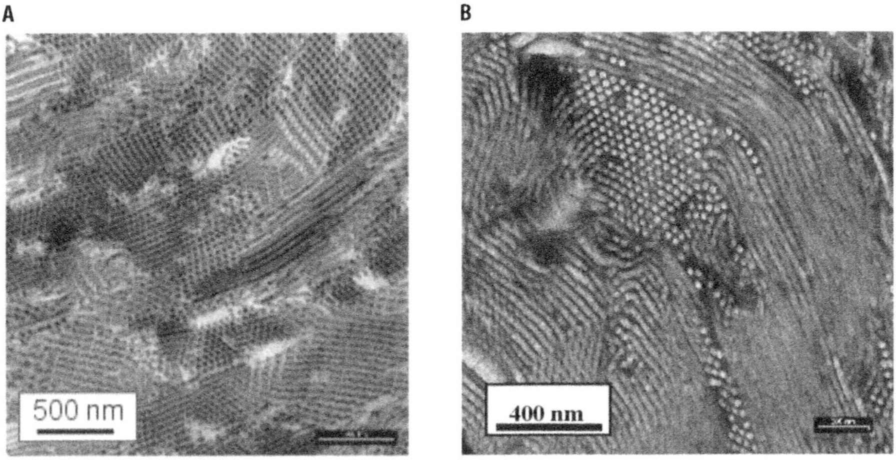

Fig. 16. **A** Lyotropic phase of hexagonally packed cylinders. **B** Mesoporous silica prepared therefrom (B) [60]

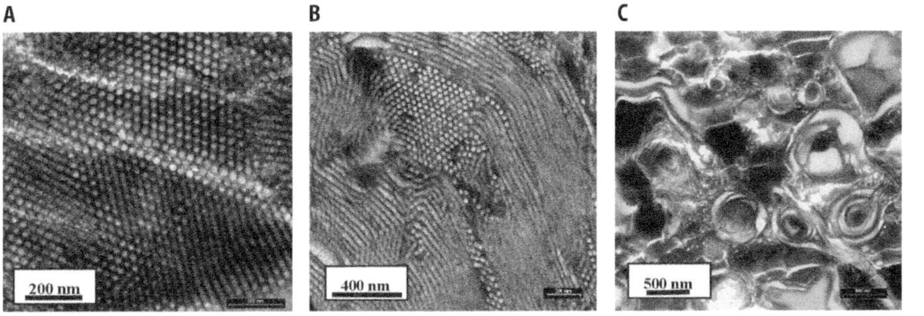

Fig. 17 A–C. Silicates with: **A** spherical pores of cubic order; **B** hexagonally ordered cylindrical pores; **C** lamellar pores, prepared from lyotropic phases of block copolymers [5, 60]

Using this procedure pore diameters and a wall thickness of more than 50 nm can be prepared which leads to mechanically stable monolithic materials. According to the structure of the lyotropic phases, porous cubic, hexagonal, and lamellar structures can be prepared (see Fig. 17) [60, 96]. By means of the sol/gel process it is possible to prepare mesoporous SiO_2 [97–108], TiO_2 [108, 109], NiO_2 [109], ZrO_2 [109], Al_2O_3 [109], Nb_2O_5 [109], Ta_2O_5 [109], WO_3 [109], HfO_2 [109], SnO_2 [109], as well as mixed oxides such as $SiAlO_{3.5}$ [109], $SiTiO_4$ [109], $ZrTiO_4$ [109], Al_2TiO_5 [109], and ZrW_2O_8 [109]. After removing the polymer by calcination the porous inorganic material is obtained.

In case of anisotropic templates it is possible to orient them macroscopically to prepare uniformly oriented casts. Macroscopic orientation can preferably be obtained by applying external shear fields. High degrees of orientation with order parameters >0.9 can be realized and afterwards be cast into oriented pore systems. Figure 18 shows ultra-small-angle X-ray (USAXS) patterns obtained

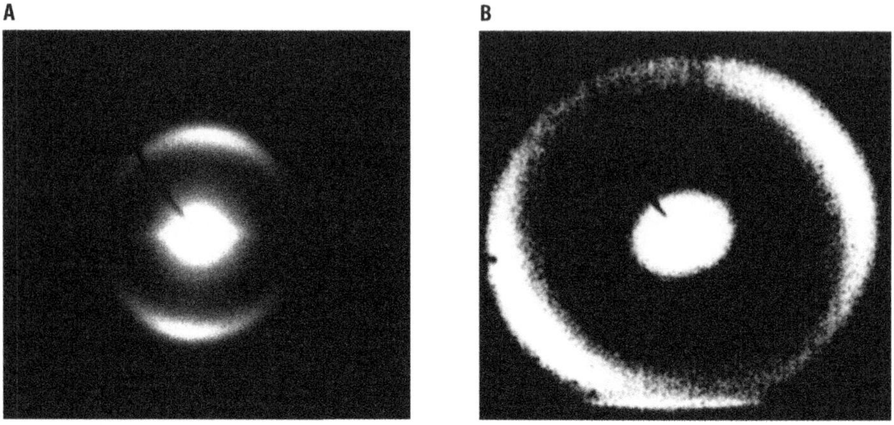

Fig. 18 A, B. USAXS-patterns of: **A** a shear-oriented nematic block copolymer phase; **B** the oriented mesoporous monolith prepared therefrom. The shift of the Debye-Scherrer ring is due to removal of solvent during the sol-gel process [110]

from a nematic lyotropic phase of cylindrical block copolymer micelles and a macroscopically oriented silica monolith prepared therefrom [110]. Also, oriented mesoporous fibers can be prepared from a lyotropic liquid crystalline phase by pulling a fiber from the solution, which after calcination remains stable [111].

Studies are currently being carried out on a large number of applications for mesoporous structures, e.g.:

- As sorbents for the removal of heavy metals from sewage and of gaseous contaminations from natural gas
- As hydrogen and methane reservoir for fuel cells
- As catalysts and catalyst supports
- As transparent thermal insulating layers
- For the storage and controlled delivery of plant nutrients (fertilizers, water)

Intensive investigations are presently being performed on how to vary the chemical composition of the silicates, to coat the surface, to vary the pore size, or to fix catalysts at the surface [112].

6
Consecutive and Hierarchical Templating

In many cases the casts obtained from templates can themselves by used as templates for a subsequent template synthesis (*consecutive templating*). This often involves a sequence of endo/exo-templating. Further, one can make use of *hierarchical templating* on different length scales. The scheme in Fig. 19 shows common pathways employing different templating steps to prepare nanoparticles, nanofibers, hollow particles, hollow fibers, and nano- and mesoporous materials.

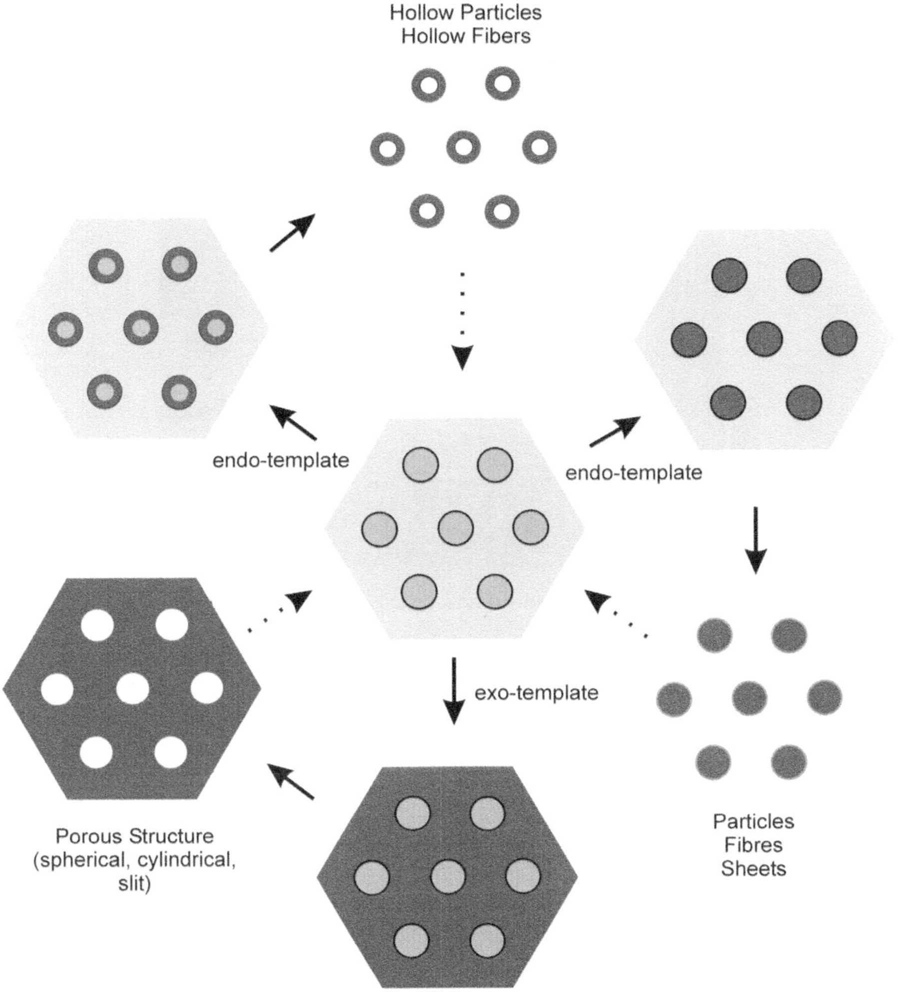

Fig. 19. Various possibilities for templating materials into solid, hollow and porous nanostructures

An example of a consecutive use of block copolymer micelles as endo- and exo-templates is the preparation of mesoporous silica with embedded Pd-nanoparticles [113]. As a first step Pd-nanoparticles are prepared in the micellar core (see Sect. 4.1). In a lyotropic phase of these micelles they are further employed as exotemplates for the preparation of mesoporous silica (see Sect. 5). After removal of the block copolymer by calcination, nanoparticles within the open mesopore structure are obtained (Fig. 20). This represents a promising way to incorporate catalytically active nanoparticles into mesoporous oxides as stable catalyst supports.

In another *consecutive templating* procedure a bicontinuous block copolymer lyotropic phase is used for the preparation of a bicontinuous porous polymer gel

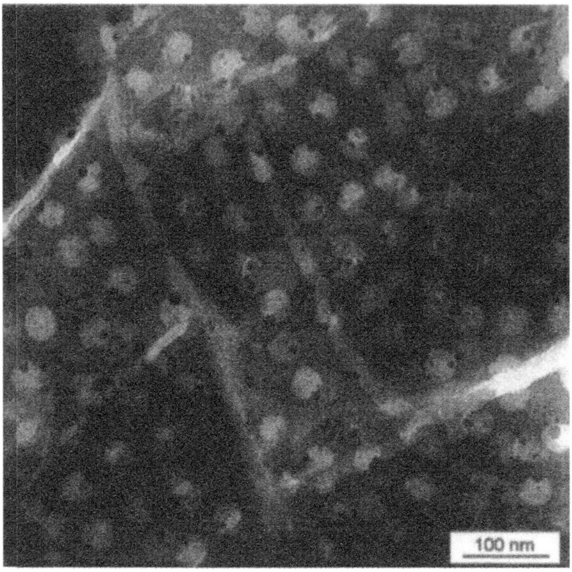

Fig. 20. Mesoporous silica with embedded Pd-particles prepared via a consecutive templating procedure [113]

[114]. First, the lyotropic phase is used as a template for the preparation of a bicontinuous silica structure, from which the polymer is removed by calcination or extraction. In the second step the porous inorganic structure is filled with monomer and crosslinker which is polymerized to form a bicontinuous organic polymer network from which the silica template is removed by treatment with hydrofluoric acid. An example for the preparation of *hierarchical structures* is the synthesis of bicontinuous pore structures by using two templates simultaneously [115]. In this case a liquid crystalline lyotropic phase of an amphiphilic block copolymer is used as a template together with suspended latex particles. The sol-gel process with subsequent calcination leads to a bicontinuous open pore structure with pores of 300 nm and 3 nm.

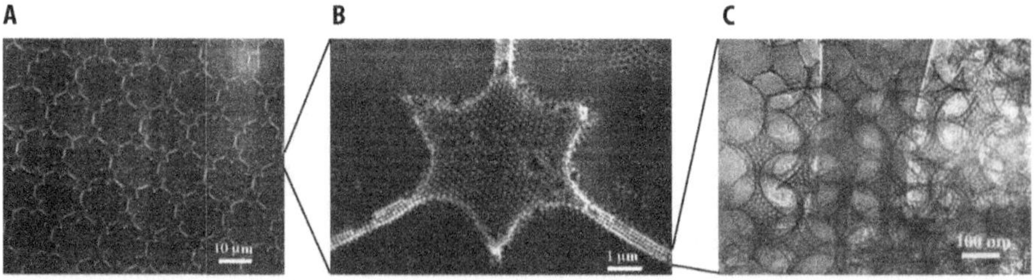

Fig. 21 A–C. Electron microscopic graph of a structure hierarchy over three organization levels [108] prepared with the aid of *microcontact-printing* and *self-assembly*: **A** microcontact sample (d=15 µm); **B** macroporous structure (d=200 nm); **C** mesoporous structure (d=2 nm)

The following example shows how *bottom-up* self-assembled templates and *top-down* prepared templates are used together for the generation of functional structure hierarchies. By microcontact-printing periodic structures (triangles linked by bridges) with a periodic length of d_1=15 μm are produced (Fig. 21). The bridges and triangles consist of porous SiO_2 with pore sizes of d_2=200 nm achieved by templating of latex particles, and of d_3=2 nm (lyotropic block copolymer phase) [108]. Microscopic images of these three structural levels are depicted in Fig. 21. Such porous silicate structures have a low refractive index ($n \approx 1.15$) and can be loaded with laser dyes. They are currently being studied as wave guides for lasers in integrated optical systems [116].

7 Conclusions and Outlook

In recent years simultaneous progress in the understanding and engineering of block copolymer microstructures and the development of new templating strategies that make use of sol-gel and controlled crystallization processes have led to a quick advancement in the controlled preparation of nanoparticles and mesoporous structures. It has become possible to prepare nanoparticles of various shapes (sphere, fiber, sheet) and composition (metal, semiconductor, ceramic) with narrow size distribution. In addition mesoporous materials with different pore shapes (sphere, cylindrical, slit) and narrow pore size distributions can be obtained. Future developments will focus on applications of these structures in the fields of catalysis and separation techniques. For this purpose either the cast materials themselves are already functional (e.g., TiO_2) or the materials are further functionalized by surface modification.

A particular drawback of many templating strategies is the waste of the template which is usually destroyed to obtain the desired cast material. In nature templating processes are necessary to cast information, e.g., in the form of DNA and m-RNA into functional molecules like proteins. The template is always reused and repaired if necessary. In this way templating becomes a replication or copying process with no loss of information or material. This concept has been used in the development of the polymerase chain reaction (PCR). This technique can be automated which enables investigators to obtain large quantities of DNA in many rounds of replication, producing billions of copies of DNA in only a few hours.

It would be desirable to develop similar strategies to cast forms from templates that can later be removed and reused. A part of this concept is already realized in the preparation and use of molecular imprinted gels [117, 118]. These gels are synthesized by a crosslinking reaction in the presence of small molecules which after removal leave an imprint in the gel. The imprinted gel is then used as the stationary phase in column chromatography for recognition, binding, and separation of similarly shaped molecules. If templates could be bound to a stationary phase, removal of the cast from the template could be achieved by, e.g., shrinking the template via an actuator function and releasing the cast into the mobile phase. In this way the template could be reused and the templating procedure would become an efficient copying or replicating process.

References

1. Antonietti M, Göltner C (1997) Angew Chem 109:944
2. Göltner C, Cölfen H, Antonietti M (1999) Chem Unserer Zeit 33:200
3. Cohen RE (1999) Curr Opin Solid State Mater Sci 4:587
4. Förster S, Antonietti M (1998) Adv Mater 10:195
5. Förster S, Plantenberg T (2002) Angew Chem Int Ed 41:688
6. Jerome R, Tong JD (1998) Curr Opin Solid State Mater Sci 3:573
7. Yagci Y, Mishra MK (1996) In: The polymeric materials encyclopedia. CRC Press, Boca Raton
8. Riess G, Hurtrez G, Bahadur P (1985) In: Mark HF, Bikales NM, Overberger CG, Menges G (eds) Encyclopedia of polymer science and engineering. Wiley, New York
9. Kennedy JP, Ivan B (1991) Designed polymers by carbocationic macromolecular engineering. Hanser, Munich
10. Matyjaszewski K (1996) Curr Opin Solid State Inteface Sci 1:769
11. Chong BYK, Le TPT, Moad G, Rizzardo E, Thang SH (1999) Macromolecules 32:2071
12. Chiefari J, Chong YK, Ercole F, Krstina J, Jeffery J, Le TPT, Mayadunne RTA, Meijs GF, Moad CL, Moad G, Rizzardo E, Thang SH (1998) Macromolecules 31:5559
13. Hawthorne DG, Moad G, Rizzardo E, Thang SH (1999) Macromolecules 32:5457
14. Krstina J, Moad CL, Moad G, Rizzardo E, Berge CT (1996) Macromol Symp 111:13
15. Tuzar Z, Kratochvil P (1993) In: Matijevic E (ed) Surface and colloid science. Plenum Press, New York
16. Gast A (1998) NATO ASI Ser E303:311
17. Chu B (1995) Langmuir 11:414
18. Alexandridis P (1996) Curr Opin Colloid Interface Sci 1:490
19. Selb J, Gallot Y (1985) In: Goodman I (ed) Developments in block copolymers. Elsevier, Amsterdam
20. Moffitt M, Khougaz K, Eisenberg A (1996) Acc Chem Res 29:95
21. Förster S, Zisenis M, Wenz E, Antonietti M (1996) J Chem Phys 104:9956
22. Qin A, Tian M, Ramireddy C, Webber SE, Munk P, Tuzar Z (1994) Macromolecules 27:120
23. Eckert AR, Webber SE (1996) Macromolecules 29:560
24. Voulgaris D, Tsitsilianis C, Grayer V, Esselink E, Hadziioannou G (1999) Polymer 40:5879
25. Voulgaris D, Tsitsilianis C (2001) Macromol Chem Phys 202:3284
26. Elias HG (1990) Makromoleküle. Hüthig & Wepf, Basel
27. Förster S, Berton B, Hentze HP, Krämer E, Antonietti M, Lindner P (2001) Macromolecules 34:4610
28. Discher B, Won YY, Ege D, Lee J, Bates FS, Discher D, Hammer D (1999) Science 284:113
29. Dimova R, Döbereiner HG, Förster S, Pouligny B, Seifert U (2002) Eur Phys J E 7:241
30. Förster S, Borchert U, Lipprandt U (in preparation)
31. Wanka G, Hoffmann H, Ulbricht W (1994) Macromolecules 27:4145
32. Förster S, Khandpur AK, Zhao J, Bates FS, Hamley IW, Ryan AJ, Bras W (1994) Macromolecules 27:6922
33. Khandpur AK, Förster S, Bates FS, Hamley IW, Ryan AJ, Bras W, Almdal K, Mortensen K (1995) Macromolecules 28:8796
34. I thank Dr. Christian Burger for providing the images
35. Nagarajan R, Barry M, Ruckenstein E (1986) Langmuir 1:210
36. Slocum SA, Kilara A, Nagarajan R (1990) In: Charalambous G (ed) Flavors and off-flavors. Elsevier, Amsterdam
37. Matijevic E (1996) Curr Opin Colloid Interface Sci 1:176
38. Tuzar Z, Kratochvil P (1993) Surf Colloid Sci 15:1
39. Tuzar Z, Bahadur P, Kratochvil P (1981) Makromol Chem 182:1751
40. Gupte A, Nagarajan R, Kilara A (1991) Biotechnol Prog 7:348
41. van Ooij WJ (1979) Rubber Chem Technol 52:605
42. Srisiri W, Lee YS, Sisson TM, Bondurant B, O'Brian DF (1997) Tetrahedron 53:15,397
43. Gray DH, Hu S, Juangt E, Gin DL (1998) Adv Mater 9:731

44. Yang JL, Wegner G (1992) Macromolecules 25:1791
45. Guo A, Liu G, Tao J (1996) Macromolecules 29:2487
46. Liu G, Qiao L, Guo A (1996) Macromolecules 29:5508
47. Liu G (1998) Curr Opin Colloid Interface Sci 3:200
48. Hillmyer MA, Lipic PM, Hajduk DA, Almdal K, Bates FS (1997) J Am Chem Soc 119:2749
49. Lipic PM, Bates FS, Hillmyer MA (1998) J Am Chem Soc 120:8963
50. Ishizu K, Fukutomi T (1988) J Polym Sci Part C Polym Lett 26:281
51. Ishizu K, Sugita M, Kotsubo H, Saito R (1995) J. Colloid Interface Sci 169:456
52. Huang H, Kowaleswki T, Remsen EE, Gertzmann R, Wooley KL (1997) J Am Chem Soc 199:11,653
53. Huang H, Remsen EE, Wooley KL (1998) Chem Commun 1415
54. Won YY, Davis HT, Bates FS (1999) Science 283:960
55. Prochazka K, Baloch MK, Tuzar Z (1979) Makromol Chem 180:2521
56. Wilson DJ, Riess G (1988) Eur Polym J 24:617
57. Saito R, Ishizu K, Fukutomi T (1992) Polymer 33:1712
58. Thurmond KB, Kowalewski T, Wooley KL (1996) J Am Chem Soc 118:7239
59. Thurmond KB, Kowalewski T, Wooley KL (1997) J Am Chem Soc 119:6656
60. Hentze HP, Krämer E, Berton B, Förster S, Antonietti M, Dreja M (1999) Macromolecules 32:5803
61. Donath E, Sukhorukov GB, Caruso F, Davis SA, Möhwald H (1998) Angew Chem Int Ed 37:2201
62. Caruso F, Caruso RA, Möhwald H (1998) Science 282:1111
63. Roos C, Schmidt M, Ebenhoch J, Baumann F, Deubzer B, Weis J (1999) Adv Mater 11:761
64. Ng Cheong Chan Y, Schrock RR, Cohen RE (1992) Chem Mater 4:24
65. Clay RT, Cohen RE (1995) Supramol Sci 2:183
66. Spatz JP, Roescher A, Möller M (1996) Adv Mater 8:337
67. Spatz JP, Mößmer S, Möller M (1996) Chem Eur J 1:1552
68. Möller M, Spatz JP, Roescher A, Mößmer S, Selvan ST, Klok HA (1997) Macromol Chem Macromol Symp 117:207
69. Antonietti M, Wenz E, Bronstein L, Seregina M (1995) Adv Mater 7:1000
70. Antonietti M, Förster S, Hartmann J, Oestreich S, Wenz E (1996) Nachr Chem Tech Lab 44:579
71. Förster S, Antonietti M (1998) Adv Mater 10:195
72. Bronstein L, Chernyshov D, Valetsky P, Tkachenko N, Lemmetyinen H, Hartmann J, Förster S (1999) Langmuir 15:83
73. Antonietti M, Förster S, Hartmann J, Oestreich S (1996) Macromolecules 29:3800
74. Bronstein L, Sidorov SN, Gourkova AY, Valetsky PM, Hartmann J, Breulmann M, Cölfen H, Antonietti M (1998) Inorg Chim Acta 280:348
75. Ng Cheong Chan Y, Craig GSW, Schrock RR, Cohen RE (1992) Chem Mater 4:885
76. Roescher A, Hempenius M, Möller M (1996) Acta Polym 47:481
77. Bronstein LM, Seregina MV, Platonova OA, Kabachii YA, Chernyshov DM, Ezernitskaya MG, Dubrovina LV, Bragina TP, Valetsky PM (1998) Macromol Chem Phys 199:1357
78. Tsutsumi K, Funaki Y, Hirokawa Y, Hashimoto T (1999) Langmuir 15:5200
79. Sidorov SN, Bronstein LM, Valetsky PM, Hartmann J, Cölfen H, Schnablegger H, Antonietti M (1999) J Coll Interface Sci 212:197
80. Platonova OA, Bronstein LM, Solodovnikov SP, Yanovskaya IM, Obolonkova ES, Valetsky PM, Wenz E, Antonietti M (1997) Colloid Polym Sci 275:426
81. Klingelhöfer S, Heitz W, Greiner A, Oestreich S, Förster S, Antonietti M (1997) J Am Chem Soc 119:10,116
82. Moffitt M, McMahon L, Pessel V, Eisenberg A (1995) Chem Mater 7:1185
83. Yue J, Cohen RE (1994) Supramol Sci 1:117
84. Möller M, Spatz JP (1997) Curr Opin Colloid Interface Sci 2:177
85. Sankaran V, Yue J, Cohen RE, Schrock RR, Silbey RJ (1993) Chem Mater 5:1133
86. Yue J, Sankaran V, Cohen RE, Schrock RR (1993) J Am Chem Soc 115:4409
87. Kane RS, Cohen RE, Silbey R (1996) Chem Mater 8:1919

88. Wenz E (1996), PhD Thesis, Freie Universität Berlin
89. Chan WCW, Nie SN (1998) Science 281:2016
90. Bruchez M, Moronne M, Gin P, Weiss S, Alivisatos AP (1998) Science 281:2013
91. Grätzel M, Brooks K, McEvoy AJ (1999) Innovative materials in advanced energy technologies In: Advanced science and technology. Faenza, Italy
92. Templin M, Franck A, Du Chesne A, Leist H, Zhang Y, Ulrich R, Schädler V, Wiesner U (1997) Science 278:1795
93. Ulrich R, Du Chesne A, Templin M, Wiesner U (1999), Adv Mater 11:141
94. Kresge CT, Leonowicz ME, Roth WJ, Vatuli JC, Beck JS (1992) Nature 359:710
95. Beck JS, Vartuli JC, Roth WJ, Leonowicz ME, Kresge CT, Schmitt KD, Chu CTW, Olson DH, Sheppard EW, McCullen SB, Higgins JB, Schlenker JL (1992) J Am Chem Soc 114:10,834
96. Krämer E (2000) PhD thesis, Universität Potsdam
97. Göltner CG, Antonietti M (1997) Adv Mater 9:1
98. Göltner CG, Henke S, Weißenberger MC, Antonietti M (1998) Angew Chem Int Ed Engl 37:613
99. Göltner CG, Weißenberger MC (1998) Acta Polym 49:704
100. Weißenberger MC, Göltner CG, Antonietti M (1997) Ber Bunsenges Phys Chem 101:1679
101. Göltner CG, Berton B, Krämer E, Antonietti M (1999) Adv Mater 11:395
102. Göltner CG, Berton B, Krämer E, Antonietti M (1998) Chem Commun 2287
103. Zhao D, Feng J, Huo Q, Melosh N, Fredrickson GH, Chmelka BF, Stucky GD (1998) Science 279:548
104. Melosh NA, Lipic P, Bates FS, Wudl F, Stucky GD, Fredrickson GH, Chmelka BF (1999) Macromolecules 32:4332
105. Yang PD, Zhao DY, Chmelka BF, Stucky GD (1998) Chem Mater 10:2033
106. Krämer E, Förster S, Göltner CG, Antonietti M (1998) Langmuir 14:2027
107. Zhao D, Huo Q, Feng J, Chmelka BF, Stucky GD (1998) J Am Chem Soc 120:6024
108. Yang PD, Deng T, Zhao DY, Feng PY, Pine D, Chmelka BF, Whitesides GM, Stucky GD (1998) Science 282:2244
109. Yang PD, Zhao DY, Margolese DI, Chmelka BF, Stucky GD (1998) Nature 396:6707
110. Förster S, Konrad M, Schnablegger H, Cunis H (2001) HASYLAB Annual Report 1:695
111. Yang P, Zhao D, Chmelka BF, Stucky GD (1998) Chem Mater 10:2033
112. Ying JY, Mehnert CP, Wong MS (1999) Angew Chem 38:58
113. Bronstein L, Kramer E, Berton B, Burger C, Forster S, Antonietti M (1999) Chem Mater 11:1402
114. Weissenberger MC (1997) PhD thesis, University of Potsdam
115. Antonietti M, Berton B, Göltner C, Hentze HP (1998) Adv Mater 10:154
116. Yang PD, Wirnsberger G, Huang HC, Cordero SR, McGehee MD, Scott B, Deng T, Whitesides GM, Chmelka BF, Buratto SK, Stucky GD (2000) Science 287:465
117. Lanza F, Sellergren B (2001) Chromatographia 53:599
118. Sellergren B (2001) J Chromatogr A906:227

Top Curr Chem (2003) 226: 29–54
DOI 10.1007/b10823

Nanocasting of Lyotropic Liquid Crystal Phases for Metals and Ceramics

Christine Göltner-Spickermann

School of Chemistry, University of Bristol, Bristol BS8 1TS, UK
E-mail: goltner@supanet.com

The utilization of ordered surfactant and block copolymer bulk phases is a versatile tool for the preparation of porous nanostructured ceramics and metals. The soft, lyotropic liquid crystal phases resemble a casting mold in which the chemical synthesis and solidification of the inorganic material takes place without altering the self-assembled structure. The result is an exact cast of the lyotropic phase, hence the name of the process, nanocasting. This chapter is aimed at outlining the general principles and scope of the method with special emphasis on the materials properties of the resulting inorganic nanostructures.

Keywords. Surfactants, Amphiphilic block copolymers, Templating, Sol-gel processing, Electrochemistry, Nanostructures

1 Introduction

1.1 Structure of Surfactant Phases

Amphiphilic molecules, such as surfactants or amphiphilic block copolymers (ABCs), are compounds in which a hydrophilic and a hydrophobic component are linked covalently. When dissolved in water, the presence of the hydrophobic moieties will give rise to supramolecular self-assembly into micellar structures or, at higher concentration, lyotropic liquid crystal (LLC) phases. All these aggregates form without external manipulation by processes of supramolecular self-assembly. Although a self-assembled structure can never be without defect, the degree of order in such systems is often surprisingly high. Depending on the molecular structure and concentration of the amphiphile, a series of different phase structures is presented in a binary phase diagram (e.g. surfactant/water) [1]. For a fictitious surfactant, the phase diagram comprises phases of all degrees and signs of curvature, going from a micellar phase through micellar cubic, hexagonal, bicontinuous cubic, lamellar, inverse hexagonal, inverse micellar cubic to inverse micellar with increasing surfactant concentration [1]. For a real compound, the phases observed in the phase diagram mainly depend on the ratio of headgroup to tail volume. The appearance of a micellar, a hexagonal and a lamellar phase are schematically illustrated in Fig. 1.

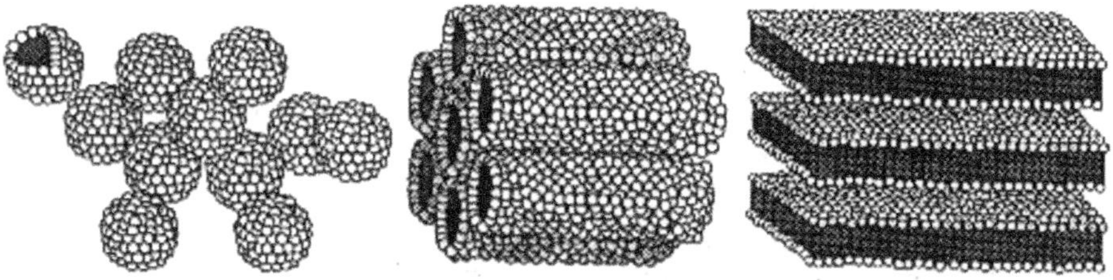

Fig. 1. Typical surfactant aggregation structures. From *left to right*: micellar, hexagonal, lamellar

1.2 Surfactant-Assisted Precipitation of Inorganic-Organic Hybrid Phases

Due to their ability to self-assemble spontaneously into regular arrays, surfactants are ideal candidates for the preparation of nanostructured inorganic-organic hybrid materials [2]. The first mention of the utilization of surfactants as structure-directing media for the preparation of nanostructured siliceous ceramics was the precipitation of an ordered inorganic-organic hybrid material in the presence of ionic surfactants [3–9]. This process is based on the electrostatic interaction between a condensing silica sol with oppositely charged surfactants or equally charged species via the mediation of counterions. As a consequence of this interaction, a surfactant-rich gel phase precipitates from homogeneous solution. The structures found in these gel phases strongly resemble those of LLC phases, for example hexagonal, cubic or lamellar, with the exception that the water is replaced to a great extent by the inorganic ceramic. Removal of the surfactant by calcination leaves behind porous, regularly structured inorganics denoted M41S. The discovery of the M41S family of mesoporous molecular sieves by researchers of the Mobil Corporation has inspired a new trend towards inorganic large-pore systems. These are expected to be a welcome complement to classical microporous zeolites, the pore diameter of which is limited due to the crystalline structure of these materials. Mesoporous materials possess large specific surface areas that can be functionalized chemically with respect to acidity, catalytic activity or specific affinity. Their relatively narrow pore-size distribution makes them promising for size-exclusion processes. Excellent reviews have been published since on the synthesis of different pore geometries and sizes, functionalities and possible applications [10–12]. However, although the discovery of M41S-type materials represents a breakthrough in porous materials, there are certain drawbacks to these materials, which mainly originate from the way they are made, namely by precipitation. First of all, the structure of such porous silicates can only be predicted a posteriori, as the synthesis takes place at surfactant concentrations above the critical micellar concentration, but below that of LLC phase formation. The structure of the precipitate is temperature- and time-dependent. Second, precipitation occurs as soon as the supramolecular inorganic-organic aggregate is too large to be soluble in the surrounding medium, usually water. Therefore the particle size of M41S materials is commonly small, indicating that the large specific surface area originates partly from the external particle surface rather than the internal pore system. This would be a drawback as soon as applications involving size selectivity are involved, as the outer particle surface can be expected to contain the same kind and density of functional groups or sorption sites.

The problem of small particle sizes due to precipitation can best be avoided by solidifying surfactant LLC bulk phases that contain an inorganic precursor. Furthermore, solidifying such media means that the degree of predictability of such syntheses can be vastly improved, as liquid crystal phases can be analyzed by a variety of simple, noninvasive analytical techniques.

1.3 LLC Phases as Structural Media

The macroscopic characteristics of LLC phases are usually best described by a soft, slimy or margarine-like consistency. Their domains undergo continuous structural rearrangement via molecular exchange of water and surfactant molecules. This is why these systems can be ruled out as structural components, e.g. for reinforcement purposes or for the creation of materials with defined diffusion pathways. Some recent applications of amphiphilic compounds in structure generation are summarized in [13]. The easiest approach to making full use of the structural potential of surfactant phases is a simple solidification process.

Shock-freezing of the phase does not alter the assembly structure, but has the disadvantage of being reversible, as upon temperature increase the system softens and regains its original properties. The polymerization of surfactants with polymerizable head groups or tails is commonly associated with phase transitions or even a complete breakdown of the structure [14–16]. The same goes for the polymerization of organic monomers used as swelling additives for LLC phases. Here the polymerization imposes osmotic and thermodynamic pressure onto the system with the consequence of structural deterioration or rearrangement.

The prerequisites to be fulfilled for a structural preservation of a preformed LLC assembly are as follows: (i) the solidification has to be irreversible, (ii) the resulting solid product should not compete with the surfactant head groups for water, as this would result in substantial changes of the composition with the consequence of phase changes, (iii) no macroscopic demixing must occur, which indicates that the solid substance has to be compatible either with the hydrophobic, or (more commonly) with the hydrophilic domains of the phase, (iv) the presence of reactants or the release of by-products should not affect the surfactant phase structure, and (v) the synthesis has to occur at moderate temperature or at least far below the boiling point of the least volatile component (usually water).

To fulfil these aspects, three synthetic pathways appear particularly promising, namely precipitation, sol-gel processing and electrochemical conversion. In fact, all three approaches have been successfully applied. The synthetic procedures, outcome and properties of the respective products are described in this chapter. The ceramic oxide most frequently synthesized in LLC phases is silica; therefore the following section is going to be focused mainly on its preparation.

2
Sol-Gel Processing of Ceramic Oxides in LLC Phases

The sol-gel process for the production of silica is an industrially widely applied procedure. The process is usually conducted in homogeneous solution, and it is possible to manufacture nano- or micro-sized particles, clear macroscopic silica objects (monoliths), such as fibres or lenses [17]. Sol-gel processing is commonly conducted at quiescent temperature, and no increased pressure is necessary, which makes this method suitable for LLC templating. The starting material for the sol-gel preparation of silica is an orthoester (alkoxide) of the general structure $Si(OR)_4$, which is hydrolysed in order to formally yield silicic acid. The latter undergoes polycondensation into a three-dimensional network of silicon

$$Si(OR)_4 + 4\,H_2O \xrightleftharpoons[\text{Esterification}]{\text{Ester Hydrolysis}} Si(OH)_4 + ROH$$

$$2Si(OH)_4 \xrightleftharpoons[\text{Hydrolysis}]{\text{Condensation}} (OH)_3Si\text{-}O\text{-}Si(OH)_3 + H_2O$$

$$Si(OR)_3OH + Si(OR)_4 \xrightleftharpoons[\text{Alcoholysis}]{\text{Alcohol Condensation}} (OR)_3Si\text{-}O\text{-}Si(OR)_3 + ROH)$$

Scheme 1

dioxide, whose surface is saturated by silanol (Si-OH) groups. The various reactions occurring simultaneously in a sol-gel mixture (see Scheme 1) represent equilibria, which can be influenced by parameters, such pH, temperature, or solvent content and which all have a substantial effect on the structure of the resulting silica gel. The most common alkoxide precursors are tetraethylorthosilicate (TEOS) and tetramethylorthosilicate (TMOS) because of their relatively low cost, easy handling, and relatively fast hydrolysis.

2.1 Surfactants Used for LLC Templating

In contrast to the surfactant-assisted precipitation of nanostructured M41S-type materials, the templating of silica in LLC phases usually employs non-ionic surfactants, such as the monoalkyl ethers of oligo(ethylene oxides) (C_nEO_m), to avoid strong interactions that would invariably lead to phase changes. The nonionic nature of the surfactant headgroup minimizes the electrostatic interaction between inorganic polymer and does not give rise to precipitation or structural rearrangement during the synthesis. The phase diagrams of these surfactants are well documented in the literature [18]. One further aspect of surfactants containing oligo(ethylene oxide) (EO) headgroups is the temperature dependence of their LLC phase behaviour, which is a consequence of a dehydration of the EO groups at temperatures above 50 °C [19].

2.2 Templating of Nonionic-Surfactant Phases

The templating of LLC surfactant phases involves the "swelling" of the LLC phase, the structure of which can be predicted using the phase diagram of the surfactant, with the inorganic precursor [20]. This mixing is easily achieved by dissolving tetramethylorthosilicate (TMOS) in the lyotropic phase at pH 2, upon which fast hydrolysis occurs, catalysed by the presence of the surfactant. The lyotropic phase order is temporarily lost as methanol, evolved during the hydrolysis, causes the mixture to become isotropic and of low viscosity. This apparent drawback (see Sect. 1.3) is overcome by the advantage that the isotropic character of the mixture allows fast and homogeneous mixing. Methanol, the least volatile component, is evaporated in a slight vacuum or by blowing air over the sample, and consequently the original LLC phase structure is re-established. This process can be easily monitored by either polarized-light optical microscopy (see Fig. 2), deuterium NMR spectroscopy (using D_2O to hydrolyse the TMOS) or small-angle X-ray diffractometry [20–23]. As the amount of water in the system is low after hydrolysis, the pH is considerably lower than 2, and the siliceous LLC phase starts solidifying. During this process the hydrophilic inorganic silica precursor (silicic acid and its oligomers) undergoes polycondensation that is confined within the aqueous domains of the liquid crystal, hence producing a cast or replica of the lyotropic system without changing the supramolecular structure. This procedure was termed "*nanocasting*", and in many ways it resembles a nanoscale analog of the lost-wax casting technique,

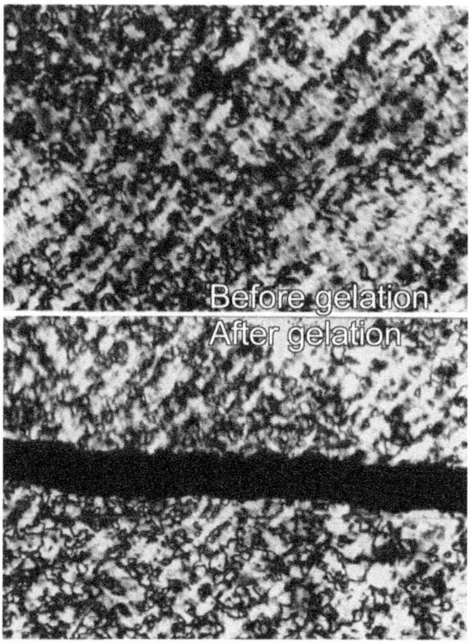

Fig. 2. Optical textures of a hexagonal siliceous LLC phase comprising the surfactant $C_{12}EO_8$. The *top picture* was taken before polycondensation, i.e. of the still liquid phase, the *bottom picture* after polycondensation was complete. The crack is a consequence of shrinkage

which is still applied for the manufacturing of bronze statues and church bells. *Nanocasting* is the method, which produces an exact cast of the lyotropic block copolymer or surfactant phase structure.

The structure of the materials obtained by *nanocasting* is similar to that of M41S-type mesoporous molecular sieves (see Fig. 3). However, since *nanocasting* is conducted in the presence of nonionic templates, which do not demand electrostatic compensation on a specific length scale, the wall thickness of the porous ceramics can be individually adjusted via the precursor content. Furthermore, the nanostructure derived from liquid crystalline bulk phases is a result of a homogeneous mixture solidifying, as opposed to a precipitation process. Therefore the dimensions of the inorganic particles are by far larger. As the interaction between a nonionic surfactant and the ceramic material is not that strong, the surfactant can even be removed by simple extraction, which allows a recovery of the usually quite expensive template.

All the mesoporous ceramic oxides obtained by this method are amorphous on the atomic level, but show periodicities on the nanometer length scale and narrow pore size distribution. The outcome of the process is very predictable, as the binary phase diagram of the surfactant can be used as a guideline towards the nanostructure design. The transmission electron microscopy images of silicas derived from three different LLC surfactant phases are shown in Fig. 3.

It is worth mentioning that the oligo(ethylene oxide) headgroups of the nonionic surfactants are firmly anchored in the solidifying inorganic matrix. There-

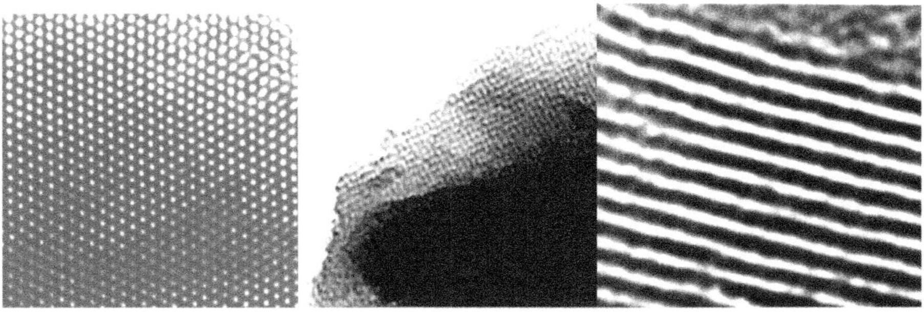

a b c

Fig. 3 a–c. TEM images of three silicas obtained by nanocasting of LLC surfactant phases: **a** hexagonal; **b** cubic; **c** lamellar

fore, nonionically templated mesoporous ceramics are microporous too, as every headgroup forms its own cavity within the inorganic wall.

Although the nanostructured particles derived by nanocasting of LLC surfactant phases are by far larger than those of the previously known M41S-type precipitates, the shrinkage occurring during the solidification of the silica sol causes extensive crack formation (see bottom photograph in Fig. 2). The preparation of monolithic mesoporous silica has, however been reported by various research groups. Ammundsen et al. [24] prepared macroscopic objects with disordered hexagonal structure from LLC surfactant phases, and more recently, Jones et al. [25] studied the effect of nonionic surfactant type, aluminium content and additives on the structure and ability of monolith formation of LLC-templated siliceous ceramics.

In order to prepare a macroscopic object of specific size and shape, it is necessary to introduce some elasticity and ductility into the inorganic-organic hybrid system to prevent crack formation due to shrinkage and allow *nanocasting* to be conducted in a precisely shaped mold. Improving the mechanical properties can be achieved by introducing a typically elastic component into the mixture, which is a polymer, and indeed, the nanocasting of amphiphilic block copolymer phases [26] has proven a versatile and successful templating procedure for the synthesis of tailor-made porous silica monoliths.

2.3 Block Copolymer Phases

The creation of ceramic nanostructures with controlled structure is a rapidly emerging field, which greatly profits from the self-assembly of amphiphilic block copolymers as well as the variety of ABCs available. The ordering characteristics of amphiphilic block copolymers can be almost continuously tuned by varying the chemical nature of the blocks, solvent content, molecular weight, block length ratios, copolymer architecture, or even by varying external parameters that influence the aggregation behaviour, such as temperature or the addition of salt.

Simplistically, amphiphilic block copolymers can be assumed to act as "large surfactants", in that they allow self-organization into larger aggregate structures than their low-molecular weight analogs. The aggregation structures of amphiphilic block copolymers exhibit decreased exchange dynamics, which go along with higher kinetic stability of the phases formed. Furthermore, amphiphilic block copolymers extend the synthetic methods for mesoporous ceramic nanostructures over the inherent limits of low-molecular weight surfactant templates. The latter are only available up to certain alkyl chain lengths (usually 22 carbon atoms at the very most). Accordingly, the pore diameters available for mesoporous inorganic nanostructures are limited to a maximum of 4.5 nm, unless inert organic auxiliaries are introduced into the system during synthesis. These would be hydrophobic oils that are successfully applied in M41S-type syntheses to increase the pore diameter, but do not work as well for LLC phases, as their utilization would involve moving on to a ternary templating system, which is notoriously more prone to phase separation.

ABCs can be made (or even purchased) with considerably higher molecular weight, so that the synthesis of larger-pore materials is possible. In particular, modern polymer chemistry provides the tools to make these block copolymers in a vast variety of shapes and sizes, allowing an equally rich variety of nanostructures to be produced.

Like the surfactant templates discussed above, the compatibilizing amphiphilic block copolymers not only stabilize the interfacial area between the inorganic and the surrounding water, but also act as porogens, leaving nanometer-sized voids behind after its removal. Amphiphilic block copolymer templates simply replace the low-molecular weight surfactants used previously, introducing mechanical stability and opening access to a wider range of pore diameters.

2.4
Nanocasting of ABC Phases

Templating of lyotropic ABC phases is as straightforward as the nanocasting of LLC surfactant phases, which stresses another analogy between ABCs and classical surfactants. The precipitation of inorganic ceramic oxides in the presence of amphiphilic block copolymers is one method of preparing large-pore mesoporous materials, which show a high degree of order [27–29]. This synthetic approach stands in analogy to the previously discussed preparation of M41S-type materials and will not be discussed here.

Nanocasting of the highly concentrated LLC phases of ABCs represents a versatile route towards inorganic-organic nanostructured hybrid materials [30–32]. Like in the templating of LLC surfactant phases, the direct utilization of lyotropic amphiphilic-block copolymer phases is predictable with respect to the structure of the final material, as again this approach can be understood as the vitrification of a microphase-separated medium, during which no changes of the phase structure occur.

Using the binary phase diagram of an amphiphilic block copolymer with water as a guideline (an example is shown in Fig. 4), the structure of the final

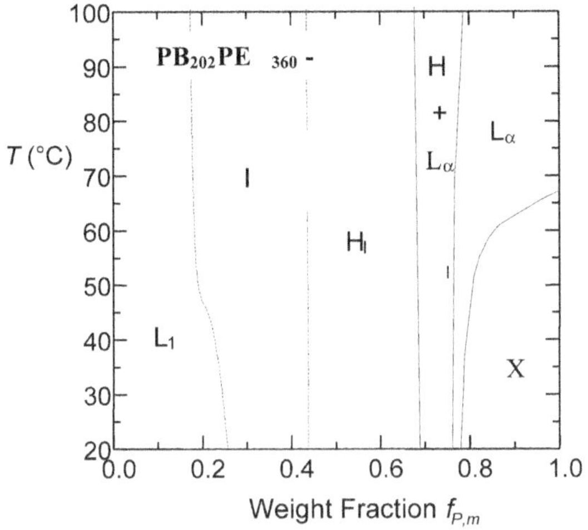

Fig. 4. Binary phase diagram of poly(butadiene-*b*-ethylene oxide) (PB_{202}-PEO_{360})

product can be predicted a priori: As the *nanocast*, an ordered hybrid material consisting of silica and the templating block copolymer, is the result of adding a component to a binary lyotropic liquid crystal phase without ultimately changing the phase structure. This method is equally applicable for nonionic [26, 31, 32] as for ionic [33] amphiphilic block copolymers, which is another strong indicator that supramolecular templating of the lyotropic mesophase governs the mechanism of *nanocasting*.

The nonionic ABC templates used for *nanocasting* commonly consist of a hydrophobic soft block (T_g below or around room temperature in order to warrant sufficient solubility at room temperature), such as poly(butadiene) [34], poly(ethylene-*co*-butylene) (Kraton Liquid) or relatively short poly(styrene) and a poly(ethylene oxide) block as the hydrophilic moiety [35].

The binary phase diagrams of aqueous amphiphilic-block copolymer solutions follow the general phase sequence shown by surfactants (Fig. 4). While at low block copolymer concentration the formation of micelles is observed, high concentrations give rise to more complex aggregates, such as cylindrical micelles or lyotropic liquid crystal phases of different structure. A rich lyotropic polymorphism is observed that can be utilized directly for porous-nanostructure design. The phase diagrams correspond to those of low-molecular weight surfactants in that they are the more defined the narrower the molecular weight distribution. Heterodisperse block copolymers often show a tendency to form ill-defined phases, whereas the phase diagrams of monodisperse amphiphilic block copolymers exhibit wide regions of high order. Consequently, heterodisperse ABC templates afford nanostructures, which appear to be affected by high defect densities where the defect sites determine the overall structure [26]. In contrast, *nanocasts* of low-polydispersity amphiphilic block copolymers are generally more ordered. The structure elucidation of the latter by imaging or scattering methods is, therefore, considerably simpler.

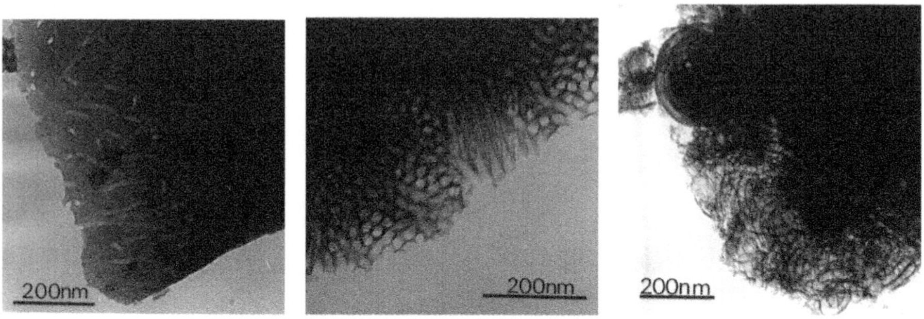

a b c

Fig. 5 a–c. TEM images of KLE-templated silicas (from [31])

After mapping the phase diagram with respect to phase composition and temperature, the variation of one experimental parameter during the synthesis, namely the template concentration, allows the precise tailoring of pore systems with respect to their shape, density, and connectivity. Figure 5 shows TEM images of three samples, all obtained with the same template (KLE 3729), but at varying concentration [31].

Isolated, bent cylindrical pores result from a system at relatively low template contents, while a hexagonally packed cylindrical pore array is achieved at higher concentration. Increasing the template contents even further gives rise to an entirely different structure, which is again in accordance with the lyotropic phase diagram of the polymer. Multilamellar vesicles are found in coexistence with isolated fragments of sheet-like structures. Calcination, that is removal of the organic scaffolding from a lamellar system, drives the structure to collapse.

Nitrogen adsorption-desorption isotherms provide valuable information not only on the structure of the materials, but also allow one to gain deeper insight into the nature of their porosity. They support the conclusions drawn from the TEM micrographs, but in addition they illustrate the fact that each template molecule delivers its individual contribution to the overall interface area. Within the regime of one particular pore structure, the specific surface area increases significantly with growing template concentration [31]. They also demonstrate that the same double templating action can be observed for the *nanocasting* of amphiphilic block copolymer phases as well as for that in LLC surfactant media. The resulting mesoporous ceramic oxides possess an undeniable microporosity, which is due to the hydrophilic poly(ethylene oxide) blocks (the "headgroups"), which are initially homogeneously distributed in the silicious microphase, displacing silica growth from their own molecular volume [36, 37].

Some phase diagrams of low-polydispersity amphiphilic block copolymers, exhibit areas of coexistence over a relatively wide range of composition (see Fig. 4) [32]. This is probably due to kinetic inertia or to the fact that at the borderline between two thermodynamically stable phases the energetic differences between two structures are marginal. Swelling these coexisting phases with a siliceous precursor affords a microphase-separated siliceous phase, which has the same structure as the binary mixture consisting of water and amphiphilic

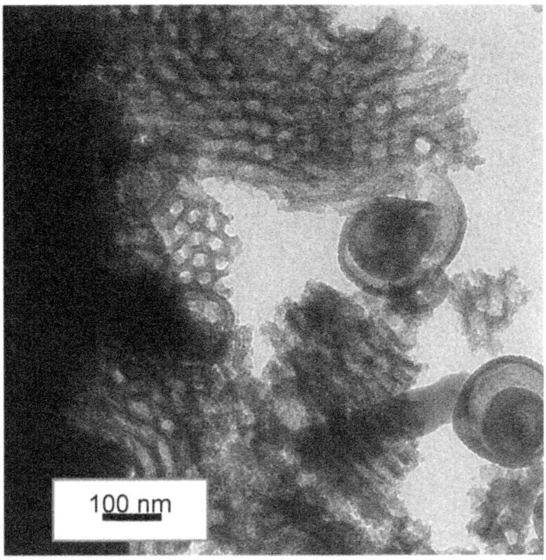

Fig. 6. TEM image of silica prepared in the hexagonal/lamellar area of coexistence of PB_{202}-PEO_{360}

block copolymer. As a result of the inorganic precursor undergoing polycondensation, the bulk phase solidifies without altering its superstructure, as shown in the TEM image displayed in Fig. 6.

The fact that the structural integrity of the binary lyotropic phase does not seem to be harmed by the *nanocasting* procedure suggests a reversal of the principle. If the nanostructure of the silica can be predicted a priori, the a posteriori analysis of a silica cast should provide valuable information on the structural composition of an unknown binary ABC phase. Assuming the noninvasive character of *nanocasting*, this method can help elucidate more complex lyotropic phase structures of amphiphilic block copolymers as a complementary technique to diffraction methods and may help to avoid time-consuming preparations (e.g. cryo-TEM, freeze-etching, etc.). The phase structure simply has to be "frozen" into a solid silica cast, which is perfectly stable in a high vacuum under the electron beam.

The applicability of *nanocasting* as an analytical tool has been demonstrated [38] by comparing the silica structures obtained from the lyotropic phase, which has been crosslinked using γ-rays, in order to provide sufficient mechanical stability to allow thin-sectioning, with those of a silica *nanocast* obtained from a lyotropic phase of the same composition (Fig. 7). The similarity between the structures is striking. A reference sample was prepared by filling the pore system of the crosslinked polymer gel with silica and subsequent calcination. The pictures prove without doubt that the sol-gel process indeed does not have any structurally disrupting effect on the liquid crystalline phase [38].

In contrast to precipitation procedures, *nanocasting*, particularly of amphiphilic block copolymer phases, allows the fabrication of large objects (monoliths) that are macroscopically devoid of cracks and defects. Despite their mechanical robustness, the porosity of these monoliths can be as high as 85%.

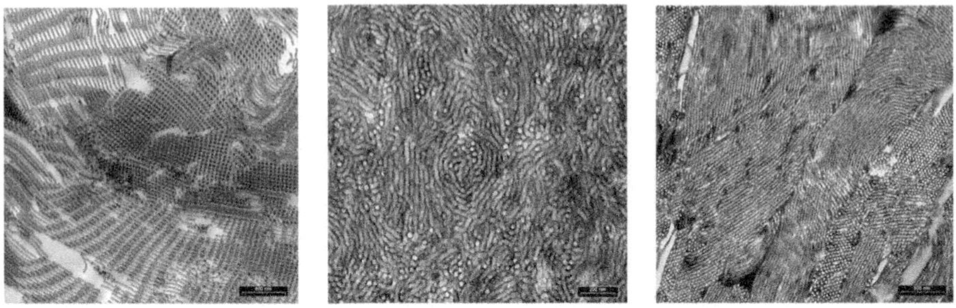

a b c

Fig. 7a–c. TEM images of: **a** the crosslinked LLC ABC phase; **b** nanocast of the LLC ABC phase; **c** nanocast of the crosslinked LLC ABC phase

Moreover, diffusion pathways can be individually designed by templating particular phase structures. Above all, the pore system of a macroscopic object is exclusively determined by the pore system, whereas particulate powders show a significant contribution to the surface area caused by the nonstructured particle surface. The direct liquid crystal templating approach was also used to prepare monolithic silica from TMOS or TEOS in block copolymer-water mixtures mixed with alcohol cosurfactants and hydrophobic swelling agents [39–41].

Nanocasting of lyotropic ABC phases allows one to design predictably the structure, size, connectivity and shape of nanoscopic pore systems in sol-gel-derived silicates. It is expected that the generation of defined diffusion pathways, combined with the possibility of shaping macroscopic objects, undeniably causes highest expectations regarding the application in separation processes. In these respects, nanocasting is a faithful method of replicating the structural characteristics of lyotropic liquid crystal phases, be it for synthetic or analytical purposes. This procedure allows for the utilization of the full potential of such self-organized media as structural rather than only functional materials.

2.5 Template Mixtures

Nanocasting of LLC phases has one further advantage over the precipitation of M41S-type materials, namely the possibility to introduce a second template into the lyotropic phase. For example, adding a polymer latex dispersion to the *nanocasting* mixture can be used to generate hierarchical pore systems: The synthesis of silica is performed in the presence of a polymer latex dispersion as the porogen, giving rise to spherical pores in the size and size distribution corresponding to those of the templating latex dispersion (not shown here). By adding an additional amphiphilic block copolymer as a second template, materials can be prepared with bimodal pore size distributions (see Fig. 8) [30], which as monolithic species would be ideal candidates as stationary phases for chromatographic separation, as they allow for convection as well as diffusion to occur in the pore system.

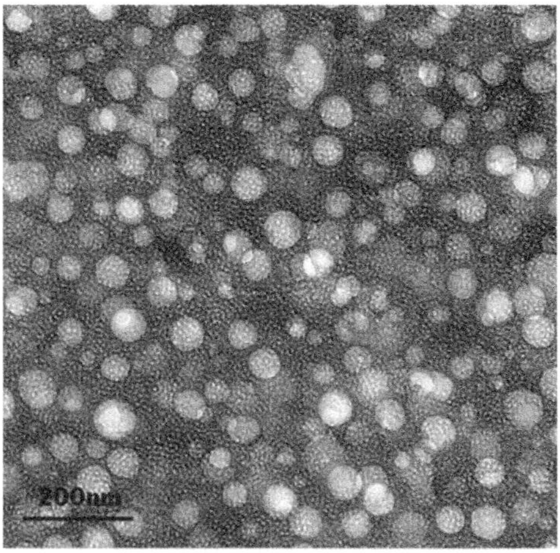

Fig. 8. TEM image of silica prepared in the presence of an ABC template and an additional latex dispersion. The bimodal pore size distribution allows for convection *and* diffusion

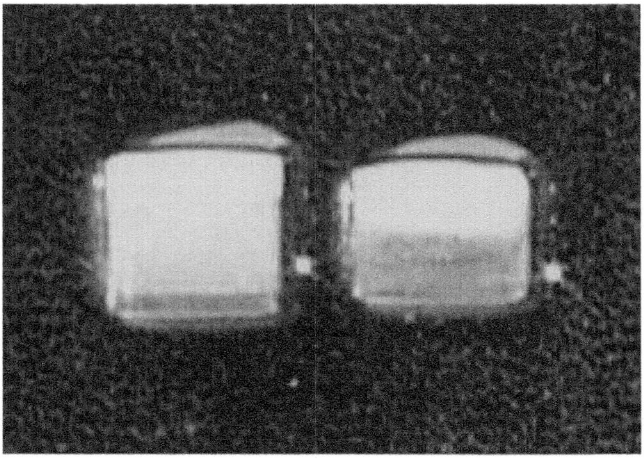

Fig. 9. Photograph of monoliths made from silica sols containing an ABC template as well as a latex dispersion. The latex, as the least dense component, floats on top of the centrifuge cell, whereas dense, larger silica clusters sediment to the bottom. The ABC-silica hybrid phase remains between the two

An additional ordering principle can be imposed on a system comprising a combination of such templates by conducting the solidification of he reaction mixture in a centrifuge field [42]. The different components partially separate according to their density to create a gradient of porosity in a monolithic material, as shown in Fig. 9. The monoliths shown here contain dense, mesoporous and macro-mesoporous domains, which makes them promising candidates for sequential separation of molecules according to size.

2.6
Nanocasting of Inverse ABC Phases: Particles

Another approach towards the synthesis amphiphilic block copolymer-templated inorganic nanostructures is the adjustment of the lyotropic phase structure in ABC phases via the content of a pre-hydrolysed mixture of aluminium and silicon alkoxides [43–45]. Again, the formation of a solid, nanostructured inorganic-organic hybrid material is a consequence of the strict microphase separation between a hydrophobic (or better: "silicatophobic") poly(isoprene) block and the "silicatophilic" poly(ethylene oxide) interacting with the inorganic sol-gel precursor.

The novelty of this process lies in the inorganic sol acting as a swelling agent, whose volume fraction determines the overall microphase structure of the hybrid material. As the microphase-separated structure forming depends on the volume fraction of each block (or microphase), the amount of inorganic "microphase" added to the amphiphilic block copolymer determines the structure of the resulting inorganic-organic hybrid material. Interestingly, the chemical nature of the inorganic phase appears to be independent of the inorganic precursor/template ratio, hence manifesting the main difference to the ABC templating processes discussed above. In ABC bulk phase templating, it is the amount of inorganic that directs the structure of the hybrid material rather than the amphiphilic block copolymer. As a consequence, the whole phase diagram of this bulk block copolymer with respect to block length ratio can therefore be traversed without having to synthesize amphiphilic block copolymers with different block length ratios. Depending on the amount of inorganic sol present in the block copolymer, a whole variety of differently structured hybrid materials can be produced (see Fig. 10).

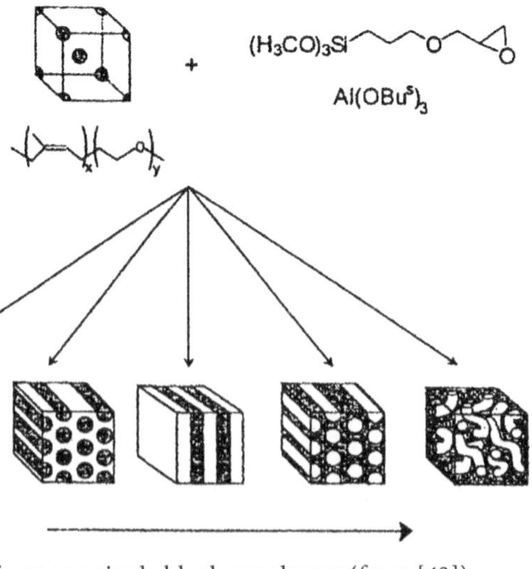

Fig. 10. Hybrid structures from one single block copolymer (from [43])

NMR spectroscopic analysis finally proved that the poly(ethylene oxide) blocks are firmly anchored in the inorganic phase rather than being located at the interphase adjacent to the hydrophobic domains. Solid-state NMR spectroscopy revealed that this anchoring leads to a substantial hindrance of the conformational mobility in the poly(ethylene oxide) chains compared with the relatively mobile hydrophobic poly(isoprene) [44].

Two possible scenarios can be envisaged for the structure of the hybrid material (see Fig. 11). The poly(ethylene oxide) block, albeit strongly interacting and partially penetrating, forms a pure PEO layer at the interface to the hydrophobic poly(isoprene) (Fig. 11, left-hand sketch) ("three-phase" system). The other possibility is the complete "dissolution" of the PEO chains in the aluminosilicate, which results in the "two-phase" system depicted in the right-hand sketch of Fig. 11. Spin-diffusion NMR experiments showed that there appears to be no dynamic heterogeneity in the poly(ethylene oxide) chains, as would be expected for a "three-phase" system, giving rise to the conclusion that the hydrophilic

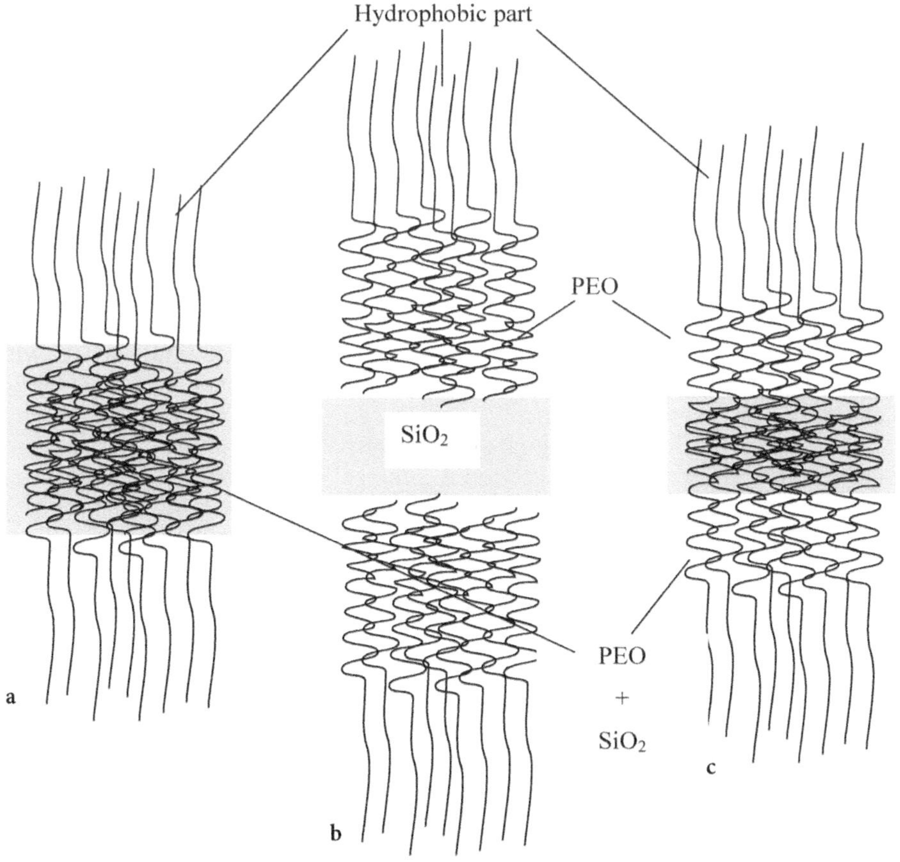

Fig. 11 a–c. Interaction between ABC "headgroups" and a siliceous ceramic polymer

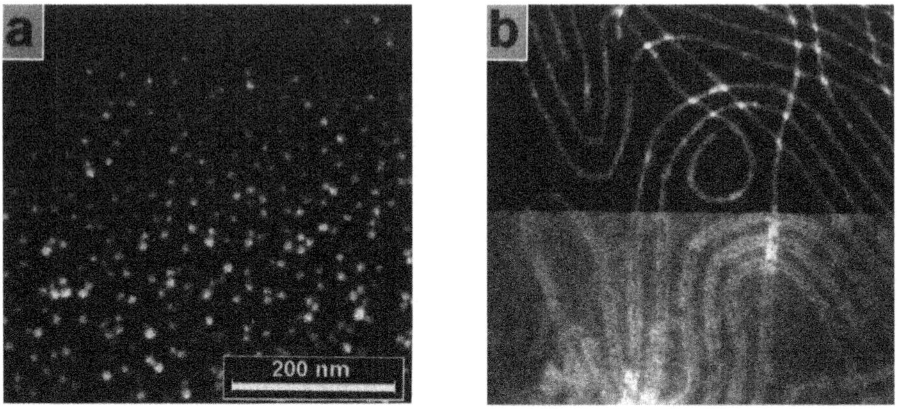

Fig. 12 a, b. TEM micrographs of nanoscopic spheres and rods obtained from inverse ABC phases

domains comprise one homogeneous hybrid phase consisting of poly(ethylene oxide) and the inorganic.

This approach towards nanostructured inorganic-organic hybrid materials is the first one to allow the synthesis of inverse-topology systems, in which the hydrophobic polymer blocks represent the outside of the microphase-separated structure. After solidification of the inorganic sol, the hydrophobic phase can be swollen with organic solvents. This procedure allows the isolation of colloidal objects, such as spheres or ceramic rods (see Fig. 12), from one another, which are sterically stabilized, because the hydrophilic block is firmly anchored in the ceramic material [45].

The position of the hydrophilic poly(ethylene oxide) (PEO) headgroups in ABCs and nonionic surfactant templates bears important implications with respect to the overall pore structure of the resulting porous material. The PEO chains can be assumed to be firmly anchored and molecularly dispersed within the silica matrix, hence giving rise to substantial microporosity of the mesoporous system. The following section is aimed at describing the pore structure of nonionically templated mesoporous ceramics more exactly.

2.7 Microporosity in Mesoporous Ceramics

The assumption of the EO chains being intimately mixed with the condensing silica has crucial consequences regarding the porosity of the solidified material after removal of the template, as each EO chain will create its own cavity or "molecular imprint" in the amorphous silica wall. The EO chains can therefore be expected to act as a porogen that creates microporosity within the walls of the mesoporous structure. This has indeed been evidenced experimentally. The combination of a complete evaluation of the small-angle X-ray diffraction data, in combination with nitrogen sorption experiments prove that mesoporous silicas, prepared in the presence of templates containing oligo(ethylene oxide) headgroups, indeed possess a substantial degree of microporosity in the wall. Prouzet et al. [46] conducted the preparation of mesoporous ceramics in two steps by adding a silica sol-gel precursor to a micellar solution of nonionic surfactant. X-ray diffraction studies, in combination with dynamic light scattering, shows clearly that the oligo(ethylene oxide) corona of these micelles is initially infused by silica oligomers. Davidson et al. [47] prove the existence of a microporous corona around the mesopores of nonionically templated mesoporous silicas. This corona can be attributed to the hydrophilic, or better: silicatophilic oligo(ethylene oxide) chains being firmly anchored in the silica wall. In short, a substantial proportion of the specific surface are of nonionically templated mesoporous silicas originates from additional microporosity.

It can be concluded that the surfactant headgroup of nonionic surfactants provides an additional tool to tailor the pore structure of silicas, in particular those derived from LLC phases. The favourable interaction between oligo(ethylene oxide) headgroups (or hydrophilic blocks in amphiphilic block copolymer templates) and growing silica gives rise to microporosity. Therefore the application of surfactant templates, the headgroup of which is less compatible with the growing inorganic polymer should give rise to purely mesoporous materials. One attempt in this direction has been reported by Antonietti et al. [48], who employed cyclodextrins as porogens. These cone-shaped polysaccharides show a different kind of amphiphilicity, as expressed by the aggregation into stacks rather than micelles. Therefore no "protruding" headgroup will be pointing towards the solidifying silica matrix, in contrast to the oligo(ethylene oxide) headgroups of the surfactant templates discussed above.

3
Precipitation in LLC Phases

Although the sol-gel process for the synthesis of siliceous ceramics is the most widespread synthetic method when the templating of LLC phases is concerned, there are other ways to imprint the structure of the soft LLC material onto an inorganic solid. One example is the precipitation of inorganics from the aqueous domains in an LLC phase. Although this approach sounds about as simple as nanocasting, the experimental conditions are much more complex. The main reason for this phenomenon is that an aqueous salt, used as a precursor for the preparation of the inorganic precipitate, will always compete with the surfactant headgroups for water. Most salts are only poorly soluble in nonionic surfactant phases, as they have a tendency to be "salted out" from the ternary mixture. Furthermore, the presence of large amounts of salt will always affect the phase behaviour of surfactants. Even nonionic surfactant phases can suffer only a limited amount of salt present before structural disintegration occurs. This phenomenon is even more pronounced for ionic surfactants. Another problem is the volume ratio of a hydrated salt to the nonhydrated solid product, which will never allow a monolith or even larger particles to be precipitated from LLC phases, as the diffusion of the salt ions is slower than the nucleation of a different particle. Finally the creation of a crystalline morphology within the nanoscopic confinement of the LLC phase implies that a substantial degree of curvature has to be introduced into the crystal lattice to follow that of the parental LLC phase. This curvature leads to a significant increase in the defect density in the crystalline product. More commonly, however, the crystal will grow regardless of the arrangement and direction of the lyotropic assembly, and simply displace the soft matter from its original location.

The most serious difficulty encountered when the precipitation of an inorganic solid in an LLC phase is attempted is the fact that two reactive components have to be mixed intimately within LLC phase before precipitation occurs. The mixing process is usually associated with difficulties, as both reactants should be dissolved in the LLC phase, which is of more or less viscous consistency, making homogenization difficult. One way around this problem is to let the reactant that causes precipitation diffuse into the LLC phase either via the gas or the solution phase. In this way nucleation can be slowed down considerably so that larger particles form instead of small ones that would be dispersed in the hydrophilic domains of the liquid crystalline phase.

These considerations make it evident that a delicate balance between rate of nucleation, crystal growth rate and stability of the surfactant assembly is necessary to warrant a successful outcome of the experiment.

3.1 Precipitation of Semiconductor Particles

The growth of cadmium sulfide superlattices has been conducted successfully by Stupp et al. [49–51] in the hexagonal LLC phase of a nonionic surfactant (see Fig. 13). Again, the structure of the resulting inorganic nanostructure depends on that of the parental LLC phase; accordingly a lamellar nanostructure was obtained from a lamellar phase, an hexagonal one from a hexagonal phase and hollow spheres from a micellar cubic phase.

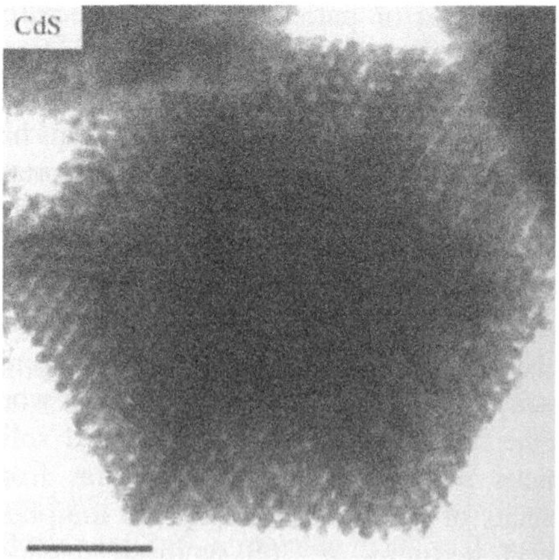

Fig. 13. TEM image of hexagonally structured cadmium sulfide

The same procedure was applied to the synthesis of cadmium selenide and zinc sulfide in LLC surfactant phases using C_nEO_m surfactants. A relationship between the covalent nature of bonds in the final product and the success of the templating procedure was established on the basis of silver sulfide, copper sulfide, mercury sulfide and lead sulfide not producing the same results. The interaction of the surfactant headgroups with the precipitated mineral and with its precursor ions are necessary for direct templating. This is also confirmed by the fact that salts that bind precursor ions prevent the formation of an ordered inorganic nanostructure within the LLC phase [51].

As expected, no monolithic structures of the semiconductor are observed in the product, as the LLC medium that immediately surrounds a crystal nucleation site is quickly impoverished of cadmium ions. However, this report clearly demonstrates that LLC phases are more widely applicable as structure-directing casting molds than shown previously.

3.2 Chemical Reduction of Noble Metal Salts

In principle, the reduction of a metal salt represents one form of precipitating an inorganic solid. This has been achieved in LLC surfactant phases of nonionic surfactants [52]. Hexachloroplatinic acid and ammonium tetrachloroplatinate were employed as the starting materials for the chemical synthesis of platinum in the hexagonal LLC phase of $C_{16}EO_8$. Hexachloroplatinic acid was reduced by the presence of large pieces of less noble metals (iron, zinc or magnesium). In contrast, the LLC phase containing ammonium tetrachloroplatinate was reacted by contacting the platinum-containing LLC phase with a hydrazine-containing one of the same structure. Immediate reaction in both cases causes the precipitation of nanostructured platinum particles from the LLC phase. The particulate nature of the resulting materials was revealed by scanning electron microscopy. The volume of the final product, the metal, is much smaller than that of the aqueous domains in the LLC phase; therefore the formation of the nanostructure only yields small, albeit porous, particles. It is further noteworthy that the particle size of the nanostructured product is much smaller than the domain size in the parental LLC phase, indicating that the domain boundaries in the self-assembled system are not the location at which particle growth necessarily stops or starts. The nanostructure of the surfactant-free metal particles is evidenced by TEM (see Fig. 14) and small-angle X-ray diffraction analysis, which both prove the existence of a hexagonally structured material. The cylindrical pores in this material are ca. 3 nm in diameter, whereas the walls are about 3 nm thick, giving an average pore-to-pore distance of 6 nm, which is slightly wider than that of the corresponding *nanocasting*-derived silica. This widening of the overall dimension is attributed to the crystalline nature of the walls.

The formation of the nanostructured metal strongly depends on the nature of the reduction agent. Slow reduction (e.g. using hydrogen) produces dendritic structures, in which the nanostructure of the LLC phase is not detectable. As the reaction dramatically alters the chemical composition of the phase at the location of reduction (or particle formation), the phase becomes destabilized. Slower

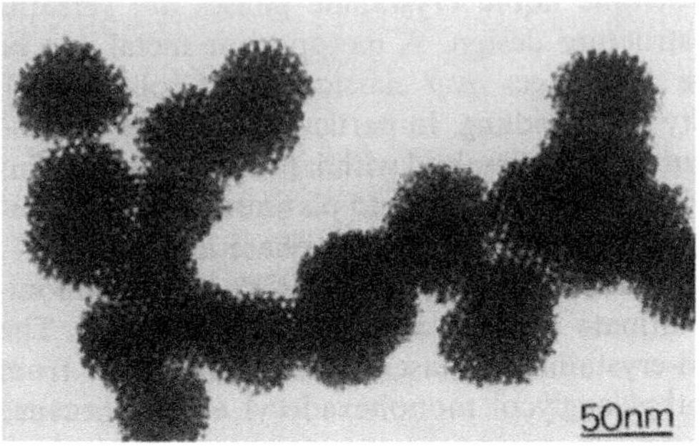

Fig. 14. TEM of hexagonally structured mesoporous platinum

reduction, therefore, provides enough time for the LLC phase to rearrange and give way to the growth of the metal crystal without inducing the curvature necessary to form the nanostructure. The platinum salt necessary for the large crystals to grow is evidently provided by diffusion through the aqueous domains of the LLC phase. This phenomenon again stresses the problems encountered whenever precipitation reactions are attempted in the soft LLC media.

The fact that no nanostructured metal can be obtained by conducting the reduction in micellar systems of the surfactant nevertheless proves that it is indeed the LLC phase that acts as the structure-directing environment, in which the growth of the metal is confined to the aqueous domains of the LLC phase. The Fourier transform of the Extended X-ray Absorption Fine Structure (EXAFS) data obtained from nanostructured, hexagonal platinum reveals a significantly reduced amplitude as compared to that of the bulk metal, which is attributed to a lower average coordination number, hence a large quantity of surface atoms.

The reduction of metal salts from LLC phases not only provides nanostructured materials with defined diffusion pathways and large surface areas, but is also suitable to deposit nanostructured alloys [53, 54] which are not accessible by any other method. The main interest in preparing nanostructured alloys lies in the field of catalysis, as most catalytic processes rely on a well-defined and reproducible surface chemistry. Furthermore, alloying can afford materials the catalytic activity which is enhanced compared with that of the constituent materials. The co-reduction of binary mixtures of metal compounds dissolved in the hexagonal phase of a nonionic surfactant leads to bimetallic alloy powders with long-range mesoporous nanostructure and large specific surface areas. In particular, hexachloroplatinic acid and ruthenium trichloride were co-deposited electrochemically from the hexagonal mesophase of the commercially available nonionic surfactant Brij 76 (mainly $C_{18}EO_8$) using formaldehyde in alkaline conditions. The resulting nanostructured powder has a large specific surface area of 86 m^2/g according to nitrogen sorption experiments. The existence of a true alloy was confirmed by EXAFS measurements in combination with wide-angle X-ray diffractometry. These techniques prove that no heterogeneous reduction of the metals has occurred, which would give rise to a physical mixture of the metals.

Because of their regular structure (defined pathways for electrons and the diffusion of ions) and large specific surface area, nanostructured metals are expected to be useful in electrochemistry and catalysis; however, many technological applications rely on the use of a coherent layer of the large-surface area material rather than on a powder, which would be mechanically less robust and prone to abrasion. In order to obtain nanostructured films of noble metals it is therefore mandatory to support diffusion of the metal ions towards the site and of by-products away from it. Depositing the metal electrochemically from an LLC phase onto an electrode surface, as will be discussed in the following sections, is currently the only way to warrant a constant supply of metal ions at the location of reduction. The resulting materials are expected to be of interest for a wide range of applications, such as in batteries, fuel cells and sensors [55–58].

4 Electrochemical Deposition of Noble Metals from LLC Phases

4.1 Noble Metal-Containing LLC Phases

Although the presence of large amounts of salts are commonly described to destabilize even nonionic, but more so ionic, LLC surfactant phases, sometimes a stabilizing effect is observed. The effect of hexachloroplatinic acid on the LLC phase boundaries of ternary systems containing a nonionic surfactant and water was reported by Attard et al. [59], who mapped complete pseudo-binary phase diagrams as a function of surfactant concentration and temperature. A dramatic effect on the phase behaviour was observed by polarized-light optical microscopy. In particular, the presence of the platinum compound was found to enhance the stability of the more highly curved phases (micellar cubic and hexagonal). Moreover, the temperature ranges over which the LLC phases are stable are dramatically increased with stabilities up to 95 °C for the ternary mixtures as compared with 58 °C for the binary system. The formation of biphasic regimes, as reported previously in the literature [60], is also observed.

The observations made in this investigation show clearly that simple dehydration of the surfactant headgroups by the solvated hexachloroplatinic acid cannot be the only reason for these changes. The inorganic component is rather suggested to interact with the surfactant headgroups, hence formally increasing their spatial demand. At the same time the electrostatic repulsion between the now formally ionic headgroups is enhanced. The net effect is to stabilize aggregates and phases with a higher degree of mean curvature than the case in the platinum-free binary system.

4.2 Electrochemical Reduction of Metal Salts

Noble metals, such as platinum and palladium films, have been successfully deposited from LLC surfactant phases onto flat electrode surfaces [61, 62]. Simplistically, a flat graphite electrode is dipped into the metal salt-containing LLC surfactant phase of a defined structure and then the reduction conducted in an electrochemical cell under potentiostatic and thermostatic control. Choosing an electrode of known diameter further allows the elucidation of the surface area of the final nanostructured deposit via comparing the integrals of the cyclic voltammogram with that obtained from a polished electrode made from the same metal. The large surface areas can also be confirmed by impedance spectroscopy [63].

The deposits show nanostructures that, apart from slightly increased dimensions of the material, are very similar to the structure of the LLC phases they are derived from. These new electrode surfaces are adherent, macroscopically smooth (see Fig. 15) and shiny and expectedly show high specific surface areas. Variations in the deposition conditions, such as temperature and deposition potential, cause dramatic changes in the specific surface area, nanostructure and macroscopic morphology of the films, providing an additional tool to tailor the surface properties of such electrodes [62].

The mass transport characteristics through nanostructured metal deposits on microelectrodes are excellent, which, in combination with the large real surface area, makes them potential candidates for electroanalysis where, for example, the amperometric detection of organic species with poor electrode kinetics is largely facilitated.

A literally nanoarchitectured electrode was reported by Owen et al. [64, 65], who electrodeposited nickel/nickel oxide, which is regarded an electrochemically extremely important material whenever applications like fuel cells, the electrolysis of water, batteries or catalysts for the electrochemical hydrogenation of organic species are concerned. Mesoporous nickel electrodes have been electrodeposited by reduction of a nickel(II) solution dissolved within the aqueous

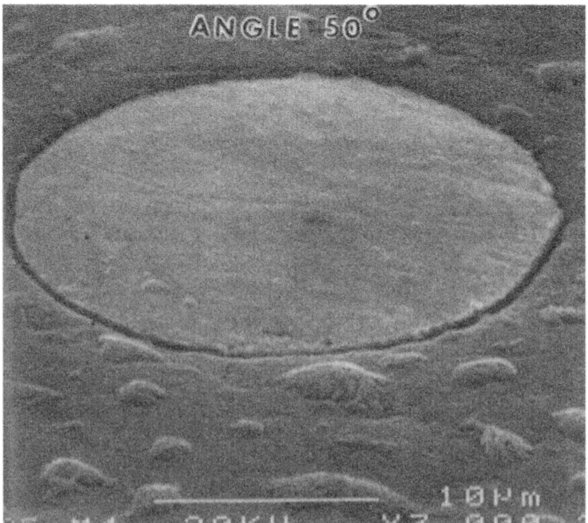

Fig. 15. Surface morphology of a microelectrode with a mesoporous platinum coating

domains of the hexagonal (H_I) LLC phases of Brij 56, a nonionic surfactant. The nanostructured films consist of well ordered, periodic arrays of channels arranged in a hexagonal lattice with nanometer-sized dimensions. Cyclic voltammetry has again been applied to demonstrate the large specific surface areas (typically surface area enhancements of 100 times over non-templated electrodes) and a pore structure that allows extremely rapid diffusion of species to active sites within the structure. These characteristics suggest application as a low cost-current collector and use in high-power devices.

Electrodeposition from aqueous systems containing a palladium and a platinum salt [66] as well as the nonionic surfactant yield a regularly nanostructured material, which was further characterized by cyclic voltammetry and X-ray diffraction measurements. The evident lack of macroscopic demixing as well as the fact that palladium and platinum are miscible over the whole range of composition gave rise to the conclusion that, again, a nanostructured alloy was deposited. A very peculiar redox behaviour was observed by cyclic voltammetry of this mixed-metal electrode in sulfuric acid (see Fig. 16). Unexpected, very sharp and intense peaks appear in the hydrogen region, which are interpreted in terms of dissolution and re-plating of the less noble component palladium during electrochemical cycling.

Mesoporous tin was deposited from LLC surfactant phases in order to provide a nanostructured negative electrode material for applications in lithium ion batteries. Whitehead et al. [67, 68] electrodeposited tin films from lyotropically liquid crystalline solutions of tin sulfate in the hexagonal phase of Brij 76 onto copper foil. The authors further employed *n*-heptane as an inert swelling agent in order to tailor the dimensions of the resulting electrode materials.

The electrodeposition of nanostructured materials from LLC phases is not restricted to noble metals, but can be extended into the field of semiconductors [69]. Mesoporous regular semiconductors have great potential for applications in optical sensors and solar cells. The successful electrodeposition of selenium was demonstrated by Nandhakumar et al., who subjected a hexagonally struc-

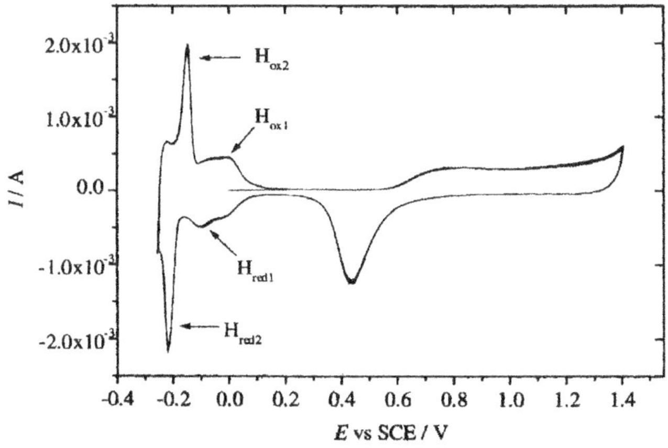

Fig. 16. Cyclic voltammogram of a mesoporous platinum-palladium electrode

tured, quaternary system comprising selenium dioxide, sulfuric acid, nonionic surfactant ($C_{16}EO_8$) and water, to reduction on either nickel or platinum electrodes under potentiostatic and thermostatic control at 65 °C. Again, nanostructured, homogeneous deposits were obtained. The selenium films were post-synthetically modified with cadmium.

The electrodeposition of nanostructured inorganics from LLC phases is a further example of the versatility provided by bulk lyotropic liquid crystal phases as templating media [70, 71]. In all these cases the structure can be predicted from that of the parental mesophase. By careful choice of surfactant, addition of a swelling agent and by changing the temperature and composition of the lyotropic phase it is possible to alter pore size and architecture of the nanostructure in a controllable manner.

5 Summary and Outlook

The three above-mentioned examples of utilizing the lyotropic supramolecular aggregates of surfactants and amphiphilic block copolymers show impressively how self-assembled systems can be applied as structure-directing media in modern materials science. *Nanocasting* of siliceous ceramics in LLC phases of surfactants and amphiphilic block copolymers is a most versatile tool to prepare porous ceramics with large specific surface areas. Suitable choice of the templating phase assists in tailoring diffusion pathways, pore diameters, pore-size distributions and allows for the manufacturing of macroscopic objects (monoliths). The precipitation of inorganic solids within the aqueous domains of surfactant phases affords nanostructured porous particles with extremely large specific surface areas, which are expected to find application in catalysis (noble metals) and solar cells (semiconductors). The electrodeposition of noble metals and semiconductors onto electrode surfaces is an elegant method of synthesizing nanostructured electrodes with excellent transport properties, good mechanical performance and, again due to the large specific surface area, high efficiency.

Most importantly, such structures form spontaneously by mechanisms of supramolecular self-assembly rather than necessitating external manipulation, for example by microlithographic processing. The utilization of LLC surfactant and amphiphilic block copolymer phases is therefore a simple method to tailor directly the nanometer-scale structure, which undoubtedly defines the properties of the final product.

Finally, it needs to be mentioned that the mechanisms of supramolecular self-assembly apply to many systems more complex than simple surfactants and amphiphilic block copolymers. Supramolecular self-assembly and pattern formation is one of the crucial principles in nature and gives rise to hierarchical structure ranging from the length scale of a few nanometers to the macroscopic domain. The generation of intricate and very regular structures by templating supramolecularly self-assembled structures in the shape of LLC phases has proven extremely successful. Therefore, the utilization of the more complex systems found in nature (e.g. protein assemblies in viruses) is at the hands of modern scientists and merely a question of time spent observing and learning from these spontaneously forming structures.

Acknowledgement. The author would like to thank the members of the Inorganic and Materials Section at the School of Chemistry, University of Bristol, for their support.

6 References

1. Collings PJ (1990) Liquid crystals, nature's delicate phase of matter. Princeton Science Library, Princeton, New Jersey
2. Göltner CG, Antonietti M (1997) Adv Mater 9:431
3. Kresge CT, Leonowicz ME, Roth WJ, Vartuli JC, Beck JS (1992) Nature 359:710
4. Vartuli JC, Kresge CT, Leonowicz ME, Chu AS, McCullen SB, Johnson ID, Sheppard EW (1994) Chem Mater 6:2070
5. Beck JS, Vartuli JC, Kennedy GJ, Kresge CT, Roth WJ, Schramm SE (1994) Chem Mater 6:1816
6. Beck JS, Vartuli JC, Roth WJ, Leonowicz ME, Kresge CT, Schmitt KD, Chu CTW, Olson DH, Sheppard EW, McCullen SB, Higgins JB, Schlenker JL (1992) J Am Chem Soc 114:10,834
7. Monnier A, Schüth F, Huo Q, Kumar D, Margolese D, Maxwell RS, Stucky GD, Krishnamurty M, Petroff P, Firouzi A, Janicke M, Chmelka BF (1993) Science 261
8. Huo Q, Margolese DI, Ciesla U, Demuth DG, Feng P, Gier TE, Sieger P, Firouzi A, Chmelka BF, Schüth F, Stucky GD (1994) Chem Mater 6:1176
9. Huo Q, Margolese DI, Ciesla U, Feng P, Gier TE, Sieger P, Leon R, Petroff PM, Schüth F, Stucky GD (1994) Nature 368:317
10. Ciesla U, Schüth F, (1999) Microporous Mesoporous Mater 27:131
11. Hüsing N, Schubert U (1998) Angew Chem 110:22
12. Ying JY, Mehnert CP, Wong MS (1999) Angew Chem 38:58
13. Antonietti M, Göltner C (1997) Angew Chem Int Ed Engl 36:910
14. Candau F (1992) In: Paleos CM (ed) Polymerization in organized media. Gordon Science, Philadelphia, p 215
15. Antonietti M, Göltner C, Hentze HP (1998) Langmuir 14:2670
16. Antonietti M, Caruso RA, Göltner CG, Weissenberger MC (1999) Macromolecules 32:1383
17. Brinker CJ, Scherer GW (1990) Sol-gel science: the physics and chemistry of sol-gel processing. Academic Press, San Diego New York Boston London Sydney Tokyo Toronto
18. Mitchell DJ, Tiddy GJ, Waring L, Bostock T, McDonald MP (1983) J Chem Soc Faraday Trans 79:975
19. Zana R (1997) Colloids Surf A Physicochem Eng Aspects 123:27
20. Attard GS, Glyde JC, Göltner CG (1995) Nature 378:366
21. Attard GS, Edgar M, Göltner CG (1998) Acta Mater. 46:751
22. Attard GS, Edgar M, Emsley JW, Göltner CG (1996) MRS Conf Proc 425:179
23. Attard GS, Edgar M, Emsley JW, Göltner CG (1996) MRS Conf Proc 425:185–189
24. Ammundsen B, Burns G, Jones DJ, Rozière J (1997) J Sol-Gel Sci Technol 8:331
25. Rozière J, Brandhorst M, Dutatre R, Jacquin M, Jones DJ, Vitse P, Zajac J (2001) J Mater Chem 11:3264
26. Göltner CG, Henke S, Weißenberger MC, Antonietti M (1998) Angew Chem Int Ed Engl 37:316
27. Zhao D, Feng J, Huo Q, Melosh N, Fredrickson GH, Chmelka BF, Stucky GD (1998) Science 279:548
28. Zhao D, Huo Q, Feng J, Chmelka BF, Stucky GD (1998) J Am Chem Soc 120:6042
29. Zhao D, Yang P, Melosh N, Feng J, Chmelka BF, Stucky GD (1998) Adv Mater 10:1380
30. Antonietti M, Berton B, Göltner C, Hentze HP (1998) Adv Mater 10:154
31. Göltner CG, Berton B, Krämer E, Antonietti M (1998) Chem Commun 2287
32. Göltner CG, Berton B, Krämer E, Antonietti M (1999) Adv Mater 11:395

33. Krämer E, Förster S, Göltner C, Antonietti M (1998) Langmuir 14:2027
34. Förster S, Krämer E (1999) Macromolecules 32:2783
35. Poly(styrene-*b*-ethylene oxide) (SEs) (for SE10/10 the average molecular weight is 1 kg/mol, for SE30/30, it is 3 kg/mol) of various block lengths and low-dispersity poly[(ethylene-*co*-butylene)-*b*-(ethylene oxide)]s (KLEs) are a product of Th Goldschmidt AG, Essen, Germany
36. Göltner CG, Berton B, Smarsly V, Antonietti M (2001) Chem Mater 13:1617
37. Smarsly B, Göltner C, Antonietti M, Ruland W, Hoinkis W (2001) J Phys Chem B 105:831
38. Hentze HP, Krämer E, Berton B, Förster S, Antonietti M, Dreja M (1999) Macromolecules
39. Melosh NA, Lupic P, Bates FS, Wudl F, Stucky GD, Fredrickson GH, Chmelka BF (1999) Macromolecules 32:4332
40. Feng P, Bu X, Pine DJ (2000) Langmuir 16:5304
41. Feng P, Bu X, Stucky GD, Pine DJ (2000) J Am Chem Soc 122:994
42. Mann S, Cölfen H, Göltner CG (unpublished results)
43. Templin M, Franck A, Du Chesne A, Leist H, Zhang Y, Ulrich R, Schädler V, Wiesner U (1997) Science 278:1795
44. De Paul SM, Zwanziger JW, Ulrich R, Wiesner U, Spiess HW (1999) J Am Chem Soc 121:5727
45. Ulrich R, Du Chesne A, Templin M, Wiesner U (1999) Adv Mater 11:141
46. Eric Broissière, Larbot E, Prouzet E, Bunton CA (2001) Chem Mater 13:3580
47. Impéror-Clerc M, Davidson P, Davidson A (2000) J Am Chem Soc 122:11,925
48. Polarz S, Smarsly B, Bronstein L, Antonietti M (2001) Angew Chem Int Ed Engl 40:4417
49. Braun PV, Osenar P, Stupp SI (1996) Nature 380:325
50. Braun PV, Stupp SI (1999) Mater Res Bull 34:463
51. Braun PV, Osenar P, Tohver V, Kennedy SB, Stupp SI (1999) J Am Chem Soc 121:7302
52. Attard GS, Göltner CG, Corker JM, Henke S, Templer RH (1997) Angew Chem Int Ed Engl 36:1315
53. Attard GS, Leclerc SAA, Maniguet S, Russel AE, Nandhakumar I, Gollas BR, Bartlett PN (2001) Microporous Mesoporous Mater 44:159
54. Attard GS, Leclerc SAA, Maniguet S, Russel AE, Nandhakumar I, Bartlett PN (2001) Chem Mater 13:1444
55. Leroux F, Koene BE, Nazar LF (1996) J Electrochem Soc 143:L181
56. Ye S, Vijh AK, Dao LH (1996) J Electrochem Soc 143:L7
57. Pell WG, Conway BE (1996) J Power Sources 63:255
58. Wang J, Agnes L (1992) Anal Chem 64:456
59. Attard GS, Bartlett PN, Coleman NRB, Elliott JM, Owen JR (1998) Langmuir 14:7340
60. Miller CA, Raney KH (1993) Colloids Surf A 74:169
61. Attard GS, Bartlett PN, Coleman NRB, Elliott JM, Owen JR, Wang JH (1997) Science 278:838
62. Elliott JM, Birkin PR, Bartlett PN, Attard GS (1999) Langmuir 15:2411
63. Elliott JM, Owen JR (2000) Phys Chem Chem Phys 2:5653
64. Nelson PA, Elliott JM, Attard GS, Owen JR (2002) Chem Mater 14:524
65. Nelson PA, Elliott JM, Attard GS, Owen JR (2002) J New Mater Electrochem Syst 5:63
66. Guerin S, Attard GS (2001) Electrochem Commun 3:544
67. Whitehead AH, Elliott JM, Owen JR (1999) J Power Sources 82:33
68. Whitehead AH, Elliott JM, Owen JR, Attard GS (1999) Chem Commun 4:331
69. Nandhakumar I, Elliott JM, Attard GS (2001) Chem Mater 13:3840
70. Attard GS, Elliott JM, Bartlett PN, Whitehead A, Owen JR (2000) Macromol Symp 156:179
71. Elliott JM, Attard GS, Bartlett PN, Owen JR, Ryan N, Singh G (1999) J New Mater Electrochem Syst 2:239

Top Curr Chem (2003) 266: 55–90
DOI 10.1007/b10824

Nanoparticles Made in Mesoporous Solids

Lyudmila M. Bronstein

Indiana University, Chemistry Department, Bloomington, IN 47405, USA
E-mail: lybronst@indiane.edu

This review with 98 references describes the synthesis of mesoporous materials containing nanoparticles, characterization of these materials and their material properties: catalytic, optical, and magnetic. All synthetic methods are grouped into seven categories depending on the way of metal compound incorporation and metal particle formation. Advantages and disadvantages of each method for each particular application are discussed. A short description of methods used for characterization of mesoporous materials with nanoparticles is presented including examples of their applications. Catalytic, optical, and magnetic properties of mesoporous solids containing nanoparticles are discussed.

Keywords. Mesoporous materials, Nanoparticles, Catalysis, Optical properties, Magnetic properties

List of Abbreviations

ASAXS	Anomalous small angle X-ray scattering
BET	Brunauer-Emmet-Teller
BJH	Barrett, Joyner, and Halenda
EXAFS	Extended X-ray absorption fine structure
FSM-41	Silica with cylindrical pores (similar to MCM-41)
HRSTEM	High resolution scanning transmission electron microscopy
HRTEM	High resolution transmission electron microscopy
MCM-41	Silica with hexagonal pore structure (Mobil recipe)
MCM-48	Silica with cubic pore structure (Mobil recipe)
MCM-50	Silica with lamellar pore structure (Mobil recipe)
SBA-15	Silica with hexagonal pore structure (Santa Barbara recipe)
STEM	Scanning transmission electron microscopy
TEM	Transmission electron microscopy
XAFS	X-ray absorption fine structure
XANES	X-ray absorption near-edge structure
XRD	X-ray diffraction

1 Introduction

After discovery of ordered mesoporous silica by Mobil scientists in 1992 [1, 2], the field of mesoporous materials developed vigorously in several directions. Originally, mesoporous silicas were classified into three groups depending on their structure: a hexagonal (MCM-41), a cubic (MCM-48), and a lamellar phase (MCM-50). Later, similar to MCM-41, FSM-41 mesoporous silica with cylindrical pores was synthesized [3]. Varying the template structure allowed developing a number of various ordered mesoporous material structures [4–7]. Numerous metal oxides with structures mimicking the structure of mesoporous silica have been synthesized [8–10]. Different kinds of functionalization have been applied to mesoporous solids including attaching organic ligands, organometallic compounds, and inorganic species. These results have been summarized in recent review articles [11, 12]. In the present review, functionalization of mesoporous solids with inorganic compounds followed by nanoparticle formation will be discussed: synthetic methods, characterization, and properties of mesoporous materials containing nanoparticles will be scrutinized and summarized. This kind of functionalization provides a vast variety of materials which draw a significant attention nowadays: each year dozens of scientific papers on this subject are published, encouraged by properties of such materials and their expected applications. Three main application areas are foreseen so far for mesoporous solids with nanoparticles: catalysis [13–15], magnetic [16, 17], and optical materials [12, 18, 19]. For catalytic applications, the most essential characteristics are an interpenetrating pore structure of the mesoporous solids and formation of small particles located within the pores. Such particles offer a huge surface area and superior catalytic properties in a variety of catalytic reactions. Yet their location in the pores results in their stabilization and modification of their properties (surface modification). For optical materials, particle size, particle size distribution, and regular particle location are crucial, so the ordered mesoporous oxides can also be perfect hosts for inclusion of optically interesting particles in 1D, 2D, and 3D structures. For magnetic applications, particle size and particle size distribution are most crucial parameters determining the type of magnetic properties. Considering these requirements, particular synthetic methods should be chosen when certain applications are targeted.

2 Methods of Synthesis

Seven general ways for inorganic functionalization of mesoporous materials have been described. Below, analysis of each method will be given taking into consideration its advantages and disadvantages from the viewpoint of particle growth control and influence on the material properties.

2.1 Addition of Metal Compounds in Sol-Gel Mixture

A very simple synthetic method is addition of inorganic compounds (metal salts or alkoxides) in the sol-gel mixture. In this way a modifying metal is introduced in the mesoporous material body (but not necessarily placed in the pores) and particles form during calcination [20–23]. Normally, to avoid negative influence of additive on the ordering of mesoporous oxides, metal compound loading is kept low. In [23] authors describe two regimes of thermal treatment for synthesis of rhodium oxide particles: 373 K for 10 days and 423 K for 48 h, while the Si/Rh value was varied within the range 200–70. Depending on the regime, the mean particle diameter was either 6 nm or below 3 nm, respectively. The authors showed that for both regimes most Rh atoms form particles and are located within the material (not on the outer surface as often occurs during impregnation), but no details have been given on precise location of nanoparticles: whether they are mainly embedded in the silica walls or positioned in the pores. Other authors [20] obtained Pd nanoparticles in the mesoporous silica by addition of H_2PdCl_4 in the sol-gel mixture. Despite the low Pd content, particle size after thermal treatment exceeds the pore size, which is evident for at least partial location of Pd species within the pore walls.

If mesoporous oxides with nanoparticles are designed for magnetic application, embedding in the mesoporous material walls can be of no concern, as it should not strongly affect the magnetic properties: they are mainly dependent on the particle size and particle size distribution. On the other hand, the latter should be narrow: otherwise particle properties will vary. However, narrow particle size distribution is difficult to achieve during uncontrolled particle growth within the silica body during calcination. If optical properties of the material are targeted, the interaction of particle surface with mesoporous oxide host will strongly influence the properties. Then the position of the nanoparticles determining the size of a contact area between a particle surface and mesoporous solid is also important. In the case of catalytic applications, embedding the nanoparticles in silica (or other mesoporous material) walls will kill a significant fraction of the nanoparticle activity, as particle surface will be partly obscured from reacting molecules. Thus, this method to prepare nanoparticles within mesoporous solids for catalytic applications has a number of disadvantages which make it inferior compared to other methods. On the other hand, this method works perfectly when silica or other oxide should be modified (doped) with suitable ions without nanoparticle formation [8, 13].

2.2 Impregnation with Metal Compounds Followed by Reduction, Thermolysis, or Other Treatment

Straightforward impregnation of preformed mesoporous solids with solutions of the desired metal compound followed by reduction, thermal decomposition, UV-irradiation, or ultrasonic treatment of metal-compound-containing material is very extensively used in preparation of mesoporous materials containing nanoparticles. Incorporation of metal compounds normally occurs due to capillary forces. Since no noticeable chemical interaction of the mesoporous oxide walls with the metal compounds or the forming metal (metal compound) particles is involved in this case, control over particle formation can be realized only due to limits imposed by the pore structure. As this is a case, conditions of impregnation and further treatment are crucial for structure of the materials synthesized. Speaking about "quality", several material characteristics should be considered: (i) preservation of mesoporous material ordered structure, (ii) uniformity of particle distribution within the pores, (iii) narrow particle size distribution, and (iv) match of the particle size to pore size. The typical impregnation of a mesoporous silica film with $[Ag(NH_3)_2NO_3]$ followed by reduction with H_2-N_2 mixture is described in [24]. As seen in Fig. 1, particle size distribution is rather bimodal which is determined by the particle formation both in the pores and in defects, such as locally coalesced pores or walls. In a similar way, Fe_2O_3 nanoparticles within mesoporous MCM-48 silica have been synthesized by wet impregnation with $Fe(NO_3)_2$ followed by drying and calcination at 673 K [25]. Thorough examination of the material has shown that iron oxide is randomly distributed within the mesoporous silica. The filling of the pores was also random, without any structural order. Though in most cases the mesoporous host stays intact after incorporation of metal salt and particle forma-

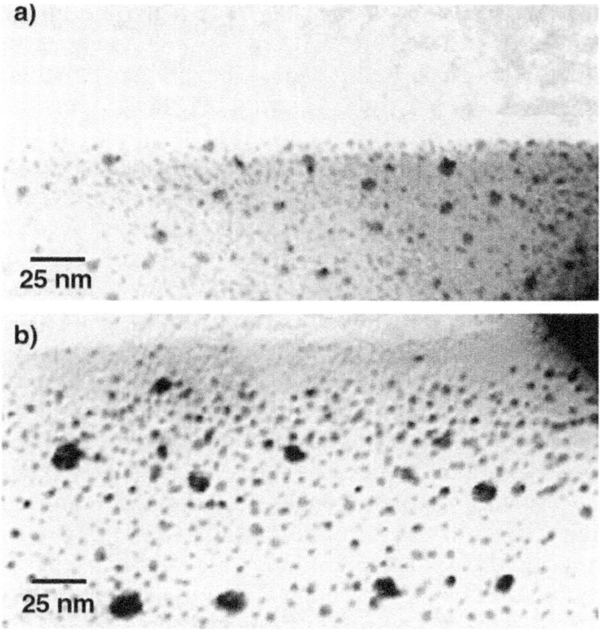

Fig. 1a,b. Transmission electron micrographs of the calcined mesoporous silica film observed in cross-section view after: **a** one; **b** three contacts with an aqueous solution of $[Ag(NH_3)_2]NO_3$ and subsequent reductive treatment at 400 °C [24]. Reproduced by permission of The Royal Society of Chemistry

tion, the material is disordered from the viewpoint of guest location and its parameters. The situation slightly improves if the mesoporous solid is a monolith [26]; then the external surface is low and the negative influence of free outer surface on particle formation is minimized. If particle formation occurs due to UV-irradiation [27] or ultrasonication [28–30] of the mesoporous slurry in the mother solution containing metal salt, the probability of uncontrolled particle precipitation on the outer surface is high. Though some authors [28–30] do not report broad particle size distribution showing only TEM or HRTEM (which characterizes a small area of the material), the lack of SAXS or ASAXS studies do not allow judging of the particle size distribution in these mesoporous materials.

To improve control over nanoparticle formation while using wet impregnation, several authors used consecutive impregnations (four to five times) with drying the material in between to insure that mesopores are completely filled with metal precursor [31, 32]. In this case, the amount of precursor is fixed and determined by the pore size and volume; the recipe is well reproducible as the final particle size is controlled by the precursor amount and in some cases by the pore size. This avenue led to a 3D quantum dot lattice inside the mesoporous silica film due to regular CdS particle formation, when impregnation was followed by precipitation with H_2S and the procedure was repeated until the film saturation was obtained [32]. It is remarkable that repetitive impregnation with tetramineplatinum(II) nitrate $[Pt(NH_3)_4(NO_3)_2]$ followed by drying-reduction cycles (four times) allowed the authors [31,33] to prepare Pt nanowires in mesoporous oxides SBA-15 and MCM-41. As found in [34,35], depending on the type of metal compound precursor and conditions of its incorporation into the silica and further transformation, either nanoparticles or nanowires can be synthesized even after a single wet impregnation. However, it is not yet entirely understood when nanowires and when nanoparticles are formed. As discussed in [34], if metal particle precursor is a prepared in situ ("ship-in-bottle") metal carbonyl cluster (for example, $[Pt_{15}(CO)_{30}]^{2-}$) in FSM-16 having cylindrical pores, metal nanoparticles with a diameter of 1.5 nm (50–60 Pt atoms) are formed after mild thermolysis (at 473 K). If FSM-16 is filled with platinic acid (H_2PtCl_6), subjected to water and 2-PrOH vapor (see Scheme 1) and then irradiated by UV, Pt nanowires are formed within the silica pores. If reduction of the same precursor is performed with H_2 at 673 K, nanoparticles are formed. Presumably, in the former case, slow nucleation at ambient temperature under UV light and presence of water and 2-propanol vapors facilitate slow nucleation and migration of Pt species providing the nanowire formation. Similarly, Pt-Rh nanowires were synthesized. It is proposed [36] that Pt nanowires are formed by the photoreduction of $[PtCl_6]^{2-}$ ions on the small Pt nanoparticles initially generated in the FSM-16 mesopores. The authors [35] found that, in order to achieve complete filling of the mesopores within SBA-15, selection of the impregnation solution is crucial. For example, if 0.2 mol/l $AgNO_3$ in EtOH-H_2O (1:1 v/v) was used, nanowires were formed (Fig. 2). If the same salt was dissolved in pure water or EtOH-H_2O (3:7 or 8:2 v/v), only spherical nanoparticles were obtained after thermal treatment. Yet the channel filling was low. Here the authors assume that it can be related to very different surface tensions of water and ethanol, so their right combination is a critical factor.

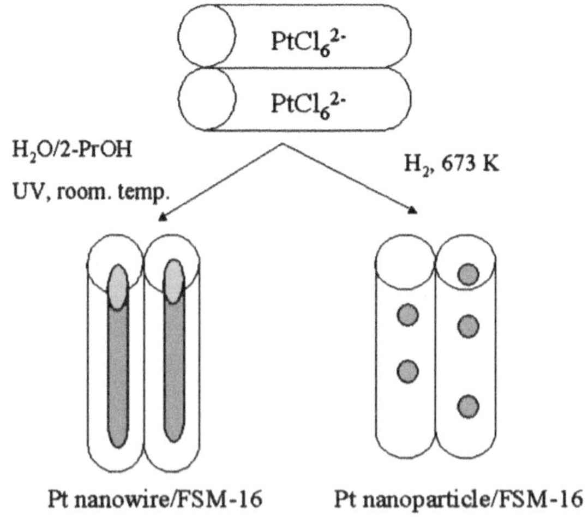

Scheme 1. Schematic representation of preparation of Pt nanowires and nanoparticles in FSM-16

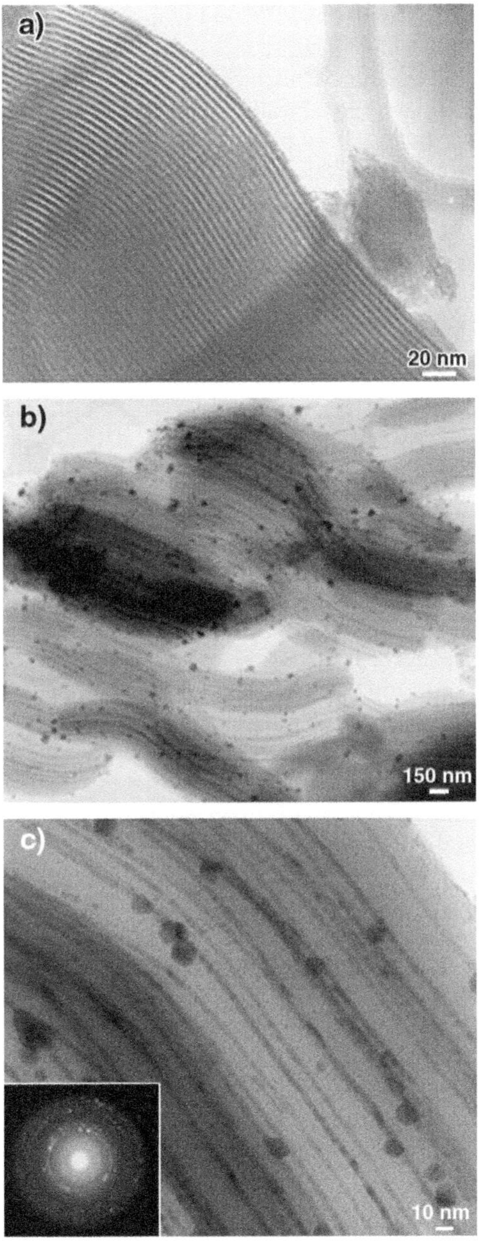

Fig. 2. **a** TEM (Topcon 002B, 200 KeV) image of the as-made SBA-15 mesoporous silica recorded along the [110] zone axis. **b** Low magnification TEM image of SBA-15 mesoporous silica after the Ag nanowire loading; Ag nanowires appear as *dark lines* in the image. **c** High-magnification TEM image of the Ag nanowires within the SBA-15. A selected area electron diffraction pattern recorded on several Ag nanowires is shown in the *inset* of c [35]. Reproduced by permission of The Royal Society of Chemistry

Impregnation of mesoporous silica with metal carbonyl clusters instead of simple metal salts leads to superior control over nanoparticle formation and can even result in ordered location of such carbonyl clusters. In [37] it was shown that encapsulated carbonylate cluster salts preserve their integrity within MCM-41 (no aggregation is found). Nanoparticles obtained by gentle thermolysis of these clusters may be attached to the interior surface of mesoporous silica. Moreover, such carbonylate salts as $[Ag_3Ru_{10}C_2(CO)_{28}C][PPN]_2$ and $[Ru_{12}C_2Cu_4(CO)_{32}Cl_2][PPN]_2$ (where PPN stands for $^+N(PPh_3)_2$) allow the obtaining of bimetallic particles the sizes of which are determined by the amount of atoms in each carbonylate cluster. This becomes possible if the interaction of the clusters with the pore surface is stronger than that between the precursor species, so that aggregation into small molecular crystallites on the surface is suppressed and the clusters after CO removal do not aggregate. The authors [37] suppose that the silanol groups belonging to the silica walls facilitate the uptake of bulky cluster anions due to hydrogen bonding. Thus, this case may rather be considered as chemical interaction with mesoporous solid walls, not as impregnation, assuming (as will be shown below) that the latter method allows higher degree of control. For two Ru salts, $[Ru_6C(CO)_{16}][PPN]_2$ and $[H_2Ru_{10}(CO)_{25}][PPN]_2$, formation of 1D ordering within the silica channels has been observed [38]. The authors believe that intrachannel ordering of the guest ions is mainly governed by a combination of molecular recognition, pore-wall interaction, and electrostatic considerations. After thermolysis monodisperse particles are formed within MCM-41.

Normally, impregnation provides the efficient incorporation of metal compounds inside the pores, but the particle growth is not controlled (particle size distribution is broad and particles are located statistically) if no special conditions are applied (see above). On the other hand, particle size is often restricted by the pore size. If silanol groups of the mesoporous silica are involved in the interaction with metal compounds as discussed in the last example, additional control can be expected although the amount of silanol groups in the calcined silica is rather low.

A very special example of successful impregnation of mesoporous material was recently reported in [39] when mesoporous carbon was used as support material. In turn, mesoporous carbon was prepared by templating over ordered mesoporous silica followed by removal of the template. As a result, an ordered graphitic framework was obtained. It is interesting to note that, despite the apparent absence of functional groups, the nanostructured carbon is able to stabilize Pt nanoparticles as small as 1 nm in diameter preventing formation of large particles outside the pores or in the pore defects. One can assume that surface of such carbon is not inert to Pt species when the impregnation takes place. Graphitic aromatic rings or unsaturated carbon-carbon bonds can play the role of "functional" groups both interacting with Pt species and Pt nanoparticles. A similar phenomenon was described in [40] when Pt nanoparticles with a mean diameter of 1.2 nm were stabilized in hypercrosslinked polystyrene. The Pt-nanocarbon materials reported in [39] can also be prepared as free-standing films by using ordered silica films as templates. This and also the enhanced electrocatalytic activity of Pt nanoparticles/mesoporous carbon in O_2 reduction makes it promising in fuel cell systems.

2.3 Chemical Interaction of Metal Compounds with Functional Groups of Mesoporous Oxides

If the mesoporous oxide walls are functionalized, the chemical interaction with metal compounds is a strong driving force for the incorporation of metal species into the pores. Functionalization with organic groups was widely explored by many authors [12]. It can be carried out both during sol-gel reaction when one of the silica precursors bears such groups [12] or as a post-synthesis via interaction of various compounds with silanol groups [41–43]. If the removal from organic template of the mesoporous solid is carried out in mild conditions (extraction instead of calcination), silanol groups on the pore walls are mainly preserved which insures successful functionalization of silica walls [44–46].

As functional groups for further interaction with inorganic compounds, thiol [47, 48] and amine [49] ones are mostly used. A functional group can play a dual role (i) being an anchor for metal compounds and/or (ii) interacting with the surface of the growing nanoparticle. In [47] the authors described the formation of CdS particles in mesoporous silica (MCM-41) functionalized with thiol groups. Despite the functional group presence, incorporation of cadmium acetate was carried out by mere impregnation. At the same time, these thiol groups served as anchors for the forming of CdS particles: the size of the CdS particles decreased with increasing a number of thiol groups, i.e., the thiol groups interacted with the particle surface stabilizing the smaller particles. Similar stabilization of ZnS nanoparticles was carried out by diamine groups present in MCM-41 [49]. The authors of [48] explored both functions. Mesoporous SBA-15 functionalized with thiol substituents hosted Au nanoparticle growth, first anchoring the gold precursor and second governing the chemical reduction. This procedure allowed avoiding the outer pore growth of nanoparticles that is a major drawback of the impregnation/calcination method and provided particle formation solely in the pores.

To replace less acidic silanol groups with more acidic Al-OH, walls were modified with sodium aluminate [17]. Subsequent ion exchange using $NiBr_2$ resulted in incorporation of Ni^{2+} ions into the AlMCM-41 host material. To remove the extra-pore Ni^{2+} ions from the host material, the resulting Ni^{2+}-AlMCM-41 was suspended in water for 1 h. If simple impregnation would be used for incorporation of metal precursor, no rinsing would be possible as such a procedure would remove metal compound from the mesoporous material. After reduction with $NaBH_4$, nickel nanoparticles with diameters of 1–2 nm were prepared.

In several cases silanol groups of silica walls can be used for direct functionalization with inorganic compounds followed by the corresponding treatment (reduction, calcination, interaction with H_2S, etc.) [12, 14, 50]. MCM-48 mesoporous silica was functionalized with tungsten and molybdenum metal centers by the anhydrous reaction of metal alkoxides with surface silanol groups [14]. Calcination resulted in metal oxide clusters which were attached via covalent M-O-Si bonds. When silica monolith vs silica powder is used, one can avoid negative influence of the outer silica surface. However, in this case a strong driving force for incorporation of metal compound inside the silica is required. If the or-

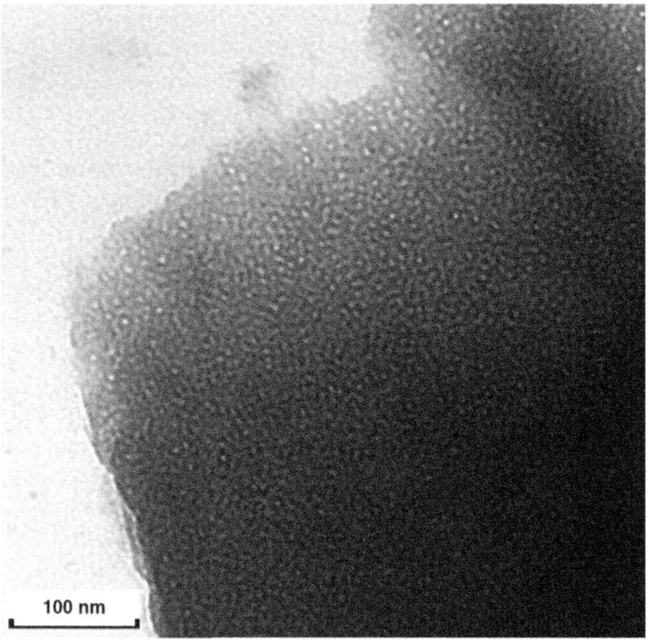

Fig. 3. TEM image of Pd-nanoparticle-containing silica monolith. Reproduced by permission of Wiley-VCH from [50]

Table 1. Characteristics of metal-free and metal containing silica monoliths ([50]-Reproduced by permission of Wiley-VCH)

Monolith	Elemental analysis data on metal		BET surface area	App. pore diameter[a]
	wt%	mol.%	m^2/g	nm
Metal-free	–	–	745	3.8
K_2PdCl_4	0.91	8.58×10^{-5}	705	3.8
$+H_2$ reduction			439	3.7
K_2PtCl_4	0.51	2.61×10^{-5}	698	3.6
$+H_2$ reduction			626	3.9
K_2PtCl_6	0.39	2.00×10^{-5}	709	3.6
$+H_2$ reduction			665	3.8

[a] From BJH data.

ganic template is removed from the silica monolith by extraction, silanol groups are mainly preserved which results in higher density of silanol groups on the silica walls. Interaction of Pt and Pd anions ($PdCl_4^{2-}$, $PtCl_4^{2-}$, $PtCl_6^{2-}$) with silanol groups (presumably via hydrogen bonding) allows efficient incorporation of Pd and Pt species evenly distributed through silica monolith [50]. Reduction with H_2 or sodium borohydride results in the regular location of sub-nanometer metal particles within the silica pores (Fig. 3). Because the metal particle size is much smaller than the pore size (Table 1), metal particle formation does not influence the mesopores structure, while the micropores (induced by the PEO chains) can be blocked during reduction. Here the metal particle size is not influenced by the type of reducing agent which is evidence of a very efficient mechanism of size restriction. Presumably the metal nanoparticles are forced to nucleate in the micropore entries, but restricted to grow beyond their cavity. This method is very simple and robust and allows synthesizing of well-reproducible materials.

2.4
Chemical Vapor Deposition of Metal Compounds for Incorporation into Mesoporous Oxides

Chemical vapor deposition (CVD) instead of impregnation seemed to be the better way to control the nanoparticle growth and particle spreading, but should be restricted only to thin silica films or silica particles (not to monolithic samples) to prevent uneven distribution of the metal compounds within the material. Another limitation of CVD is that to prevent the formation of too large or uneven particles, some additional restrictions should be applied. The authors of [51] synthesized ~1 nm diameter photoluminescent silicon clusters in the surfactant-containing channels of free-standing mesoporous silica film using disilane CVD at 100–140 °C. The hydrophobic tails of the swollen surfactant enable disilane adsorption and interaction with SiOH groups. The formation of Pd nanoparticles using sublimation of the volatile organometallic complex [Pd(η-C_5H_5)(η^3-C_3H_5)] under reduced pressure through the porous Nb-MCM-41 (niobium modified MCM-41) is described in [52]. After the palladium complex had been deposited onto the walls, the material was reduced under a stream of hydrogen at 350 °C for 3 h, yielding an air-stable black powder. Following the procedure described above, materials with typical palladium loadings of 20–25 wt% were obtained. However, this method did not lead to uniform distribution of palladium species. To decrease loading and to increase the metal surface due to obtaining the small particles, chemical modification of the pore surface was still necessary [52]. This additional restriction allowed obtaining sub-nanometer palladium particles evenly distributed through the silica.

Iron oxide nanoclusters were synthesized within mesoporous MCM-41 doped with aluminum using evaporation-condensation of volatile $Fe(CO)_5$ [53]. Subsequent calcination in an O_2 flow resulted in amorphous γ-Fe_2O_3 particles with diameters of 2–3 nm evenly distributed through the well-defined hexagonally packed cylindrical pores. These results were solidly confirmed by combination of Mössbauer spectroscopy, XRD, TEM, and STEM (Fig. 4).

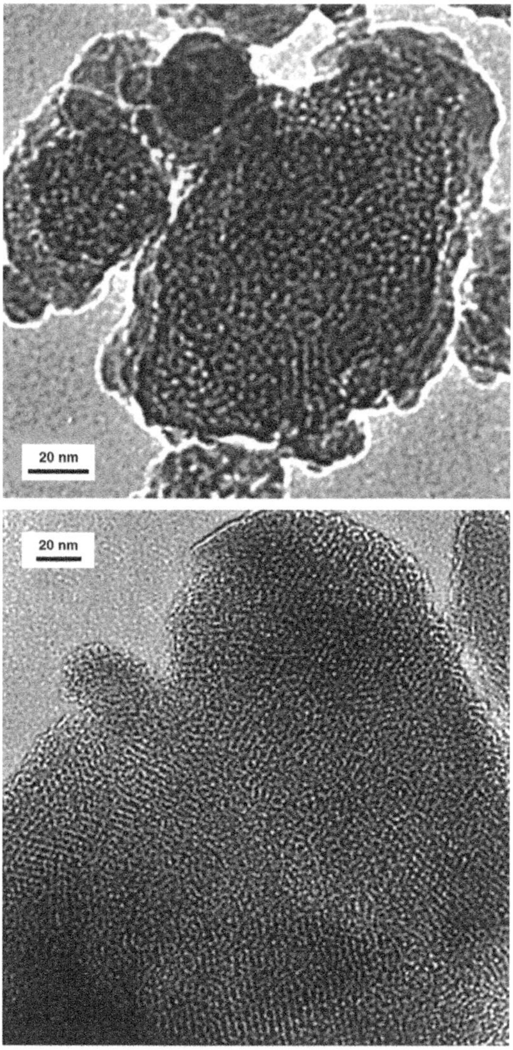

a

b

Fig. 4a,b. TEM micrographs of Fe_2O_3/AlSi-25 nanocomposite: **a** as-prepared; **b** calcined at 500 °C. Reprinted with permission from [53]. Copyright (2001) American Chemical Society

GaAs/MCM-41 heterostructures were synthesized by deposition of GaAs into the channels of MCM-41 using organometallic chemical vapor deposition with *tert*-butylarsine and trimethylgallium as volatile sources [54]. As found, GaAs nanoparticles reside both inside the pores where particle size distribution is narrow and outside the pores where particle growth was rather uncontrolled. To decrease the amount of particles growing outside the pores, the deposition time was varied. This led only to partial success and some particles grew outside the pores. Apparently, the lack of specific interactions between silica pore walls and volatile compounds did not allow circumventing the particle growth outside the pores.

An original synthetic method for functionalization of mesoporous silica is described in [55]. Silaferrocenophanes were incorporated into well-ordered mesoporous silica MCM-41 by CVD and chemically attached to the silica walls via SiOH groups. Subsequent sublimation allowed removing the monomer which was not chemically attached to the silica walls. Depending on the silaferrocenophane structure, both oligomer and polymer can form spontaneously along with ring-opening the monomer during residing in the silica pores. Thus, silica pores play role of nanoreactors already during deposition. Subsequent pyrolysis at 900 °C in nitrogen atmosphere leads to Fe particles which reside inside the silica channels. However, as for a number of materials prepared by this technique, particles grow larger than the pore diameter (ca. 3–4 nm) and the exact particle location and particle size distribution remain unclear. One can surmise that high temperature treatment (900 °C) leads to aggregation and particle location in the interpore defects. Additional studies with SAXS and ASAXS methods might allow the better characterization of these materials.

2.5
Template Ion Exchange with Transition Metal Cations

Another interesting method of inorganic functionalization of mesoporous solids is the surfactant replacement by using transition metal cations [56–59]. In this case, metal cations (Mn^{2+} [56], $[Co(en)_3]^{3+}$ [57], Cd^{2+} [58], Zn^{2+}, Mn^{2+}, Al^{3+}, etc. [59]) line the interior pore surface replacing the cationic surfactant cation. The ion-exchange reaction is normally driven by replacement of monovalent cation (surfactant) with divalent (or trivalent) metal cations, and thus entropy is responsible for an efficient ion exchange. A unique feature of this ion-exchange technique is that the metal ions are transported solely inside the pores, while the external surface can be capped with inert hydrophobic phenyl groups (Scheme 2) [58]. This ensures CdS particle formation (after treatment with H_2S)

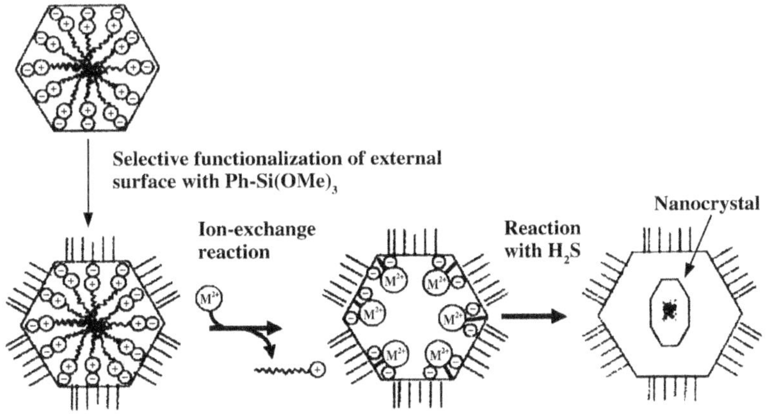

Scheme 2. Schematic diagram of transportation of metal ions into mesopores and space confined growth of nanocrystals. Reprinted with permission from [58]. Copyright (2001) American Chemical Society

takes place only inside the pores. This method seems to be superior to direct impregnation or CVD as no precipitation on outer surface takes place (if outer surface is capped), but its use was very limited (only a few papers have been published so far).

2.6 Incorporation of Prefabricated Nanoparticles in Mesoporous Solids

Apparent advantage of the prefabricated nanoparticles lies in the opportunity to use well-developed procedures to control particle size and particle size distribution and to characterize the particles by all possible means. Prefabricated particles can be introduced in the mesoporous materials in two ways. One avenue is direct incorporation of particles in sol-gel mixture as was recently reported in [60]. Here mesostructured nanoparticle-silica monoliths have been synthesized by dispersing prefabricated Au or zeolite (silicate) nanoparticles in sol-gel precursors containing $SiCl_4$ and a Pluronic triblock copolymer template. To improve the compatibility of the Au nanoparticles with the poly(ethylene oxide) block, the nanoparticle surface was modified with a SiO_2 layer. Therefore, this technique requires good compatibility of nanoparticles with the components of the sol-gel reaction, as it influences the distribution of nanoparticles through the material. Another limitation of this method is encapsulation of particles within mesoporous material, but not necessary location in the pores. As was discussed above, this limitation is especially crucial for catalytic applications.

A different approach is described in [61–63]. CdS [62, 63] or Au (Ag) [61] nanoparticles were synthesized in reverse micelles and then incorporated into mesoporous silica (MCM-41 and MCM-48) modified with thiol groups (3-mercaptopropyltrimethoxysilane was used as one of the silica precursors). The authors [62] observed a sieving effect, i.e. discrimination by particle size: (i) only particles of certain size (smaller than pore diameter) could be incorporated and (ii) the smaller the particles, the easier their incorporation inside the pores occurred. In [61] the authors compared the uptake of gold particles (size 35 Å) with silver particles (70 Å), but no correct conclusion on sieving effect can be made. In the cited papers no evidence of even distribution of nanoparticles within the mesoporous silica was presented, though this characteristic should be important from the viewpoint of optical properties. In [63] the authors used an additional CVD treatment with TMOS and water which were supplied alternately with N_2 carrier gas at 423 K. This allowed decreasing the mesopores size and stabilizing the CdS nanoparticles against heat treatment; however, this procedure is expected to obscure the particle surface so the photocatalytic activity (discussed in the paper) should suffer.

2.7 Templating over Metal-Containing Templates

This approach was first described in [64, 65]. When metal nanoparticles are located in the block copolymer micelle cores [64] or in microgels [65] and these polymeric systems are used as templates for silica casting, both pore size and metal particle growth control can be governed by a template. In [64] nanoparticle formation was carried out in the micelle cores of poly(ethylene oxide)-*block*-polybutadiene (PEO-*b*-PB) micelles in aqueous solutions. Complexation of metal precursor (Pt, Pd, Rh) with the double bonds within the PB core (π-complex formation) followed by metal nanoparticle growth after reduction was controlled by reaction conditions. Silica templating over metal-nanoparticle-containing block copolymer aggregates results in an even distribution of particles within the silica. After standard calcination at 500 °C the particles remain intact and no aggregation is observed (Fig. 5).

A similar approach described in [65] combines the use of spherical microgels as nanosized exotemplates for synthesis of metal colloids [66] and as endotemplates during casting the mesoporous sol-gel silicas, along with a block copolymer (polystyrene-*block*-poly(ethylene oxide), PS-*b*-PEO) template. After calcination the latter provides the interpenetrating mesopores structure and good access to macropores formed due to microgels (Fig. 6). This method does not ensure the regular location of metal nanoparticles through the mesoporous solid, but this disadvantage is of no importance for catalytic application. As seen in Fig. 7, particles remain discrete after calcination and access to nanoparticles surface is well ensured by the mesopore system. Using metal-particle-containing templates seem to be promising for robust control over nanoparticles growth in mesoporous solid.

Cupric oxide (CuO) nanoparticles have been prepared in amorphous SiO_2 matrix using a complex of Cu(II) with poly(vinyl alcohol) (CuPVA) as template [67]. The authors report the formation of CuO particles within the porous silica after calcination. In this case, the template was not ordered and was well compatible with forming silica, so the material formed must be fully disordered. Moreover, copper oxide particles can be present both in the pores and in the silica body; the latter will obscure particle surface from participation in catalytic reactions which authors consider being a suitable application.

Recently, dendrimers loaded with Cu^{2+} ions were used as template for silica formation [68]. As is well known, poly (propylene)imine dendrimer (DAB-Am-64) serves as a nanoreactor for well controlled nanoparticle formation [69]. However, when such dendrimers alone are used as silica templates [68], it does not lead to high porosity (BET surface area is 290 m^2/g) and does not provide ordering (Fig. 8). The latter factor is important for optical properties, but of no concern for catalytic applications. However, as silica cast over Cu-DAB-Am-64 lacks interpenetrating pore system and easy access to CuO particles, this destroys any hopes for catalytic applications as well.

The mentioned disadvantages seem to be overcome in cyclodextrin-based porous silica [70]. As recently discovered [71], cyclodextrins are able to self-assemble in concentrated solutions forming regular structures. When used as templates for silica casting, they form interpenetrating "worm-type" pore structure resembling one obtained after block copolymer templating. If hydrophobic inclusion compounds (*p*-xylene) are incorporated within the cyclodextrin cavities, the ability of cyclodextrin-inclusion complexes to self-assemble is even enhanced. An analogous idea was used in [70] when *p*-xylene was replaced with metal complexes containing cyclooctadiene (cod) ligand: $Pt(cod)Cl_2$, $Pd(cod)Cl_2$,

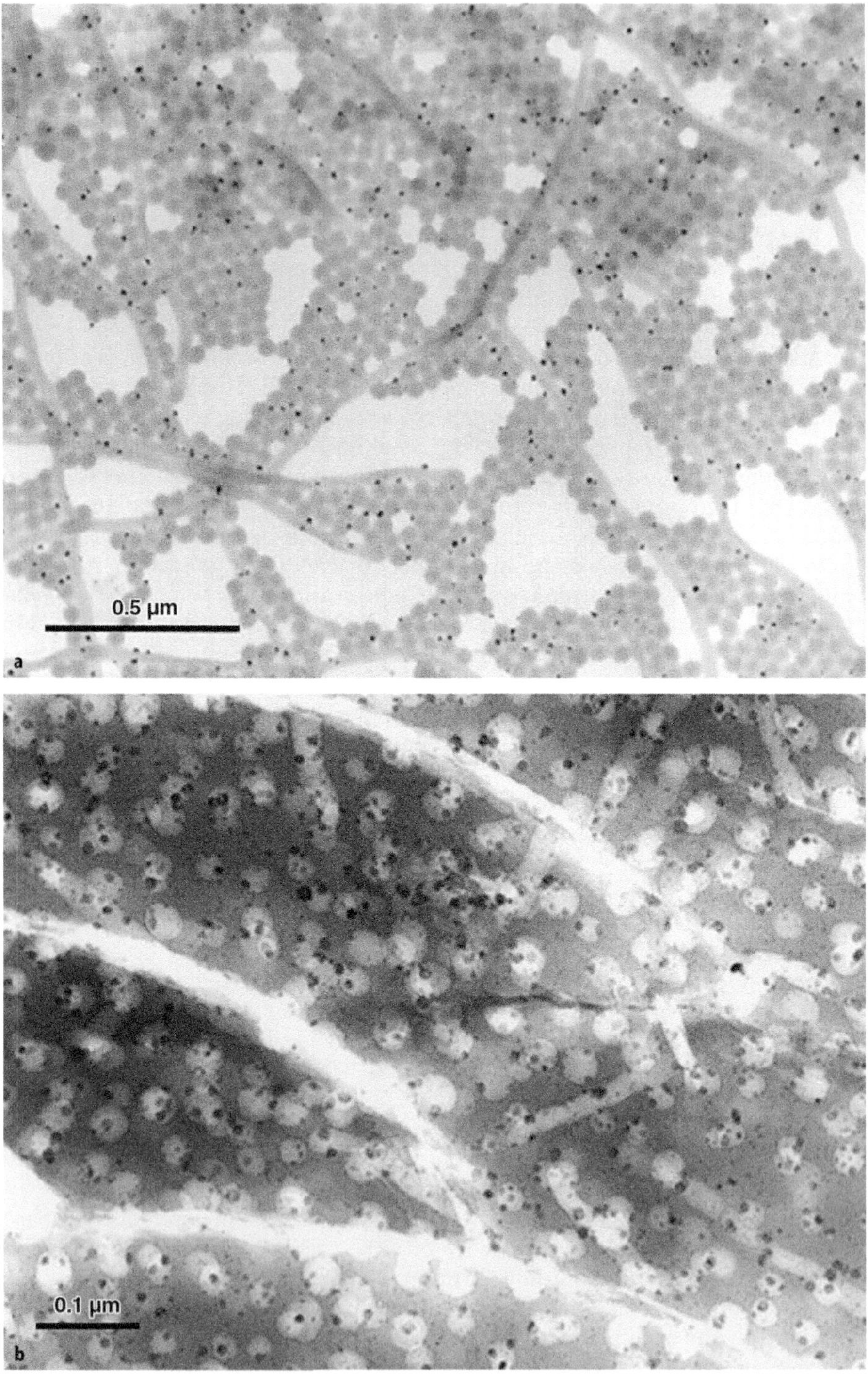

Fig. 5a,b. TEM images of: **a** the Pd colloids formed in the PB-*b*-PEO-I block copolymer micelles; **b** mesoporous silica obtained from PB-*b*-PEO-I-Pd(0) as the template after calcination. Reprinted with permission from [64]. Copyright (1999) American Chemical Society

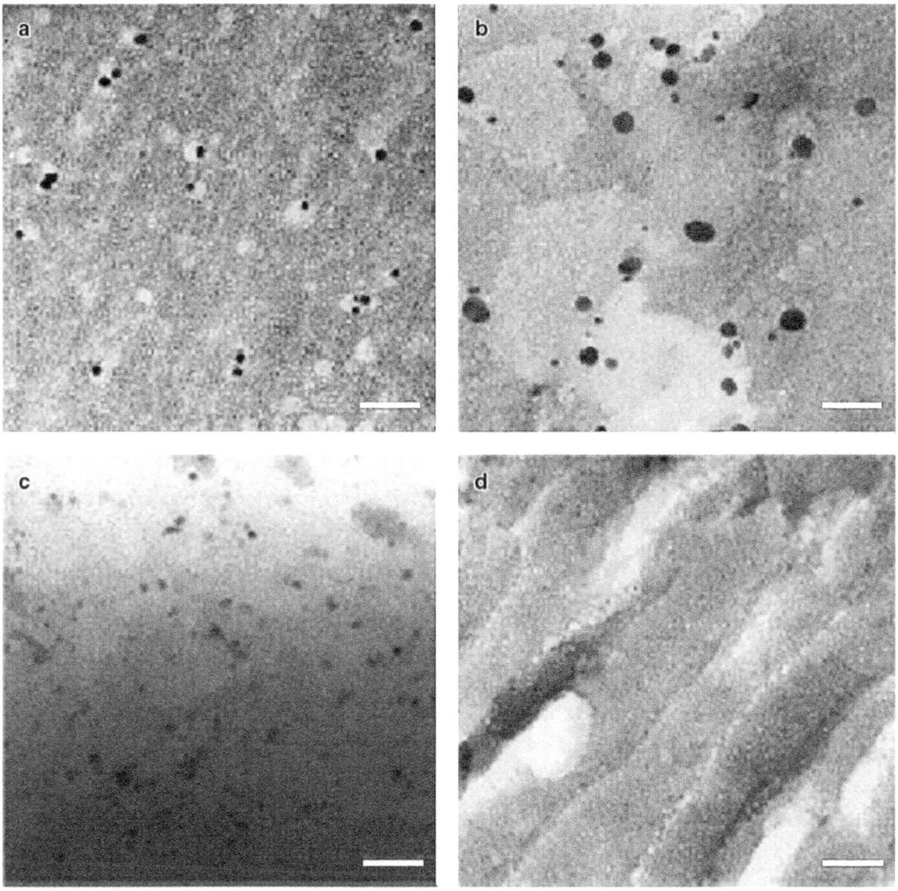

Fig. 6a–d. Transmission electron micrographs of porous silicas templated from mixtures of tetramethoxysilane, SE1010, and the respective metal microgel dispersions: **a,b** silicas obtained with platinum-loaded polymer microgels generated following reductions with $NaBH_4$ in 0.1 mol/l NaOH solution at resultant pH values of 8.0 and 10.0 respectively; **c,d** silicas obtained for platinum- and rhodium-loaded microgels, respectively; both generated under very fast reduction with an aqueous solution of $NaBH_4$ at the natural pH of the reaction mixtures. In each case the polymer microgel had a crosslinking density of 1/40. *Scale bar*=100 nm. Reproduced by permission of Wiley-VCH from [65]

$[Rh(cod)Cl]_2$. The authors showed that if the silica structure is preformed (walls are well-condensed), metal nanoparticles formed after calcination are restricted by the pore size and reside inside the pores. If the silica structure is "loose", calcination results in particles overgrowing the pores: this is caused by aggregation of the growing particles at high temperature calcination (Fig. 9). One might suggest that formation of particles prior calcination by H_2 reduction in mild conditions, would ensure the particle stability during calcination as was observed in [64].

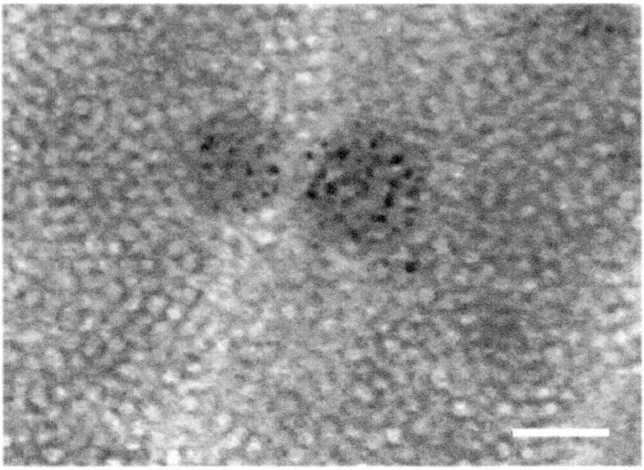

Fig. 7. Transmission electron micrograph of porous silica templated from a mixture of platinum-loaded microgels, SE1010, and tetramethoxysilane. Micrograph illustrates the retention of discreet platinum colloids following silica formation and calcination under elevated temperatures. *Scale bar*=50 nm. Reproduced by permission of Wiley-VCH from [65]

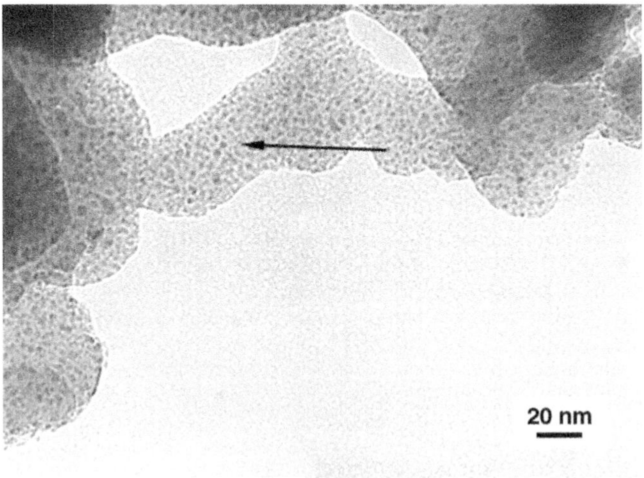

Fig. 8. TEM image of calcined silica templated over DAB-Am-*n* dendrimers with Cu^{2+} ions. Reprinted with permission from [68]. Copyright (2002) American Chemical Society

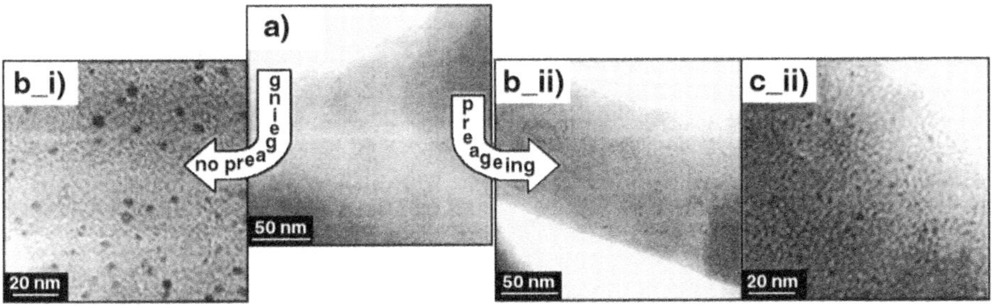

Fig. 9a–c. Evolution of Pd metal nanoparticles under different conditions for Pd(cod)Cl_2: **a** TEM image of the first nucleation at a final temperature T_e=250 °C of the in situ transformation process; **b,c** behavior for the nanoparticles in a sample not pretreated at T_e=350 °C (b_i) and the pre-aged sample (chemical confinement) at T_e=350 °C (b_ii) and T_e=500 °C (c_ii). Reprinted with permission from [70]. Copyright (2001) American Chemical Society

As can be seen from the data presented above, a variety of available methods allow synthesis of nanoparticles in mesoporous solids using various approaches. One should choose the method of synthesis taking into consideration the requirements towards particle characteristics generally determined by the possible application of such materials.

3 Characterization of Nanoparticles in Mesoporous Materials

Characterization of these materials normally includes both analysis of the mesoporous structure and assessment of nanoparticle parameters. It can be done by combination of methods used for mesoporous solid study (TEM, N_2 BET adsorption, SAXS, etc.) and additional methods allowing estimation of particle size, particle size distribution, particle positioning, and interface interactions.

3.1 Electron Microscopy

3.1.1 *Use of TEM and STEM*

Transmission electron microscopy (with magnification of up to 100,000–150,000) is most commonly used in study of mesoporous solids with or without guest particles. This method enables visualization of the mesoporous material structure (kind of packing), description of the positioning of metal particles, estimation of their size, and size distribution after processing the TEM images with designated software. If the electron contrast between mesoporous solid (normally metal oxide) and nanoparticles is strong (for example, for metal nanoparticles), information obtained from TEM images can well be descriptive and even quantitative. When metal compound particles with weak electron contrast (sulfides or similar) are small, they often cannot be distinguished in TEM image. In this case, TEM images can be used to find out whether mesoporous structure is intact or not, and what kind of changes (for example, of the pore size or wall size) occurred.

One of the key advantages of STEM compared to TEM is that the beam damage is limited to scanned area since only part of the image is exposed to electrons. Other advantages are a high resolution of particle imaging, possibility of using sub-nanometer probe and high angle annular dark field (Z-contrast) imaging. According to [72], the latter allows analysis of the distribution and homogeneity of the particles within channels more readily than any other method. In [72] examples of the STEM application are shown for Ag-Ru and Cu-Ru bimetallic particles in MCM-41. When STEM is combined with EDAX (energy-dispersive analysis of X-rays) [73], elemental analysis through the image allows locating the nanoparticles and evaluation of their chemical nature.

3.1.2
Use of HRTEM and HRSTEM

High Resolution TEM allows one to obtain more detailed information on the guest nanoparticles, since the structure of the metal particles can be elucidated [74, 75]. HRTEM, under optimal conditions, allows the locating and identifying of metal clusters in mesoporous solids when their diameters are not less than 10–15 Å. It gives information on both particle size and particle location in the ordered mesoporous solids.

On the other hand, a number of authors claim superiority of HRSTEM which using subnanometric probes provides supplementary structural and ultramicro analytical information and electron spectroscopic imaging. HRSTEM images have slightly diminished resolution compared to HRTEM, but two unique features make it really a method of choice: microanalytical capabilities of X-ray emission and Auger and electron-energy loss spectroscopy. In addition, all these signals may be collected simultaneously, together with the backscattered electrons. Since HRSTEM readily yields high-angle annular dark-field (HAADF) images, as well as both bright-field (BF) images formed from transmitted electrons, and ordinary dark-field (DF) images, formed from Bragg scattered beams, extra benefits can be gained in locating particles consisting of atoms. Chemical analyses using electron probe sizes of a few nanometers or less are now possible in modern HRTEM/STEM, and they yield chemical composition at high spatial resolution. A good example is shown in Fig. 10 where mapping the Pd and Rh species in the material gives information on location of bimetallic Pd_6Ru_6 clusters.

3.2
Nitrogen Adsorption

Nitrogen BET (Brunauer-Emmet-Teller) adsorption is a reliable method for determining the specific surface and pore volume [76]. Evaluation of the specific surface area using the BET method is based on several assumptions that deviate from the behavior of adsorbed molecules on the real surfaces. Despite the cer-

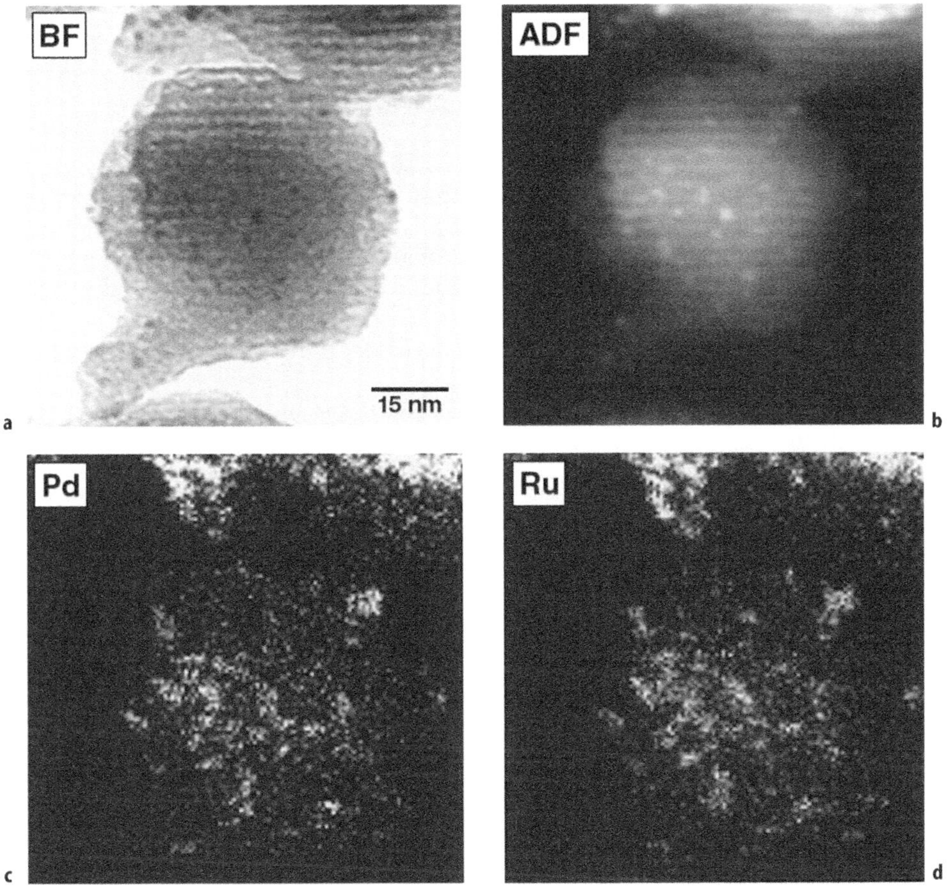

Fig. 10. **a** Bright-field (BF) images of mesoporous silica with Pd_6Ru_6 particles. **b** HAADF (high-angle annular dark field) images of mesoporous silica with Pd_6Ru_6 particles. **c,d** Electron-stimulated X-ray emission images prove that particles are intact. Reprinted with permission from [74]. Copyright (2001) American Chemical Society

tain limitations, this method is a standard and commonly used technique for specific surface evaluation. Calculations of a mean pore diameter of mesoporous materials are commonly done on the base of the BJH-approach (Barrett, Joyner, and Halenda) [77] and some other methods [76] which are based on an adsorption-desorption process. Adsorbed N_2 molecules in the mesopores of a given size are considered as multilayer adsorption followed by capillary condensation (filling the pore core) at a relative pressure determined by pore diameter. Desorption is presented as capillary evaporation (emptying the pore core with retention of the multilayer foil) at a relative pressure related to the pore diameter, followed by thinning the multilayer. These calculations are based on the Kelvin equation and underestimate the mean pore diameter of the small mesopores (in the magnitude of at least 1 nm) [78]. Additional theoretical calculations using density functional theory (DFT) often allow a more accurate estimation of pore size [79]. Despite probably absolute inaccuracy in determining the pore size distribution, differences between pristine and modified (containing guest molecules or particles) mesoporous materials can be well evaluated.

The typical observations from the BET N_2 adsorption measurements are the following: pore volume and pore size decrease after incorporation of metal compounds and nanoparticle formation within the pores (Table 1). These findings unambiguously confirm presence of the host species in the pores. Specific surface area obtained from BET measurements also normally decreases, but it is not necessarily the case. New surface of nanoparticles may add to the total surface available for N_2 adsorption, so if particles are small (large surface area) and numerous, decrease in BET surface is not a necessary indication of the guest particle formation.

The shape of N_2 adsorption-desorption isotherms and of hysteresis loop also provide important qualitative information on the presence or absence of micropores and type of mesoporosity (Fig. 11) [76].

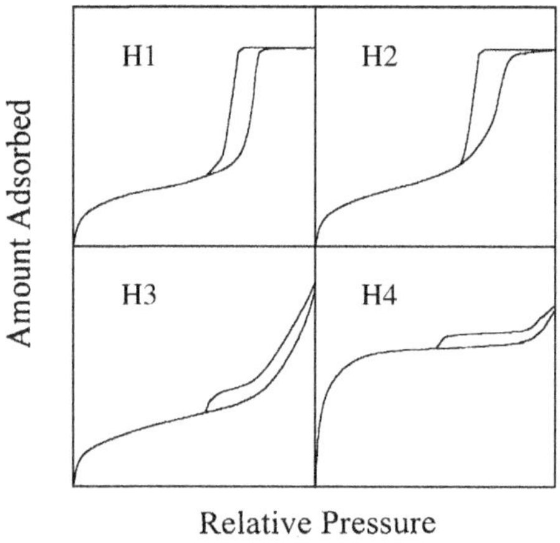

Fig. 11. Classification of adsorption-desorption hysteresis loops. H1 are typical for materials containing agglomerates and for materials with cylindrical pore geometry and a high degree of pore size uniformity. The type H2 is characteristic of materials with relatively uniform channel-like pores with the pore connectivity effects. H3 type is attributed to aggregates (loose assemblages) of platelike particles forming slitlike pores. The type H4 loop is proposed to be attributed to large mesopores embedded in a matrix with pores of much smaller size. Reprinted with permission from [76]. Copyright (2001) American Chemical Society

3.3 X-Ray Diffraction

Powder X-ray diffraction is one of the convenient methods for ordered mesoporous materials with guest nanoparticles [78]. The presence of guest species within the mesoporous structure can be detected by a reduction of the intensities typical for mesoporous silica Bragg reflections in XRD [25, 80]. This method allows rather qualitative characterization of the material if nanoparticles are too small or amorphous. When nanoparticles are crystalline, the corresponding Bragg peaks appearing in the XRD profile allow quantitative assessment of mean particle size, but no reliable information can be obtained on particle size distribution though some authors tend to consider shape of Bragg peaks as indication of broad (irregular shape) or narrow (Gaussian) particle size distribution.

3.4
Small Angle X-Ray Scattering

Small angle X-ray scattering (SAXS) can be used to characterize the type of mesoporous structure and to follow deviations of the structure when particles are formed within the mesoporous host [81]. Recently SANS (small angle neutron scattering) was used to estimate a pore size and pore size distribution in comparison with BET adsorption data [82, 83]. However, up to now SAXS (or SANS) and especially ASAXS were not used for characterization of nanoparticles in mesoporous solids. Recently a few papers reported successfully using ASAXS for estimation of size and size distribution of particles grown in nanostructured polymeric systems [84, 85] or deposited on SiO_2 [86]. Here the host matrix structure does not influence the scattering of the particles as the difference scattering (obtained via subtraction of the scattering curves at the different energies near the adsorption edge of the corresponding metal) is used [84]. So apparently the ordered structure of mesoporous solid should not obscure the assessment of particle characteristics similar to other ordered systems. Hopefully, this method will become more widely used, especially as instruments for the ASAXS measurements are available at a number of synchrotron facilities.

3.5
XAFS as Method of Choice for Studying the Nanoparticle-Containing Mesoporous Solids

Mesoporous solids (silica, aluminosilica, etc.) containing other heteroatoms (or nanoparticles) are ideally suited for study by X-ray absorption fine structure (XAFS). X-ray absorption near-edge structure (XANES) and extended X-ray absorption fine structure (EXAFS) provide information of nearest environment of metal species (first and second coordination spheres) which is especially useful for bimetallic nanoparticle characterization [87]. It allows determination of the structure of metal oxide nanoparticles [78] and interaction of particles with mesoporous material walls [87]. EXAFS provides important information about the local structure around X-ray-absorbing atoms. It can help (i) to identify the atoms and to find (ii) interatomic distances and (iii) coordination number of atoms in the neighboring atomic shells surrounding the central absorbing atoms. For bimetallic nanoparticles obtained after gentle thermolysis of $[Ru_6C(CO)_{16}Cu_2Cl]^{2-}$, analysis of both Ru and Cu *K* XAFS results showed agreement with $Ru_{12}Cu_4C_2$ structure, but only in combination with theoretical calculations was it possible to identify the lower energy free-space clusters and to solve the puzzle. In the case of metal oxide particles [78], EXAFS and XANES allow one to establish the differences between bulk oxides and nanoparticles positioned in mesoporous solids and also presence of metal atoms in different degree of oxidation. XANES and EXAFS can be also used to follow transformation of the precursor material before and after calcination. In [68], such a study showed that Cu^{2+} atoms in the precursor (before calcination) have both N and O in the first shell within DAB-Am-32 dendrimer used as template for silica casting. In a calcined sample containing CuO species, semiquantitative second shell analysis allowed the estimation of particle sizes when particles are very small and poorly detectable by other techniques [68].

Thus, one can conclude that no single method allows collecting of comprehensive information on mesoporous solids with nanoparticles. However, reasonable combination of a number of modern techniques (TEM, BET N_2 adsorption, XRD, ASAXS, etc) can provide sufficient data for correct characterization of these materials.

4
Properties of Mesoporous Materials With Nanoparticles

Properties of mesoporous solids with embedded nanoparticles can be grouped into three major categories outlining the possible applications of the materials: catalysis, optical properties, and magnetic properties. "Assignment" to the particular category is mainly determined by the nature of the guest particle. To illustrate this, one should recall that noble and other transition metal (or metal oxide) nanoparticles formed in any other environment (microemulsions, block copolymers, dendrimers, etc) show catalytic properties in a number of catalytic reactions. Some metal oxide (TiO_2) or sulfide or selenide (for one, CdS or CdSe) nanoparticles display optical properties which differ from those of bulk materials due to confinement of electron properties. Some metal oxide (Fe_2O_3) or metal (Co) nanoparticles show magnetic properties also depending on the particles size. When embedded in the pores of mesoporous materials, nanoparticles impart their properties to the materials but in a number of cases this influence is mutual: (i) host mesoporous solid affects the particle properties through particle-wall interactions and (ii) spatial organization of nanoparticles or even their close location result in the change of the properties of nanoparticle-containing mesoporous solid compare to properties of the "naked" nanoparticles (one should understand that nanoparticles are never "naked" and their surface is always stabilized by various molecules or groups depending on the way of their synthesis). Requirements for mesoporous materials with nanoparticles depend on the area where they are supposed to be used. For catalysis, one would prefer to have small nanoparticles (to ensure high surface area), but there is no clear understanding yet as to whether amorphous or crystalline nanoparticles have any advantage. One might speculate that edges of tiny crystals may be especially favorable sites for catalytic reactions [88], though surface of amorphous particles can also have many defects which are often considered as most reactive catalytic sites [89]. There are neither strong restrictions for narrow particle size distribution nor for spatial 2D or 3D arrangement if catalytic properties are of concern, although relatively narrow particle size distribution (lack of large particles where catalyst will be "wasted" inside the particle) is important. Another key trait of catalytically active porous materials is an interpenetrating pore structure and easy access of reagents to catalytic particles. As mentioned above, if metal particles are encapsulated in the silica (or other material body), no efficient catalyst can be expected. For optical or magnetic properties encapsulation is of no concern, but strong control over particle size and particle size distribution are very important, otherwise optical and magnetic properties will be not sufficiently tailored. Another essential requirement for successful application of some optical materials is regular spatial arrangement of nanoparticles within pores resulting in 1D, 2D, or even 3D ordering. Such ordering can influence absorption, luminescence, and non-linear optic properties. Below, particular examples will illustrate the success obtained so far in studying properties of mesoporous materials with embedded nanoparticles.

4.1
Catalytic Properties

The catalytic properties of mesoporous materials with embedded nanoparticles are mainly determined by the type of the inclusion (particle). All catalytic reactions, which are normally known for the particular metals or alloys, can be carried out with mesoporous solids containing nanoparticles. The important advantage of mesoporous oxides is their stability at high temperatures. Due to this feature, mesoporous oxides with nanoparticles can be successfully used as catalysts in such reactions where nanoparticles embedded in polymeric systems cannot be employed. Another probable advantage of mesoporous catalysts is an appropriate use of pores as nanoreactors of certain size. This can be applicable to large molecules or to cyclization reaction where pore size and shape will influence the reactive path [90]. However, for mesoporous solids with nanoparticles such applications are not reported so far.

In a recent review article [13] the authors described a number of catalytic reactions which were carried out with metal-deposited mesoporous silicates. The isomerization of *n*-hexane performed with Pt nanoparticles obtained in MCM-41 aluminosilicate was reported in [91]. The authors showed some improvement of the catalyst performance compared to disordered aluminosilica which might be caused by better control over nanoparticle growth. The authors of [52] demonstrated a very efficient catalyst for the Heck reaction: it was synthesized by CVD of a Pd complex within the MCM-41 pores. Using additional chemical modification of the pores, they were able to obtain sub-nanometer Pd particles. The resulting catalysts showed a catalytic activity superior to that of other heterogeneous Heck catalysts and even exceeded the catalytic activity of many homogeneous catalysts. The probable explanation of this phenomenon lays in a sub-nanometer size of nanoparticles and correspondingly strong influence of the support on their properties.

Hydrogenation of various substrates with Pd or Pt (or various bimetallic) nanoparticles embedded in mesoporous materials was studied by a number of authors (for example [92, 93]), but even if superior catalytic properties compared to traditional heterogeneous catalysts are reported, it is still questionable whether they can also compete with polymer supported catalysts. In polymeric systems Pd or Pt nanoparticle surface can be easily, efficiently, and permanently modified with organic groups inherent to polymers, yet hydrogenation is normally carried out at ambient temperatures, and the main advantage of mesoporous solids (stability at high temperatures) is not realized. One can expect that hydrogenation of polymers or other large or complex molecules can be favorably performed in nanoparticle-containing mesoporous solids but no such works have been published yet.

One of the sensible applications of mesoporous solids with metal nanoparticles is hydrocracking of vacuum gasoil [94]. Here reaction temperatures exceed 400 °C and no polymer or other organic catalysts can be used. NiMo nanoparticles were formed in MCM-41 via impregnation of molybdenum and Ni compounds followed by calcination. Despite the comparatively high metal loading (12 wt% MoO_3 and 3 wt% NiO), the surface area of MCM-41 was only slightly reduced which might be credited to increase of the surface area due to particle surface. The catalytic activity of NiMo-MCM-41 catalyst was compared with that of amorphous aluminosilica and USY zeolite with the same Ni and Mo loading. The former catalyst showed higher hydrodesulfurization and hydrodenitrogenation activity than the other heterogeneous catalysts in one-stage operation using untreated gasoil. The authors think that combination of large surface area, uniform pore size distribution with pores sufficiently large to allow diffusion of large molecules, along with mild acidity and high stability made NiMo-MCM-41 a superior catalyst.

The authors of [27] reported photocatalytic activity of titania-modified MCM-41 containing Pd in photoreduction of Cr(VI) to Cr(III) which was much greater than that of corresponding supports. As Pd clusters were not distinguishable in TEM image (Fig. 12), the authors concluded that Pd is grafted on the surface of titania. As article [27] provides no proof of such grafting (formation of thin Pd layer), one can assume that clusters formed are of a sub-nanometer size and are invisible in conventional TEM. During photodeposition of Pd

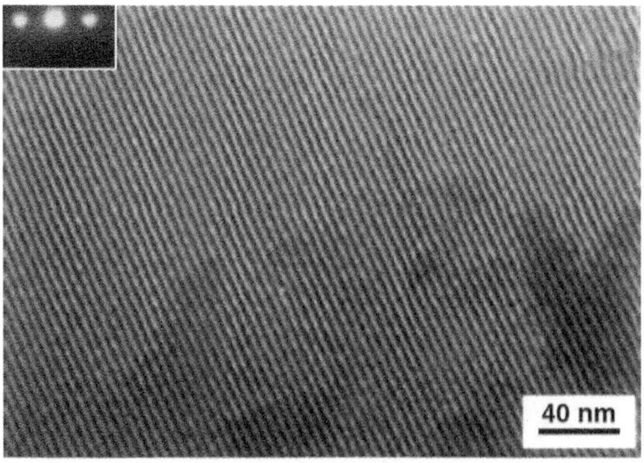

Fig. 12. TEM image of Pd/MCM-TiO_2 (1% Pd). Reproduced with permission from [27]. Copyright (2001) Academic Press

amorphous TiO_2 transforms into anatase phase which can be a cause of increased photocatalytic activity while Pd clusters rather do not participate in the reaction.

Pt and Rh nanowires and nanoparticles prepared in FSM-16 (synthesis is described above) display interesting catalytic properties in water-gas shift reaction [34] and hydrogenolysis of butane [36]. In hydrogenolysis of butane, Pt wire/FSM-16 demonstrates higher catalytic activity than Pt particle/FSM-16. The XANES and XPS studies show that the Pt wires are more electron deficient than particles which may cause a different catalytic activity. A Pt-Rh wire/FSM-16 displayed a high activity in butane isomerization. Here the use of mesoporous solid is highly justified as it ensures nanowire formation and modification of its properties due to host influence.

A recent review article on nanopore and nanoparticle catalysts [75] describes the enhanced selectivity of the catalysts with bimetallic nanoparticles with diameter of 1–2 nm formed in mesoporous silica. The authors especially underline the advantage of bimetallic particles Ru_6Pd_6, Ru_6Sn, $Ru_{12}C_2Cu_4$, and Ru_5Pt [95] formed from the corresponding carbonyl clusters (see discussion above) in the hydrogenation of (i) benzoic acid to cyclohexane carboxylic acid, (ii) dimethyl terephthalate (DMT) to 1,4-cyclohexanedimethanol (CHDM), and (iii) naphthalene in a highly selective manner to *cis*-decalin. Authors report an exceptional selectivity and activity of these catalysts although no comparison with traditional bimetallic catalysts or bimetallic nanoparticles prepared in different media is given. On the other hand, it is quite probable that the reported tiny bimetallic particles stabilized by mesoporous host and protected from leaching and contamination can display exceptional catalytic properties.

Catalytic activity of MCM-41 with rhodium oxide nanoparticles prepared by addition of $RhCl_3 \cdot 3H_2O$ in sol-gel mixture were studied in the high-temperature NO-CO catalytic reaction [22]. Catalyst containing RhO_x nanoparticles with diameters less than 3 nm exhibited a novel promotional effect in the amount of N_2 and N_2O formation with excess O_2, while most of Rh catalysts become poisoned with the O_2 excess. The authors also claim that catalysts with 6–8 nm RhO_x nanoparticles in the MCM-41 had drastically retarded the formation of target molecules. Since the Rh precursor is added to the sol-gel mixture and particles can be located in the silica body, the comparison of solely particle size in determination of catalytic properties can be misleading, as particle location (accessible for gases or not) may be a crucial feature.

4.2 Optical Properties

All publications on mesoporous solids with nanoparticles discussing optical properties can be generally divided into two large groups. The first group consists of works on different kinds of semiconductor particles (metal sulfides, selenides, oxides, etc.) where such particles are obtained in mesoporous solids and their UV-visible absorption and photoluminescence spectra are recorded. As a rule, these papers state the change of optical properties of nanoparticles compared to bulk, similar to semiconductor nanoparticles prepared in any other environment, but rarely consider the specific interactions or ordering within the mesoporous solids. However if particles are prepared via CVD or high temperature treatment, mesoporous oxides are only matrixes of choice for stabilization of such particles. One of the examples is GaAs/MCM-41 nanostructures (GaAs particles were formed via CVD) which show blue-shifted absorption and visible photoluminescence (PL) even at room temperature [54]; however a particle location both inside and outside the channels results in broad PL. ZnS nanoparticles confined in ordered mesoporous silica also showed a blue shift which is consistent with the size quantization effect [49]. However, when MCM-41 functionalized with ethylenediamine groups was used, the strong observed PL was explained by nanoparticle surface modification [49]. In_2O_3 nanoparticles formed by thermal decomposition of indium sulfate showed a significant blue shift of absorption edge when annealing temperature did not exceed 773 K [96]. The complicated luminescence was not clearly understood, but was tentatively attributed to defects or oxygen deficiencies in the In_2O_3 particles within the pores, or interactions with pore walls. The authors of [63] reported a change in the absorption spectra after incorporation of CdS nanoparticles (formed in reverse micelles) into mesoporous silica containing thiol groups. Probably, replacement of surfactant with thiol groups change modification of the particle surface and optical properties (Fig. 13).

In a review article [97], the authors describe a strong luminescence of ZnO and In_2O_3 nanoparticles in mesoporous silica which is governed by preparation conditions. Moreover, the luminescent properties differ strongly for mesoporous silica bearing nanoparticles from merely corresponding nanoparticles. Both ZnO- and In_2O_3-containing materials display strong enhancement of luminescence. The abnormal luminescence effects in the assembly systems are explained by their microstructures, both by specific properties of particles formed within pores and by interaction with the pore walls. Another important characteristic of mesoporous assembly systems described in [97] which is not inherent to conventional composites is a possibility to modulate the optical absorption edge and the absorption band by heat treatment and loading conditions which should influence the particle size and particle surface properties.

Gold nanoparticles prepared within mesoporous silica monolith showed red-shift of the Mie-resonance absorption band with decreasing the Au particle size. Comparison with theoretical calculations of Mie-resonance for particles of 5–20 nm embedded in nonabsorbent matrix showed that no particle size dependence should be expected. To explain the phenomenon observed, authors suggest that the interface interactions between particles and pore walls result in the red-shift of resonance. So the optical behavior of Au nanoparticles in the mesoporous matrix was ascribed to the charge transfer between the Au nanoparticles and the pore walls.

Silver nanoparticles obtained by thermal decomposition of $AgNO_3$ within mesoporous silica pores show interesting optical properties [26]. Ag doping results in a large red shift of the adsorption edge which is attributed to the interband absorption of Ag in the Ag-silica system and dipole interaction between Ag particles. The lack of a plasmon band typical for Ag nanoparticles is attributed to a small particle size (2.4–3.6 nm) and interaction at the interfaces between

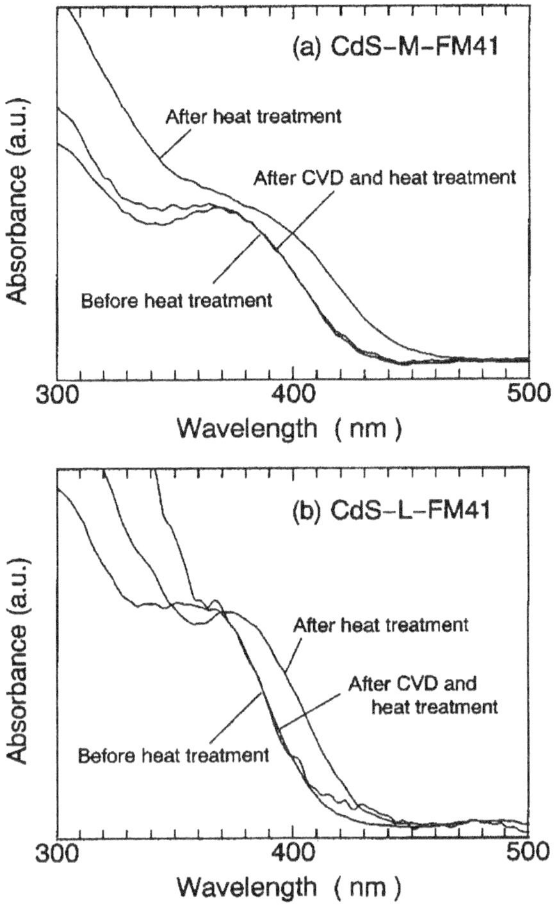

Fig. 13a,b. Diffuse reflectance spectra for CdS nanoparticles prepared in reverse micellar solution when incorporated into: **a** 0.15 g of M-FM41 (medium pore functionalized MCM-41); **b** 0.05 g of L-FM41 (large pore functionalized MCM-41). Also shown is the effect of CVD treatment for CdS-FM41 on the stability against heat treatment at 523 K for 30 min. Reproduced with permission from [63]. Copyright (2001) Academic Press

particles and pore walls. The position of the absorption edge is controlled by the amount of doped silica (or rather by the particle size).

A second group of papers on mesoporous materials with semiconductor nanoparticles fully explores the possible advantages of such systems: option of regular location of nanoparticles within mesoporous solid which strongly modifies the optical properties compared to irregular positioning. In [51] authors report a soft chemistry preparation of ~1-nm silicon clusters in free-standing oriented hexagonal mesoporous silica film. The film displays yellow-orange photoluminescence and nanosecond luminescence lifetimes, unlike milli- and microsecond lifetimes for porous and nanocrystalline silicon.

A recent example of extremely successful use of mesoporous silica films for 3D ordering of quantum dots is reported in [98]. The authors were able to fill a 300 nm thick mesoporous film with CdS nanoparticles (Fig. 14). The size and the 3D ordering of nanoparticles were directly controlled by the starting pore structure. However, the control of nanoparticles size and organization did not provide an intense PL. This could be explained by strong interactions with the silica walls, especially as the particles grew to the size of pores. So the optimal balance is required between organization and surface modification to provide the best control over optical properties.

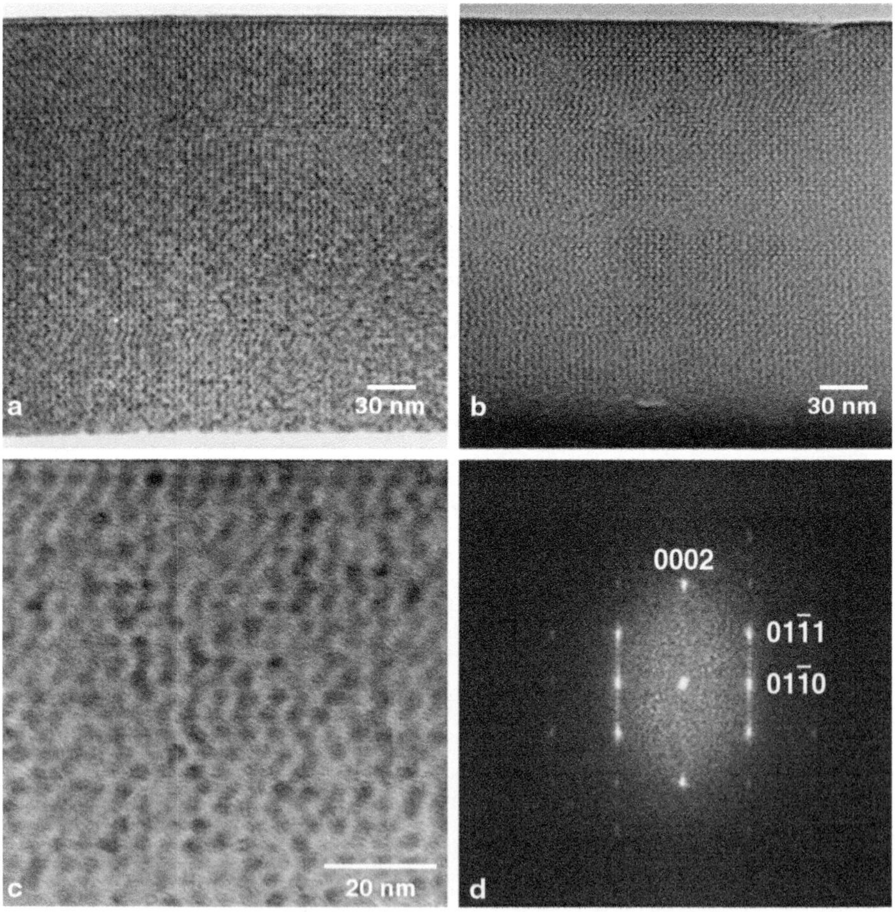

Fig. 14. **a** HRTEM image in cross section of the CdS saturated film. **b** HRTEM image in cross section of a calcined mesoporous film before impregnation. **c** Enlargement of the cross section (a) showing the periodic arrangement of the CdS particles. **d** Power spectrum of the image (a) showing the 3D hexagonal structure. Reprinted with permission from [98]. Copyright (2002) American Chemical Society

4.3 Magnetic Properties

It is well known that magnetic properties of particles strongly depend on particle size. Moreover, with decrease of particle diameter, the nature of magnetism changes: ferromagnetic particles become superparamagnetic (do not show hysteresis) and these properties are temperature-dependent.

Magnetic measurements are often used for materials containing magnetically active particles to elucidate their composition and structure. In [16], for example, magnetic measurements helped to confirm the existence of a hydroxo layer on iron oxide particles formed in mesoporous alumina. Important magnetic properties in characterization of iron oxide particles are type, strength, and direction of remnant magnetization. The values obtained give information on the domain structure and particle size distribution of the magnetic particles in the sample. Additional information on the magnetic domain structure can be obtained from magnetization curves of the samples in absence or presence of the magnetic field. It allows one to find the blocking temperature (T_b) of superparamagnetic particles. Ni metal nanoparticles formed in AlMCM-41 were 1–2 nm in diameter and showed superparamagnetic behavior with a blocking temperature of 5 K [17].

Iron nanoparticles prepared by pyrolysis of poly(ferrocenylsilanes) inside periodic mesoporous silica displayed the absence of room-temperature hysteresis in the magnetization curves which shows their superparamagnetic behavior [55]. However, magnetic properties cannot always be easily interpreted. For example, for this material data analysis of magnetization curves resulted in the ambiguous conclusion that either particle size distribution is bimodal, or iron particles have an oxide layer which behaves as small superparamagnetic nanoparticles. So magnetic measurements should be combined with other techniques (probably, in this case, EXAFS may be useful) to allow more accurate evaluation of particle structure.

Iron oxide nanoclusters (γ-Fe_2O_3) synthesized within mesoporous MCM-41 aluminosilicate matrices using volatile $Fe(CO)_5$ as precursor (the synthesis is described above) had both spherical and elongated morphologies due to influence of hexagonally packed cylindrical pore structure [53]. These materials showed superparamagnetic behavior at room temperature and ferrimagnetic nature at 5 K (according to magnetization studies and Mössbauer spectroscopy), yet magnetization properties depended on calcination conditions (Fig. 15). Fe_2O_3 nanoparticles in this nanocomposite demonstrated high stability against $\gamma \rightarrow \alpha$ phase transition. The authors suggest that the high thermal stability and increased magnetization from calcination make this material a good candidate for high-temperature magnetic applications.

However, to the best of my knowledge, no magnetic materials based on mesoporous solids are ready for any reliable application similar to magnetic fluids or magnetic polymer films although thin magnetic mesoporous films can be obtained similar to mesoporous films with semiconductor particles.

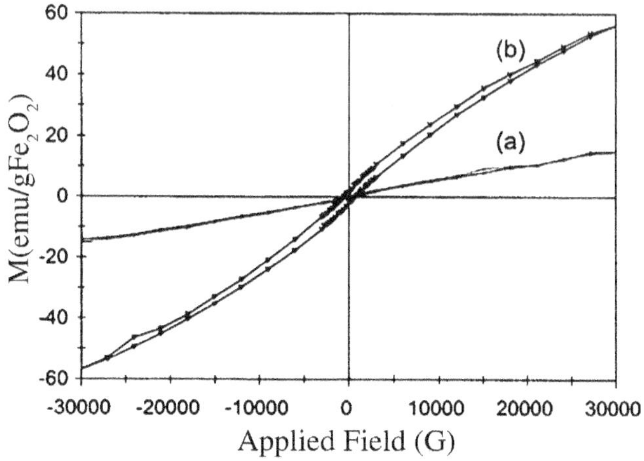

Fig. 15. Plots of magnetization vs applied field for as-prepared Fe_2O_3/AlMCM-41 obtained at (*a*) 300 K and (*b*) 5 K. Reprinted with permission from [53]. Copyright (2001) American Chemical Society

5
Summary and Outlook

Mesoporous solids containing nanoparticles of various nature became an exciting field of nanoscience and nanotechnology. This field is actively developing and new accomplishments allowing better control over particle size or regular particle location mean that "smart" applications can be expected. A vast variety of synthetic methods for obtaining such materials allows choosing of the best approach for the particular application. If the material is expected to have catalytic properties, the synthetic technique should ensure good access of nanoparticles for reagents and controlled particle size to provide a high particle surface area. Among the methods described above, incorporation of metal compounds in sol-gel mixture does not guarantee these features, as the particle surface can be obscured by the silica (or other material) body. In the same way, synthesis of nanoparticles using high temperature treatment often results in particles growing over the pore size and embedding in the intrachannel defects, so these methods seems to be inferior and should rather be avoided. For optical material synthesis, any method of choice is acceptable though techniques allowing one to control particle size distribution and positioning seem to be preferable. Another important consideration for optical materials is avoiding extensive contact with mesoporous material walls as it may quench photoluminescence or make the luminescence too broad. As for magnetic applications, no clear indications can be found yet as to which technique is superior because no magnetic materials based on mesoporous solids are yet reported which might be considered for magnetic applications. These studies are yet to come. Apparently, monolithic mesoporous material films with embedded magnetic particles should be promising for magnetic films especially as tailoring the particle size and particle positioning might be influential for magnetic properties.

As may be concluded from screening the literature, the combination of various physicochemical techniques (TEM, STEM, HRTEM, HRSTEM, XRD, BET N_2 adsorption, XANES, EXAFS) allows quite comprehensive characterization of mesoporous solids with nanoparticles. Nevertheless, when particle size distribution is questionable and important, X-ray and neutron scattering techniques (SAXS, ASAXS, SANS) should be more widely used for better understanding the material properties.

Further research will probably develop in the direction of improving the control over particle size, size distribution, and particle positioning and preparation of freestanding films or well made monolith bulk samples with guest particles. So both "guest" and "host" components of such materials need to be optimized and further developed.

References

1. Kresge CT, Leonowicz ME, Roth WJ, Vartuli JC, Beck JS (1992) Nature 359:71
2. Beck JS, Vartuli JC, Roth WJ, Leonowicz ME, Kresge CT, Schmitt KD, Chu CT-W, Olson DH, Sheppard EW, McCullen SB, Higgins JB, Schlenker JL (1992) J Am Chem Soc 114:10,834
3. Inagaki S, Fukushima Y, Kuroda K (1993) Chem Commun 68
4. Goltner CG, Henke S, Weissenberger MC, Antonietti M (1998) Angew Chem Int Ed 37:613
5. Zhao D, Feng J, Huo Q, Melosh N, Fredrickson GH, Chmelka BF, Stucky GD (1998) Science 279:548
6. Tanev PT, Pinnavaia TJ (1995) Science 267(5199):865
7. Caruso RA, Giersig M, Willig F, Antonietti M (1998) Langmuir 14:6333
8. Sun T, Ying JY (1997) Nature 389:704
9. Tanev PT, Chibwe M, Pinnavaia TJ (1994) Nature 368(6469):321
10. Lee B, Yamashita T, Lu D, Kondon JN, Domen K (2002) Chem Mater 14:867
11. Fryxel G, Liu J (2000) Surf Sci Ser 90:665
12. Moller K, Bein T (1998) Chem Mater 10:295
13. Ying JY, Mehert CP, Wong MS (1999) Angew Chem Int Ed 38:56
14. Morey MS, Bryan JD, Schartz S, Stucky GD (2000) Chem Mater 12:3435
15. Liu J, Feng X, Fryxel GE, Wang L-Q, Kim AY, Gong M (1998) Adv Mater 10:161
16. Schnider JJ, Czap N, Hagen J, Engstler J, Ensling J, Gütlich P, Reinoehl U, Bertagnolli H, Luis F, de Jongh LJ, Wark M, Grubert G, Hornyak GL, Zanoni R (2000) Chem Eur J 6:4305
17. Jung J-S, Chae W-S, McIntyre RA, Seip CT, Wiley JB, O'Connor CJ (1999) Mater Res Bull 34:1353
18. Shi H, Zhang L, Cai W (2000) J Appl Phys 87:1572
19. Powers KW, Hench LL (2000) Ceram Trans 101:253
20. Wu Y, Zhang L, Li G, Jiang C, Huang X, Zhang Y, Song G, Jia J, Zhixiang C (2001) Mater Res Bull 36:253
21. Mulukutla RS, Asakura K, Kogure T, Namba S, Iwasawa Y (1999) Phys Chem Chem Phys 1:2027
22. Mulukutla RS, Shido T, Asakura K, Iwasawa Y (2001) Scripta Mater 44:1695
23. Mulukutla RS, Asakura K, Namba S, Iwasawa Y (1998) Chem Commun 1425
24. Plyuto Y, Berquer J-M, Jacquiod C, Ricolleau C (1999) Chem Commun 1653
25. Fröba M, Köhn R, Bouffaud G (1999) Chem Mater 11:2858
26. Weiping C, Lide Z (1997) J Phys Condens Matter 9:7257
27. Zheng S, Gao L, Zhang Q, Sun J (2001) J Solid State Chem 162:138
28. Chen W, Cai W, Wang G, Zhang L (2001) Appl Surf Sci 174:51
29. Chen W, Cai W, Zhang L, Wang G, Zhang L (2001) J Colloid Interface Sci 238:291
30. Gedanken A, Tang X, Wang Y, Perkas N, Koltypin Y, Landay MV, Vradman L, Herskowitz M (2001) Chem Eur J 7:4546

31. Liu Z, Terasaki O, Ohsuma T, Hirada K, Shin HJ, Ryoo R (2001) Chem Phys Chem 229
32. Besson S, Gacoin T, Ricolleau C, Jacquiod C, Boilot JP (2001) Nano Lett
33. Liu Z, Sakamoto Y, Ohsuma T, Hirada K, Terasaki O, Ko CH, Shin HJ, Ryoo R (2000) Angew Chem Int Ed 39:3107
34. Fukuoka A, Higashimoto N, Sakamoto Y, Sasaki M, Sugimoto N, Inagaki S, Fukushima Y, Ichikawa M (2001) Catal Today 66:23
35. Huang MH, Choudrey A, Yang P (2000) Chem Commun 1063
36. Fukuoka A, Higashimoto N, Sakamoto Y, Inagaki S, Fukushima Y, Ichikawa M (2001) Microp Mesop Mater 48:171
37. Shepard DS, Mashmeyer T, Johnson BFG, Thomas JM, Sankar G, Ozkaya D, Zhou W, Oldroyd RD, Bell RC (1997) Angew Chem Int Ed 36:2242
38. Zhou W, Thomas JM, Shepard DS, Johnson BFG, Ozkaya D, Mashmeyer T, Bell RC, Ge Q (1998) Science 280(5364):705
39. Joo SH, Choi SJ, Oh I, Kwak J, Liu Z, Terasaki O, Ryoo R (2001) Nature 412(6843):169
40. Sidorov SN, Volkov IV, Davankov VA, Tsyurupa MP, Valetsky PM, Bronstein LM, Karlinsey R, Zwanziger JW, Matveeva VG, Sulman EM, Lakina NV, Wilder EA, Spontak RJ (2001) J Am Chem Soc 123:10,502
41. Zhang W-H, Shi J-L, Wang L-Z, Yan D-S (2000) Chem Mater 12:1408
42. Lebeau B, Fowler CE, Mann S, Farcet C, Charleux B, Sanchez C (2000) J Mater Chem 10:2105
43. Zhang L, Sun T, Ying JY (1999) Chem Commun 1103
44. Hua Z-L, Shi J-L, Wang L, Zhang W-H (2001) J Non-Cryst Solids 292:177
45. Antochshuk V, Jaroniec M (1999) Chem Commun 2373
46. Tanev PT, Pinnavaia TJ (1996) Chem Mater 8:2068
47. Wellmann H, Rathousky J, Wark M, Zukal A, Schulz-Ekloff G (2001) Microporous Mesoporous Mater 44/45:419
48. Guari Y, Thieuleux C, Mehdi A, Reye C, Corriu RJP, Gomez-Gallardo S, Phillipot K, Chaudret B, Dutartre R (2001) Chem Commun 1374
49. Zhang W-H, Shi J-L, Chen H-R, Hua Z-L, Yan D-S (2001) Chem Mater 13:648
50. Bronstein LM, Polarz S, Smarsly B, Antonietti M (2001) Adv Mater 13:1333
51. Dag O, Ozin GA, Yang H, Reber C, Bussiere G (1999) Adv Mater 11:474
52. Mehnert CP, Weaver DW, Ying JY (1998) J Am Chem Soc 120:12,289
53. Zhang L, Papaefthymiou C, Ying JY (2001) J Phys Chem B 105:7414
54. Srdanov VI, Alxneit I, Stucky GD, Reaves CM, DenBaars SP (1998) J Phys Chem B 102:3341
55. MacLachlan MJ, Ginsburg M, Coombs N, Raju NP, Greedan JE, Ozin GA, Manners I (2000) J Am Chem Soc 122:3878
56. Yonemitsu M, Tanaka Y, Ywamoto M (1997) Chem Mater 9:2679
57. Badiei AR, Bonneviot L (1998) Inorg Chem 37:4142
58. Zhang Z, Dai S, Fan X, Blom DA, Pennycook SJ, Wei Y (2001) J Phys Chem B 105:6755
59. Iwamoto M, Tanaka Y (2001) Catal Surv Jap 5:25
60. Khushalani D, Hasenzahl S, Mann S (2001) J Nanosci Nanotech 1:129
61. Mukherjee P, Sustry M, Kumar R (2000) Phys Chem Comm 3
62. Hirai T, Okubo H, Komasawa I (1999) J Phys Chem B 103:4228
63. Hirai T, Okubo H, Komasawa I (2001) J Colloid Interface Sci 235:358
64. Bronstein L, Krämer E, Berton B, Burger C, Förster S, Antonietti M (1999) Chem Mater 11:1402
65. Whilton NT, Berton B, Bronstein L, Hentze H-P, Antonietti M (1999) Adv Mater 11:1014
66. Antonietti M, Groehn F, Hartmann J, Bronstein L (1997) Angew Chem Int Ed 36:208
67. Eliseev AA, Lukashin AV, Vertegel AA, Heifets LI, Zhirov AI, Tretyakov YD (2000) Mat Res Innovat 3:308
68. Velarde-Ortiz R, Larsen G (2002) Chem Mater 14:858
69. Zhao M, Sun L, Crooks RM (1998) J Am Chem Soc 120:7355
70. Han B-H, Polarz S, Antonietti M (2001) Chem Mater 13:3915
71. Polarz S, Smarsly B, Bronstein L, Antonietti M (2001) Angew Chem Int Ed 40:4417
72. Ozkaya D, Thomas JM, Shephard DS, Maschmeyer T, Johnson BFG, Sankar G, Oldroyd R (1997) Inst Phys Conf Ser 153:403

73. Shimizu K, Thompson GE, Wood GC (1981) Thin Solid Films 81:39
74. Thomas JM, Terasaki O, Gai PL, Zhou W, Gonzalez-Calbert J (2001) Acc Chem Res 34:583
75. Thomas JM, Raja R (2001) Chem Rec 1:448
76. Kruk M, Jaroniec M (2001) Chem Mater 13:3169
77. Anwander R, Nagl I, Widenmeyer M, Engelhardt G, Goeger O, Palm C, Roser T (2000) J Phys Chem B 104:3532
78. Köhn R, Fröba M (2001) Catal Today 68:227
79. Jaroniec M, Kruk M, Olivier JP, Koch S (2000) Stud Surf Sci Catal 128:71
80. Marler B, Oberhagemann U, Vortmann S, Gies H (1996) Microporous Mater 6:19
81. Enke D, Janowski F, Heyer W, Gille W, Schwieger W (2000) SixthInternational Congress on Applied Mineralogy, vol 1. A.A. Balkema, Rotterdam, Netherlands, p 135
82. Smarsly B, Goeltner C, Antonietti M, Ruland W, Hoinkis E (2001) J Phys Chem B 105:831
83. Littrell KC, Khalili NR, Campbell M, Sandi G, Thiyagarajan P (2002) Chem Mater 14:327
84. Svergun DI, Kozin MB, Konarev PV, Shtykova EV, Volkov VV, Chernyshov DM, Valetsky PM, Bronstein LM (2000) Chem Mater 12:3552
85. Svergun DI, Shtykova EV, Kozin MB, Volkov VV, Dembo AT, Shtykova EV Jr, Bronstein LM, Platonova OA, Yakunin AN, Valetsky PM, Khokhlov AA (2000) J Phys Chem B 104:5242
86. Polizzi S, Riello P, Balerna A, Benedetti A (2001) Phys Chem Chem Phys 3:4614
87. Thomas JM, Sankar G (2001) J Synchrotron Rad 8:55
88. Wang ZL, Ahmad TS, El-Sayed MA (1997) Surf Sci 380:302
89. Nagamura T, Inoue H, Grieser F, Urquhart R, Sakaguchi H, Furlong DN (1999) Colloids Surf A 146:265
90. Vartuli JC, Shih SS, Kresge CT, Beck JS (1998) Stud Surf Sci Catal 117:13
91. Del Rossi KJ, Hatzikos GH, Huss A Jr (1993) US Pat 5245101, Chem Abs CAN 119:274,893
92. Chatterjee M, Iwasaki T, Onodera Y, Nagase T (1999) Catal Lett 61:199
93. Shephard DS (2000) Stud Surf Sci Catal 129:789
94. Corma A, Martinez A, Martinez-Soria V, Monton JB (1995) J Catal 153:25
95. Raja R, Khimyak T, Thomas JM, Hermans S, Johnson BFG (2001) Angew Chem Int Ed 40:4638
96. Zhou H, Cai W, Zhang L (1999) Mat Res Bull 34:845
97. Zhang L, Cai W, Mo C (1999) Prog Natur Sci 9:401
98. Besson S, Gacoin T, Ricolleau C, Jacquiod C, Boilot JP (2002) Nano Lett 2:409

Top Curr Chem (2003) 226: 91–118
DOI 10.1007/b10826

Nanocasting and Nanocoating

Rachel A. Caruso

School of Chemistry, The University of Melbourne, Victoria 3010, Australia
E-mail: rcaruso@unimelb.edu.au

The use of organic templates for the controlled structuring of inorganic materials is being widely explored. Two processes, nanocasting and nanocoating, will be discussed in this chapter for the formation of porous metal oxide structures with an emphasis on silica and titania. The difference between the two techniques is that casting is a filling of the porous structure of the organic material whereas coating results in a layer of the inorganic substance on the polymer structure. Following formation of the hybrid, the organic template can be removed, yielding a structured inorganic material. Either an *inverse* replica of the initial structure is obtained after *nanocasting* or a *hollow* replica of the organic results with the *nanocoating* procedure. The initial organic molds considered here are "rigid", preformed matrices or discrete entities that do not require the presence of the inorganic material for producing or maintaining their structure. These range from monolithic materials, including porous polymer gels and colloidal crystals, to organic fibers, crystals, and latex particles. The fabrication methods discussed that are used to obtain the metal oxide structures include sol-gel procedures and the use of preformed metal oxide nanoparticles.

Keywords. Template, Metal oxide, Silica, Porous structures, Sol-gel, Nanoparticles

List of Abbreviations

PDMS	Poly(dimethylsiloxane)
PMMA	Poly(methylmethacrylate)
PS	Polystyrene
PVP	Poly(vinyl pyrrolidone)
SEM	Scanning electron microscope
TBOT	Titanium (IV) tetrabutoxide
TEM	Transmission electron microscope
TEOS	Tetraethylorthosilicate
TIP	Titanium (IV) isopropoxide
TMOS	Tetramethylorthosilicate

1 Introduction

Morphological control of materials is required to obtain unique, often advantageous, and enhanced material properties. This chapter reviews recent research on the use of organic templates for the "transcriptive synthesis" [1] of inorganic materials, in particular silica and metal oxides, with controlled structures.

Templating generally requires three steps. First, the choice or preparation of the template, then the formation of a composite consisting of the template and final material, and third the removal of the template, giving the final structured material. Materials formed at step 2 often already have interesting and useful properties making the removal of the template unnecessary:

Step 1	→	Step 2	→	Step 3
Template		Composite		Structured material

Discussing each of these steps in turn, we start with the template. The template itself is a structured object within which, or around which, the second material is formed. The templates discussed in this chapter are organic and of fixed form, that is, the morphology of the template is apparent without the addition of the inorganic coating or casting substance and there is little change in its structure, except possibly for some expansion and shrinkage, during the templating process.

Templates require certain properties:

1 They have to have a structure that is of interest for the final material.
2 They need to be able to maintain their structure during the templating procedure.
3 If the final material is to be without the template, they need to be easily removed without disrupting the structure.

Polymeric materials are ideal for this application. Numerous morphological variants can be obtained which are of interest, different polymers can withstand a wide variation of conditions, including extreme pH and harsh solvents, and the polymer can be removed by decomposition using solvent, plasma treatment or heating procedures.

The organic templates that will be examined have been categorized according to the final inorganic structure. First, monolithic materials and films composed of both disordered and ordered three-dimensional porous structures are discussed. The formation of these materials requires templates such as polymer gels and membranes that have a bicontinuous pore morphology, and colloidal crystals and membranes, with fixed pore domains (in both size and structure). Then the fabrication of inorganic threads and hollow fibers is considered where polymeric fibers, biological threads, and organogelators act as scaffolds on or in which the inorganic material forms. This is followed by a section on templates composed of discrete entities, which result in hollow spheres and tubular materials. The organic templates in this section are spherical polymers (latex particles) or elongated organic materials such as nanotubes and crystals.

These templates can also be combined with other porogens (such as self-assembled polymers) or techniques to obtain hierarchical pore systems or structured materials on a number of length scales. Examples demonstrating the extension of simple templating to more complex structural control will also be given.

Composite materials can be formed by numerous methods. Two modes in which incorporation of the inorganic material in the template can be achieved will be discussed: sol-gel processes or nanoparticle infiltration. They are both solution methods that can be processed at low temperatures, hence allowing the use of polymeric templates. In the first method the sol-gel chemistry is performed after the incorporation of a metal oxide precursor in the polymer matrix or around the template entities. The second method makes use of preformed metal oxide nanoparticles, which are infiltrated into the organic scaffold or suspended in solution with the individual structures for controlled adhesion.

The sol-gel process involves hydrolysis and condensation reactions of the precursor, where in general (especially for the non-silica precursors) both reactions occur simultaneously [2], forming a metal oxide "polymer" and low molecular weight species (water and alcohol). The shrinkage observed during the sol-gel process is due in part to the removal of these species [3]. The overall reaction can be summarized as follows:

$$M(OR)_4 + 4H_2O \rightarrow M(OH)_4 + 4ROH$$
$$M(OH)_4 \rightarrow MO_2 + 2H_2O$$

where M represents a group 4 metal and R the alkyl group.

The use of nanoparticles requires the initial synthesis of the nanoparticles, which generally adhere to the template via electrostatic and hydrogen bonding interactions, or on processing of the nanoparticles and template the nanoparticles are physically confined within the template.

For templating in general, the final material can be achieved in a number of ways, including casting, coating, curing, and directed "assembly". This chapter will concentrate on the first two techniques applied on a submicron scale, that is either the template itself or the coating applied to the template is of nanometer proportions. Casting involves the complete filling of the porous system around the template material compared to coating, which results in a thin layer formed on the mold; see Scheme 1.

These processes are conducted with the purpose of generating materials with properties that are dependent on either a combination of the mold and the templated materials, or the structural properties obtained during the templating procedure. For the two techniques focused on in this chapter, nanocasting and nanocoating, the materials obtained on removal of the template resemble the initial template either as structural negatives (casting – gives pores where there

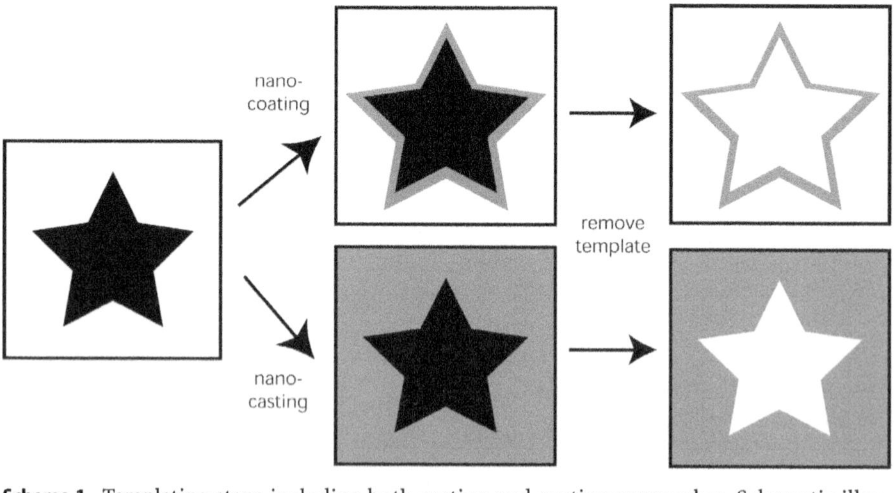

Scheme 1. Templating steps including both casting and coating approaches. Schematic illustration demonstrating the casting and coating of a star-shaped template. The casting technique gives a composite in which the second material fills the area around the mold so that on removal of the mold a structured material is obtained, which is an inverse replica of the initial template. In contrast, coating of the template results in a layer of the second material around the mold, resulting in a hollow replica on removal of the template

was originally template, and solid where there were originally pores) or as hollow structures with similar shape to the initial mold (coating – pores from the original mold are retained, plus additional pores are obtained on removal of the mold). This is depicted in Scheme 1.

The formation of such structured materials for various applications including separation, catalysis, and photonics has led to various activities in the fields of chemistry, engineering, materials science, and physics conducting detailed research on the controlled fabrication and the final properties of such materials. The use of templating procedures, including nanocoating and nanocasting techniques, allows the formation of porous structures with high-level control of the morphology (pore size and structure, as well as over all material shape and size) by a rather simple and efficient process.

Porous materials have been classified by the IUPAC (International Union of Pure and Applied Chemists) to be microporous if the pore size is below 2 nm, mesoporous if the pores have a diameter between 2 and 50 nm, and macroporous if the pore diameters are above 50 nm [4]. This chapter focuses on porous materials with pores in the meso and macro-regime.

2 Monolithic Materials and Films

Large hybrid inorganic/organic materials or inorganic structures can be formed by the use of monolithic templates or stacking of discrete entities into a monolithic structure. The monolithic structures can be disordered systems, such as polymer gels and membranes, consisting of random porous structures that are continuous and therefore can be penetrated by the inorganic material, or they can have ordered, regular pore structures such as the cylindrical pores found in polycarbonate filter membranes or the void spaces remaining on stacking of monodisperse polymer spheres. Ordered structures can also be formed with biological cells and used successfully as templates. The order or disorder of the porous structure and the overall porosity is of importance for their application. The final material retains the outer structure of the mold giving inorganic blocks or films.

2.1 Disordered Structures

The template materials discussed in this section are polymer gels and membranes. These organic matrices show that bulk hybrid or porous inorganic materials can be obtained using sol-gel or nanoparticle infiltration methods.

2.1.1 *Polymer Gels*

Polymer gels are an excellent example of an organic system that can be morphologically tailored. The synthesis of numerous polymer gels with a wide variation in the pore size and structure, and overall porosity has been achieved

[5, 6]. The choice of surfactant, monomer, and cross-linker gives a range of gels that are stable over significant pH values and that are easily handled with tweezers during the templating procedure. The sponge-like, bicontinuous pore structure makes impregnation procedures possible.

Hybrid metal oxide/polymer monoliths are formed when sol-gel reactions are performed within the pore system of the organic gel [7, 8]. The alcohol soaked polymer gel is simply placed in the liquid precursor and then into an alcohol/H_2O solution for hydrolysis and condensation, which gives an amorphous metal oxide layer on the polymer surface. Heating the hybrid at 450 °C results in removal of the organic template and crystallization of the inorganic material.

"Coral-like" TiO_2 structures have been formed which clearly demonstrate coating of the initial template, an acrylamide/glycidylmethacrylate polymer gel [7]. Highly porous inorganic networks with pore sizes from 100 nm to microns were obtained. The walls, with a thickness of 100–150 nm, are composed of TiO_2 nanoparticles, as can be seen in a thin slice of the sample viewed by TEM in Fig. 1. Variation of the initial polymer gel gives various pore sizes and different surface areas in the final titania structures, as shown in Table 1 [8]. The morphology of the final inorganic material and its properties are strongly dependent on the structure of the initial template. The crystallinity of the nanoparticles in the network (anatase or rutile phase) was determined by controlling the temperature of calcination.

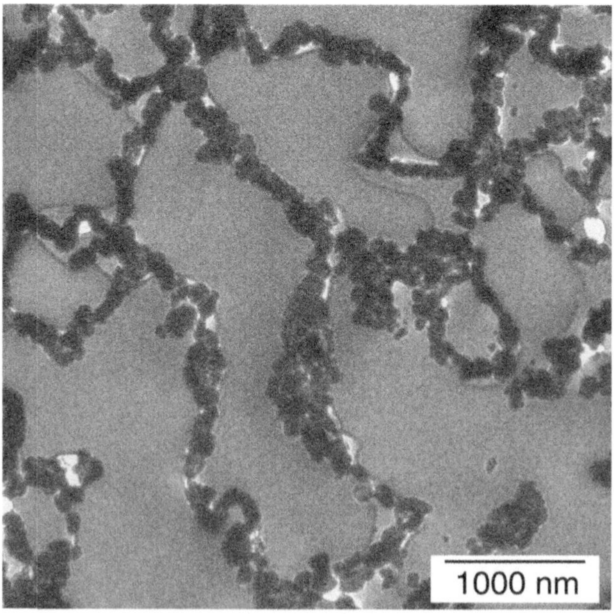

Fig. 1. TEM image of an ultramicrotome of the titania network constructed using a polymer gel template, showing the individual titania nanoparticles of which the structure is composed. Reprinted with permission from [7]. Copyright 2001 American Chemical Society

Table 1. Properties of the titania networks obtained by using polymer gel templates: calculated porosity, surface area obtained from nitrogen adsorption, pore, and titania nanoparticle diameters. Adapted with permission from [8]. Copyright 2001 American Chemical Society

Polymer gel template	Calculated porosity (vol.%)	Specific surface area[a] ($m^2 g^{-1}$)	Pore diameter range[b,c] (nm)	TiO_2 colloid diameter range[b] (nm)
B561	98	20	70–900	25–175
B562	97	11	100–3000	15–165
T601	99	39	10–325	5–40
B581	98	4.9	200–3000	25–175
B582	98	9.3	25–2300	20–110
CTA1	94	62	5–60	5–30
CTA2	94	76	–	–
SE1	99	49	3–37	10–40
SE2	99	59	5–20	7–20
EK1	–	99	2–100	4–14
EK2	–	54	5–175	7–50
EK3	–	82	3–24	8–45

[a] BET,
[b] TEM,
[c] SEM.

Monolithic zirconia networks can also be formed using a similar procedure giving porous ZrO_2 structures [9]. As the titania and zirconia precursors are miscible, binary inorganic networks of various Ti:Zr ratios could be produced [9]. The crystallinity and photocatalytic properties of the mixed material were studied: X-ray amorphous materials were produced for Ti:Zr ratios of 2:8 to 7:3, and the binary material containing 10% zirconia (the presence of which inhibited crystal transformation to the rutile phase) showed the highest photocatalytic activity for the photodecomposition of salicylic acid and 2-chlorophenol [9].

Similar templates with solvent swelling properties have been used for nanoparticle infiltration [10]. The acrylic acid/2-hydroxyethylmethacrylate copolymer was soaked in a colloidal sol of 4.5 wt%, 8 nm titania anatase particles for 7 weeks. Thermal treatment to remove the organic scaffold and induce condensation of surface hydroxyl groups between contacting nanoparticles, gave porous titania monoliths.

For the production of bulk materials monolithic structures of a desired shape are applied; however, in a number of applications, such as photocatalysis and photovoltaics, film structures are more practical. The next topic discusses the use of membranes with a similar "sponge-like" pore morphology for fabricating inorganic films.

2.1.2
Organic Membranes

Membranes such as cellulose acetate or other cellulose derivatives have porous structures that vary depending on the material composing the membrane and its particle retention diameter. The membranes used as templates have a thickness of ~100 μm. For cellulose acetate membranes (viewing the inner morphology perpendicular to the flow direction (plane) of the membrane) the pore size is relatively homogeneous and the membrane is quite dense. Polyamide membranes, on the other hand, show a distinct gradient in pore size when viewed in the same manner, and have less compact structures. Examining either membrane in the direction of flow shows a substantially different morphology to their respective perpendicular structures. The membranes can be handled with tweezers and due to their overall morphology – very thin compared to the polymer gel blocks – are processed during the templating more rapidly.

Simply soaking the membrane in the liquid precursor and then conducting sol-gel reactions by transferring the precursor soaked membrane into a water/alcohol solution results in an amorphous inorganic coating of the membrane [11]. Alternatively the liquid precursor can be filtered through the membrane, on a Büchner funnel with applied vacuum, followed by the water/alcohol solution giving a similar result. Heating was used to remove the cellulose acetate membranes to give porous metal oxide films of ~80 μm thickness. Such films, made up of crystalline nanoparticles, have been obtained for titania and zirconia. By choosing different particle retention diameters from 450 to 200 nm for the cellulose acetate membranes, the surface area of the resulting titania films was found to increase from 22 to 74 $m^2 g^{-1}$. (The initial surface areas of the 450 and 200 nm diameter particle retention membranes were 5 and 12 $m^2 g^{-1}$, respectively.) This again shows that the morphology of the organic template has an influence on the properties of the final inorganic material.

Polyamide membranes can also be successfully used as templates. Due to the more "open" structure of the polyamide membrane the final inorganic film obtained by nanocoating retains this open structure giving large pores. Nanocasting, however, gives a much denser structure.

Macroporous silica films have been obtained by casting either polyamide or cellulose acetate membranes [12]. Soaking the membranes in a tetramethylorthosilicate (TMOS)/H_2O/HCl solution for 6 min followed by drying between glass plates gave silica filled membranes. The casting is obvious on removal of the membrane. Scanning electron microscopy, SEM, images showed the porous structure in the silica to be of similar morphology to the organic material in the initial membrane; see Fig. 2.

By combining the use of membrane templates with surfactant and polymer self-assemblies as porogens, distinct macro- and mesopore structures can be obtained in the final inorganic films [12]. Mesopores of ~2.7 or 10 nm were produced when using polyoxyethylene(10) lauryl ether (C12E10) or SE30/30 (a block copolymer composed of poly(ethylene oxide) and polystyrene with an average molecular weight of 3000 g mol^{-1} per block) along with the macropores created on removal of the membrane. Again a casting of the membrane is attained. The surface area of the silica film is enhanced by the additional mesopores to surface areas of 849 and 517 $m^2 g^{-1}$ for the C12E10 and SE30/30 porogens, respectively, compared to the surface area of a macroporous silica without mesopores obtained using the same membrane template, 39 $m^2 g^{-1}$.

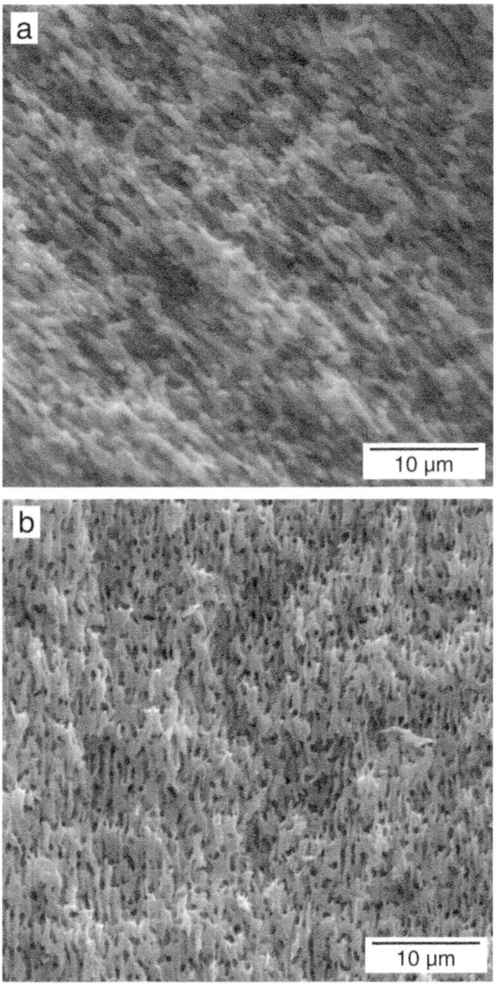

Fig. 2a, b. SEM images of: **a** the membrane template (cellulose acetate); **b** the final silica film. Adapted with permission from [12]. Copyright 2002 Wiley-VCH

2.2
Ordered Structures

Here three template systems are discussed: one that is achieved by assembling monodisperse spheres into a colloidal crystal, the second a membrane (polycarbonate) consisting of cylindrical pores of set dimensions, and the third a biological structure composed of ordered spherical yeast cells. These molds have been cast in most examples, unless blockage of smaller pores prevented the complete filling of larger pores.

2.2.1
Colloidal Crystals

Colloidal crystals result from the ordered assembly of particles (see Scheme 2). Polymer spheres with diameters of 100 nm up to microns have been packed using different methods to form films and monolithic structures with quite large, ordered domains. Monodisperse particles give good long range ordering. Casting of these crystals results in inverse colloidal crystals with ordered spherical voids in the inorganic material (Scheme 2). For a review that encompasses the formation of colloidal crystals and their use as templates for the fabrication of a range of porous materials including silica, metal oxides, metals, and polymers see [13], for macroporous metal oxides in particular the reader is referred to [14].

A lot of research has been conducted in the area of ordered macroporous metal oxides obtained from colloidal crystal templates [15–28]. Starting in 1997, Velev et al. [15] showed that such assemblies could be used as templates for the formation of an inverse silica replica. Charged polystyrene (PS) microspheres (200–1000 nm in diameter) were filtered through a smooth membrane forming a film about 10 μm [15] or 20 μm [16] thick composed of closely packed ordered layers of the spheres. The polymeric material was then functionalized with hexadecyltrimethylammonium bromide before a silica solution was filtered through the film. After gelation of the inorganic and drying, the material was calcined to give silica flakes containing uniform pores in an ordered array. The use of different sized latex particles controlled the final pore size (from ~150 nm to 1 μm) with about 20 to 35% shrinkage in the final material.

Using a surfactant-free method [17], PS particles of an average diameter of 470 nm were deposited on filter paper in a Büchner funnel forming a 1 mm thick layer. After soaking the deposit with alcohol the metal oxide precursor (titanium ethoxide, zirconium *n*-propoxide or aluminum tri-*sec*-butoxide) was dripped onto the close-packed assembly of spheres while vacuum was applied. On removal of the organics, the powder particles of titania (anatase phase), zirconia (baddeleyite), or alumina (amorphous) had spherical voids of 320–360 nm in diameter. Shrinkage was 26–32% measured by center-to-center separation between cavities compared with the original latex diameter. The sphere voids were interconnected with "windows" due to the close packing (i.e., points of particle-particle contact) of the organic spheres. Areas of hundreds of microns showed cubic close packing while hexagonal close packing and less regular packing were also obtained. Increased ordering could be obtained with a faster alkoxide flow rate and decreased humidity.

Using a different sedimentation procedure to form the crystals [18] – centrifuging the colloidal suspensions in glass capillaries (0.3 mm thick, 3 mm wide) – polycrystalline sediments larger than 10 mm in length were formed. The template was penetrated with tetrapropoxy-titanate in ethanol under N_2 before being exposed to atmospheric moisture and drying. The procedure was repeated a number of times depending on the titania precursor to ethanol ratio. The PS spheres (with diameters between 360 nm and 3 μm) were removed by heating at 450°C, giving a hexagonal pore pattern in the titania with lattice parameters about 33% less than the original colloidal crystal.

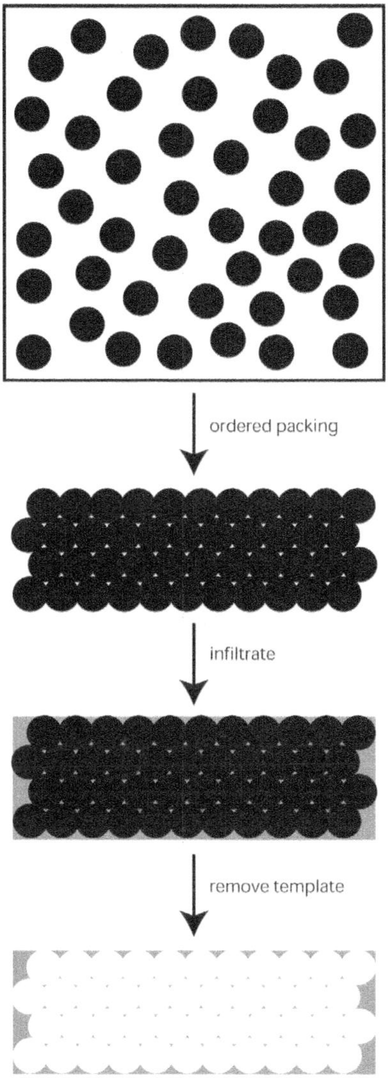

Scheme 2. Formation of colloidal crystals and their use as templates. A colloidal dispersion containing monodisperse particles undergoes controlled filtration, centrifugation, dip coating, sedimentation, or physical confinement, which results in ordered packing of the particles with void spaces between them. By infiltrating these spaces with precursor solution or preformed nanoparticles the hybrid material is formed. Removal of the polymer template (using solvent (toluene) or heating techniques) gives an inverse replica with air-filled, interconnected voids of monodisperse size, which is dependent on the initial particle size

Colloidal crystals have been combined with amphiphilic block copolymers and micromolding (the use of molds that are micron-scaled) to produce patterned silica with hierarchical pore structures [19]. The structure control displayed in this example was on three discrete length scales: 10 nm from the block copolymer (Pluronic F127 or P123), 100 nm due to the latex spheres, and the micromolding gave order in the micron domain. A poly(dimethylsiloxane), PDMS, mold with open ends was placed on a substrate so that a drop of the latex colloidal solution placed at one end could fill the channels, where upon drying the particles organized into a close-packed array. The sol-gel precursor and block co-polymer were mixed and filled the latex voids in the micromold channels by capillary action. After condensation of the silica precursor the PDMS mold was removed and the hybrid calcined at 450°C for 2 h in air to remove the latex and the block copolymer. Isolated patterned structures could also be obtained by placing the mixed sol-gel/block copolymer and latex suspension on a substrate before applying the PDMS mold.

Xia and co-workers have constructed a cell for ordering PS spheres [20]. This consists of two glass substrates framed with photoresist which contains outlets that allow drainage of solvent while retaining the PS spheres which pack to a highly ordered array with applied gas pressure and sonication. Repeated precursor/alcohol solution infiltrations were conducted to achieve filling – as complete as possible – of the voids [29]. Etching of the PS by dissolving in a toluene bath gave 3D porous titania and silica membranes with long-range periodic structure. Although organic inverse replicas with open pores on the membrane surfaces (due to template contact with the substrates) are achieved on removal of the template and substrates [30, 31], the inorganic membranes were too fragile to be taken from the cell support [29].

Holland et al. extended the possible oxide structures to include not only silica, mesoporous silica, titania, zirconia, a yttria stabilized zirconia, and alumina but also oxides of W, Fe, V, and Sb [21]. These latter transition metals formed less ordered structures, containing areas of non-porous material. Different dilutions of alkoxide in alcohol resulted in various inorganic loadings, and moderate control in the wall thickness and window sizes between spherical voids [21]. SEM images of a series of macroporous titania structures obtained with different alkoxide dilutions in ethanol are shown in Fig. 3. Gundiah and Rao have also prepared macroporous materials of ternary mixed oxides, $PdTiO_3$ and $Pb(ZrTi)O_3$ [22].

Further extension of the sol-gel and colloidal crystal templates include control of the outer shape of the colloidal crystal by assembling the PS particles in an aqueous droplet at the air/oil interface [23]. The assemblies then have regular shape (on the length scale of a few millimeters): spheres, ellipsoids, and concave disks. This was controlled by the addition of surfactant and an applied electric field. The cubic close-packed, ordered macroporous titania or silica obtained had similar outer shape as the template.

Preformed metal oxide nanoparticles have also been used in the formation of inverse colloidal crystals [24–26]. Slurries of titania nanocrystals and the PS spheres are dropped onto a glass substrate and dried slowly (over 24 h) [24]. After pressing (cold isostatic press) the film is slowly heated to 520 °C to remove the PS and produce a titania matrix of ≥10 μmx10 mmx2–3 mm with ordered

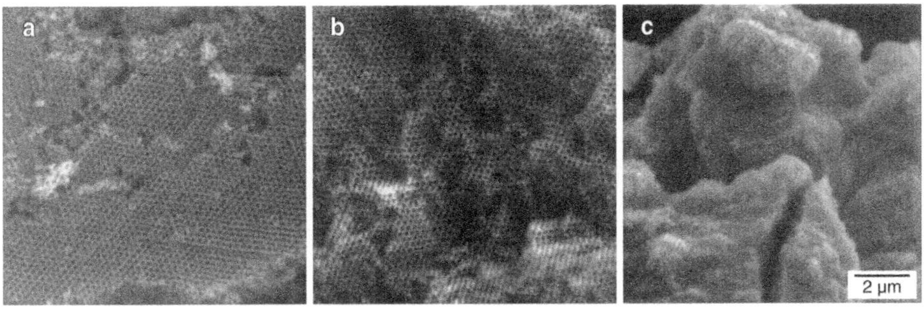

Fig. 3a–c. SEM images of macroporous titania samples synthesized from alkoxides at various dilution levels. The fractions of titanium ethoxide (TET) in ethanol are: **a** 100% TET; **b** 43% TET; **c** 14% TET. The *scale bar* shown corresponds to all of the images. Reprinted in part with permission from [21]. Copyright 1999 American Chemical Society

domains extending from 50 to >100 μm. Windows between sphere voids are also obtained using this method, indicating close packing of the PS, thereby forming an interconnected porous structure in the final inorganic material. Substantially less shrinkage was observed using the preformed nanoparticles with a linear shrinkage of about 6% (compared with 20–35% with the sol-gel approach).

To obtain larger samples Subramania et al. [25] produced samples by slowly drying mixed PS and silica or titania nanoparticles in a glass vial, which results in pieces about 2.5 mm×2 mm×6 mm composed of three regions: a top layer of ordered PS with inorganic particles in the voids, a thin middle layer of ordered PS, and a bottom layer of inorganic nanoparticles. Silica or anatase TiO_2 macroporous structures have been obtained. With further heat treatment (above 850°C) of the titania sample, conversion of anatase crystals to the rutile phase and the resulting crystal growth leads to disruption of the macropores, as is commonly observed.

Highly ordered SiO_2 and TiO_2 macroporous materials can be obtained on a substrate by using a solvent evaporation or "vertical deposition technique" [27, 28]. The PS spheres (406 nm) and SiO_2 (7 nm) or TiO_2 (13 nm) nanoparticles are mixed at a concentration ratio of 1:1 in a vial into which a glass substrate is immersed. Leaving the vial in a constant temperature (50°C) and humidity (30%) chamber leads to the cooperative assembly of the PS and nanoparticles on the substrate with solvent evaporation. Films of 0.5 cm^2 are formed over 24 h. Removal of the PS leaves highly ordered (extending more than 100 μm) porous inorganic materials.

Polymer spheres have also been used for the formation of non-ordered porous systems in inorganic structures and such structures are being tested in applications such as bone tissue implants [32]. Below are examples of casting of discrete entities to form inorganic monoliths or film structures using either sol-gel or nanoparticle infiltration.

Suspensions of polymer spheres of various diameter and surface functionality were templated to produce monolithic silica containing dispersed spherical voids [33]. The pH of the aqueous latex solution was decreased to 2 before the addition of the silica precursor (TMOS). After removal of the methanol and aging, the latex was pyrolyzed by calcination at 450°C, forming a monolith with spherical pores "possessing a liquid-type interparticular order". Bimodal pore sizes have also been prepared in the silica by the simultaneous templating of latex spheres and a liquid crystalline surfactant phase [33]. Hence, silica materials with spherical pores and a range of pore sizes and density could be obtained. The surfactant assembly controls the size and connectivity of the mesopore system, while the additional spherical pores are obtained by the presence of the polymer particles.

Macroporous titania films have been prepared on titanium surfaces [32] using PS spheres (0.5, 16, and 50 μm in diameter) with TiO_2 particles (39 nm). A slurry of the mixed PS and TiO_2 particles was spread on the titanium substrate and dried slowly (1 day). After heating to 450–950°C, titania with a disordered pore structure was obtained with a thickness of 1 to 0.1 mm on the titanium.

Quite a different approach is to use an inorganic (silica) colloidal crystal to form a polymeric material with ordered spherical voids that can then act as a template for the fabrication of other inorganic materials. Such macroporous polymer materials with ordered spherical voids can be obtained by templating monodisperse silica spheres [34]. For example, a poly(methylmethacrylate), PMMA, mold infiltrated with liquid precursors can be used for the fabrication of monodisperse (5%) inorganic spheres of titania, zirconia, and alumina. The inorganic oxide does not adhere to the PMMA surface but comes away, and therefore complete filling of the voids in the macroporous polymer can be achieved by successive precursor infiltrations. Using PS molds the inorganic material coats and adheres to the template surface, therefore giving hollow spheres of TiO_2, ZrO_2, and Al_2O_3. The wall thickness of the sphere is controlled by the number of coatings. Using this method it is also possible to construct binary metal oxide hollow spheres by successive deposition of different metal oxide precursors, such as Al_2O_3 followed by ZrO_2 [34]. The macroporous polymer template can be deformed by heating the mineral oil filled mold above the glass transition temperature of the polymer, stretching the polymer, and then cooling quickly while in the deformed state. This produces different shaped template voids, such as oblate or ellipsoidal, which can be templated in a similar fashion.

2.2.2 *Polycarbonate Membranes*

Martin and coworkers have demonstrated the use of cylindrical-pored templates for the preparation of tubes and fibers composed of metal oxides, metals and polymers [35, 36]. Track-etching of polycarbonate films gives membranes with cylindrical pores that are randomly distributed across the membrane. The pore diameters are monodisperse and, in the example described here, are 600 nm. Lakshmi et al. have used these membranes for the preparation of vanadium oxide fibers [36]. The pores of the organic filter were filled with vanadium(V) triisopropoxy oxide in an argon atmosphere. Exposure to air at 60°C induces hydrolysis of the precursor before an oxygen plasma is used to remove the polycarbonate. The crystalline alpha phase vanadium oxide (V_2O_5) fibers obtained by this

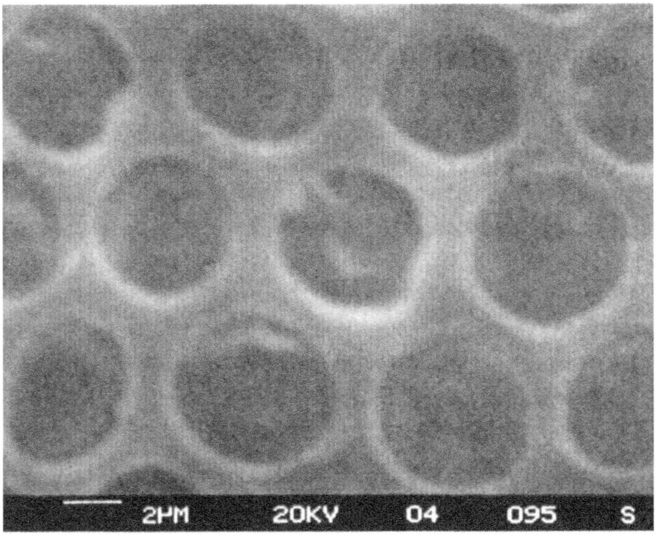

Fig. 4. Scanning electron microscopy image of the silica film templated by spherical cells. Reprinted with permission from [39]. Copyright 2000 American Chemical Society

casting method are monodisperse and protrude from a surface V_2O_5 film. These films have been assessed for reversible lithium ion intercalation [37, 38].

2.2.3
Biological Assembly

Self-organization of cell aggregates can result in ordered templates for replication. An example of spherical yeast cells is given here. The yeast cells can be coated on substrates to give films containing ordered packing of the cells, similar to that obtained in the preparation of colloidal crystals.

As for the preparation of colloidal crystals using polymer spheres, the monodispersity of the cells strongly influences the order of the material. Hence yeast cells were carefully grown to form spherical cells of similar diameter. These cells were dip coated with a silica sol on a microscope slide [39]. A monolayer of the cells arranged in a hexagonal close packing form on the microscope slide. The interstitial sites between the cells contained silica, an SEM image of the film is shown in Fig. 4. The cells remain alive and such films have potential applications in catalysis and as sensors.

3
Hollow Fibers and Threads

Fibrous templates give rise to hollow fiber materials when the nanocoating technique is applied. This is demonstrated below for polymer fibers, a bacterial assembly, and organogelators.

3.1
Synthesized Fibers

Here the use of polymer fiber templates prepared using an electrospinning technique [40, 41] is described. The polymer melt, or a polymer solution in dichloromethane, was expressed through a syringe while a potential was applied between the metal syringe tip and a plate under the collection substrate. The fibers formed using this method can be varied widely in their composition, diameter, and surface morphology. Either nanocoating or nanocasting could be achieved by the sol-gel method, depending on the concentration of the precursor solution.

3.1.1
Polymeric Fibers

Two polymer fibers are discussed: poly(L-lactide) fibers with diameters of 1–3 μm (the majority were about 1–1.5 μm) that had oval indentations on the surface (Fig. 5a), and nylon fibers with diameters of 600 nm and comparatively smooth surfaces. During the electrospinning process the fibers were formed on top of each other in a fibrous mat.

The use of such micron-sized polymeric fibers as templates with the sol-gel templating technique has proved successful in showing the ability to mimic even nanoscale surface morphologies [42]. Applying a dilute titania precursor solution (TIP:isopropanol ratio of 1:19, by volume) and drying in a vacuum desiccator, followed by calcination, afforded TiO_2 tubes of slightly decreased diameter than the original fibers. This indicated successful coating of the tubes. Transmission electron microscopy, TEM, analysis of the inorganic product showed the tubes to have nodules of similar size to the original indentations in the polymeric fiber (Fig. 5b).

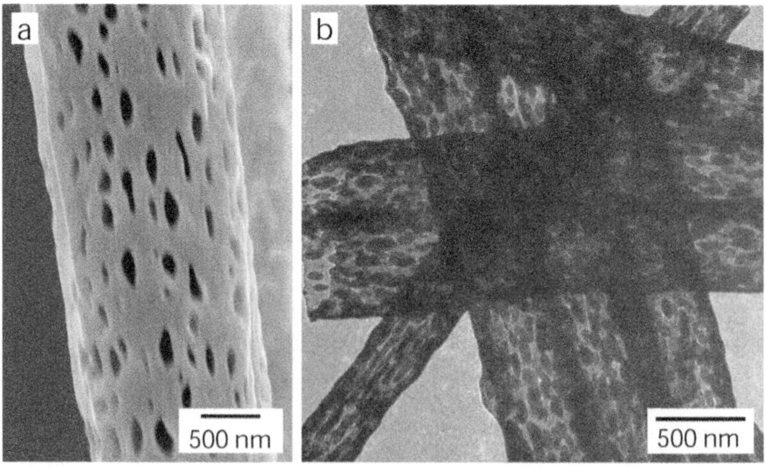

Fig. 5. **a** SEM image of an initial polymer fiber. **b** TEM image of the titania tubes. Adapted with permission from [42]. Copyright 2001 Wiley-VCH

Applying a similar method to the nylon fibers with a more concentrated precursor resulted in films containing a hollow morphology on removal of the fibers. Hence in this case a casting of the fibrous mat was attained. It was noted that the film containing the fiber had relatively few cracks compared with the flakes formed after calcination by the deposition of the titanium precursor on a glass slide without the fibers [43].

3.2 Assemblies

Here the templating of an assembly of biological cells that have a fiber-like structure with an ordered inner arrangement and gels of organogelators, which also have a fibrous structure, are discussed. Both nanocasting and nanocoating can be achieved in the first example, whereas the use of organogel templates showed controlled nanocoating, giving inorganic hollow fibers upon heating.

3.2.1 *Biological Threads*

Cylindrical bacterial filaments, formed by the growth of cylindrical shaped cells of *Bacillus subtilis*, are drawn from solution to form this template. The structure is composed of the filaments aligned along the direction of the thread and packed in a hexagonal fashion. This bacterial superstructure has been infiltrated with preformed silica nanoparticles [44]. The negatively charged particles could infiltrate the swollen macroscopic threads as binding with the anionic cell walls is inhibited. (The charge on the nanoparticles plays an important role in the use of this template, since positively charged titania and alumina particles did not infiltrate the fiber but coated its outer surface with a uniform layer [44].) On drying the silica particles are retained with 50 and ~200 nm thick walls formed between the filaments and in the voids, respectively. Aggregation and condensation of hydroxyl groups on removal of the organic template gave a fiber structure with an inverse morphology to the template: ordered macroporous channels (0.5 μm diameter) were found to run parallel along the direction of the thread.

This cellular assembly has also proven a good template for the fabrication of hierarchical pore structures by penetrating the thread with an aqueous tetraethylorthosilicate (TEOS)/hexadecyltrimethylammonium bromide/base solution. The resulting material on removal of the filaments and added porogen gave ordered macropores with periodic mesoporous silica walls [44]. Here the final structure indicated a nanocoating of the individual filaments composing the thread compared to the previous example where a nanocasting of the structure was obtained.

3.2.2 *Organogelators*

Organogelators are low molecular weight organic compounds that can gel solvents at low concentrations. Terech and Weiss [45] have recently reviewed organogelators and their properties. In general, the gel is formed by heating the gelator in an organic liquid and then cooling, which leads to fibrous aggregation of the organic compound. Hybrid fibers and, on removal of the organic materials, hollow inorganic fibers have been formed by the addition of inorganic precursor molecules to the organogel and solvent. The mechanism proposed is initially the formation of the gel, followed by electrostatic attraction of the metal oxide oligomers and continued polymerization of the metal oxide to form the coating; see Scheme 3. Hence the organic superstructures that are held together by noncovalent interactions can be permanently fixed in inorganic materials.

Ono et al. [46] showed that silica coated organic fibers could be obtained when TEOS was polymerized in the presence of a cholesterol-based gelator that contained quaternary ammonium groups. The absence of this functionality did not give the same results, as the cationic groups on the fibrils allowed the ad-

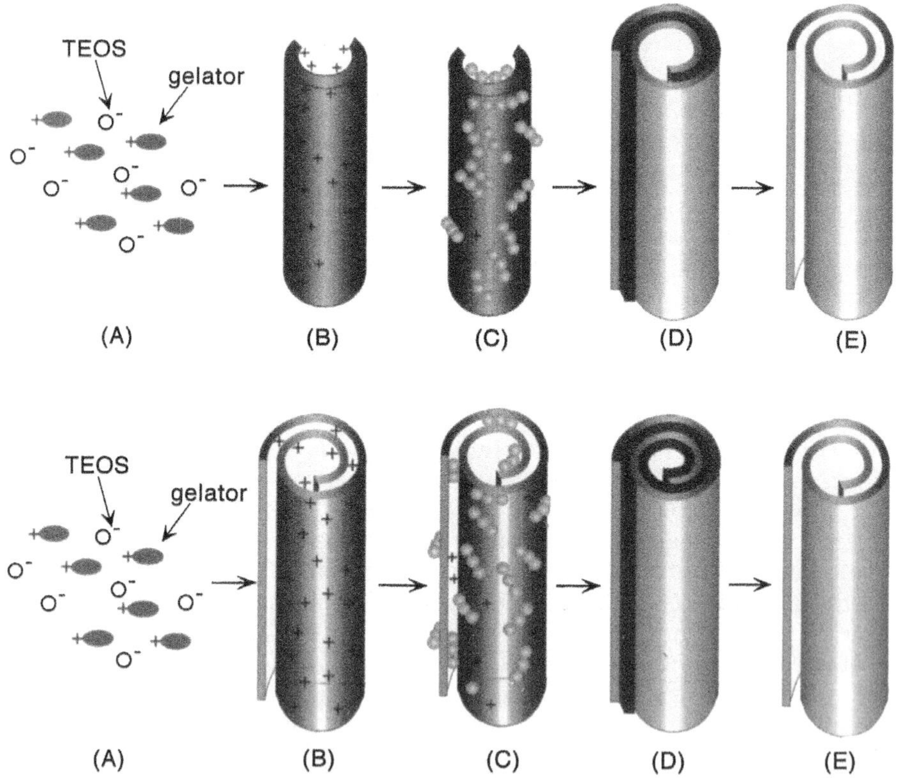

Scheme 3A–E. Mechanism of inorganic coating of organogelators. Schematic representation for the creation of paper-like roll silica by sol-gel polymerization of TEOS in the organogel state of 4: **A** mixture of gelator and TEOS; **B** gelation; **C** sol-gel polymerization of TEOS and adsorption onto the cationic gelator; **D** before calcination; **E** paper-like roll multilayer structure of the silica formed after calcination. Reprinted with permission from [49]. Copyright 2000 American Chemical Society sorption of the anionic silica oligomers where polymerization could then continue along the fibril. Hollow silica fibers were formed on removal of the gelators by calcination. The addition of an azacrown moiety to the cholesterol based gelator required high salt concentrations to produce a hollow silica material as the final product, again due to electrostatic coupling between the template and condensing silica [47]. Variation of the final silica structure, i.e., wall thickness, surface texture, and a tubular or "rolled paper-like" structure could be obtained by changing the salt concentration and the type of crown ether added to the gelator, which effected the organogel superstructure [48, 49]. Hydrogen bonding interactions [50] between gelator superstructures and silica were also successful in the formation of hollow silica structures.

Titania hollow fibrous structures can be formed as well by the use of organogel templates through either electrostatic or hydrogen bonding interactions between the template and the titanium compound [51, 52]. Calcination at 450°C removes the organogel giving hollow anatase or anatase/rutile titania fibers with lengths of up to 200 μm and outer diameters between 150 and 1200 nm. The titania materials consist of crystalline nanoparticles (diameter 15–30 nm) compared with the silica structures, which remain amorphous.

Chirality can be induced into the silica structure by the use of either a cholesterol based gelator with an azacrown moiety in the presence of metal cations [53] or organogels comprised of chiral diaminocyclohexane derivatives [54]. Helical ribbons and double layered nanotubes (diameter ~500 nm) composed of titania can also be obtained [55].

The sol-gel transcription of organogels has been extended to sugar integrated gelators showing structural versatility [56], monodisperse inner diameters [57], and control of the inner diameter [58]. Single fibers, "lotus-type" fiber structures, or spherical superstructural aggregates can be formed. These structures can then be transcribed into silica by sol-gel polycondensation giving single or multiple hollow fiber structures or spherical structures [56]. Inorganic hollow fibers with rather monodisperse inner diameters of 5 or 9 nm (outer diameter ~50 nm) can also be obtained [57]. Control over the inner diameter of the final inorganic tube was dependent on gelator concentration, with inner diameters of 20–25 nm for 0.1 wt% and 450–600 nm for 3 wt% obtained [58].

Diazocrown appended cholesterol gelators formed vesicular aggregates that have also been templated [59]. The spherical silica structures obtained had two distinct diameters, ~200 nm (interconnected spheres) and ~2500 nm (isolated spheres) which closely resembled the initial gelator assembly. Poly(L-lysine), which also forms spherical aggregates (in the presence of a hydrophobic base additive), can be templated to give, as the end product, hollow silica spheres of various diameters [60]. As yet no attempts have been made to control the aggregate size.

The formation of hierarchical pore structures in the silica has also been achieved by the use of a gelator 2, 3-di-*n*-decyloxyanthracene in methanol [61]. Hollow fibers with micron-sized diameters are obtained on removal of the gelator, which contain mesopores from smaller gelator aggregates. Changes in the mesopore diameter (5–12 nm) and shape (ink-bottle or cylindrical) occur for different gelator concentrations.

4 Discrete Entities

Individual submicron entities can be cast to form monolithic structures with voids that maintain the initial organic shape (as was discussed above), or they can be coated to give individual hybrid entities that form hollow inorganic structures on removal of the template. The particles, tubes, and crystals are generally dispersed in solution for templating.

4.1 Spherical Particles

A number of inorganic materials have been coated on polymer spheres, which can be obtained with a variety of diameters and compositions. The inorganic coating results in significant changes of the sphere properties. The organic core can also be removed to form hollow inorganic spheres. Porous polymer spheres have also been templated to produce non-agglomerated, porous inorganic beads.

4.1.1 *Solid Latex Particles*

As polymer spheres are easily accessible with low polydispersity, they are attractive templates. Coating in non-aqueous solvents is usual as most metal alkoxide precursors rapidly undergo hydrolysis and condensation in the presence of water.

Silica coatings on PS spheres (2.3 μm) were obtained in ethanol [62]. The microspheres that were prepared in the presence of poly(vinyl pyrrolidone), PVP, were dispersed in ethanol, to which water, base, and TEOS were added. Seeded polymerization of TEOS on the surface of the spheres (as well as free silica particle formation in solution) led to a coating composed of silica nanoparticles (30–40 nm). After centrifugation and drying the coating step could be repeated. An increase in the silica coating from 7.8 to 18.5 wt% was obtained after one and three coatings, respectively. Hollow silica spheres could be obtained after burning off the PS at 600°C.

Shiho and Kawahashi [63] showed that amorphous titania can be coated on PS spheres. Titanium tetrabutoxide (TBOT) is hydrolyzed in the presence of anionic PS spheres (diameter 420 nm) and PVP as a protective agent by aging at 100°C. Centrifugation and redispersion cleaned the system to give coated spheres with a coating thickness of ~60 nm. If there was an excess of TBOT, separate TiO_2 particles formed in solution. When the dried samples were calcined at 600°C, hollow anatase titania spheres were obtained. Using higher temperatures during calcination (900°C) resulted in phase transformation to the rutile crystal and these crystal particles no longer maintained the hollow spherical structure.

Zhong et al. [64] have shown that the cell designed for fabricating colloidal crystals and their metal oxide inverse can also be used for the formation of amorphous inorganic coated polymer spheres and hollow inorganic spheres. The crystalline PS array formed between the glass substrates is dried and then infiltrated with dilute precursor solution (for example, TIP:isopropanol, 1:19 v/v or tin-tetraisopropoxy:ethanol 1:19 v/v), which leads to separation of the spheres. On exposure to atmospheric moisture a homogeneous, dense, thin layer of amorphous material (TiO_2 or SnO_2) is coated on the PS beads. Immersing the hybrid material in toluene leads to dissolution of the PS giving hollow amorphous spheres that can be released from the substrate by sonication. The wall thickness was found to be dependent on the polymer sphere diameter and the concentration of the precursor. Hollow spheres were not formed when silica precursors were used due to the longer gelling time, which meant that solvent evaporation occurred before complete gelation, and hence a 3D porous structure was formed. Egg-shaped PS particles, which were prepared by deformation of monodisperse spheres embedded in an elastic matrix, were also coated with titania using a similar set-up [65]. Wall thickness (between 30 and 100 nm) was again dependent on the precursor concentration. Removing the PS by immersion in toluene gave ellipsoidal TiO_2 shells.

Titania can also be coated on cationic PS spheres (diameter 378 nm) in solution by a one-step process [66]. The PS spheres dispersed in ethanol with PVP are rapidly stirred during the addition of the titania precursor in ethanol. The slight negative charge on the titania species ensures their rapid capture by the positive PS spheres, preventing the formation of TiO_2 particles in solution. Thicker coatings can be obtained by varying the PS to titania precursor ratio or by carrying out repeated titanium isopropoxide additions. After centrifugation and redispersion to remove the PVP, smooth titania coatings on homogeneous spheres are obtained. Removing the core using either solvent extraction (toluene) or calcination leads to the formation of hollow TiO_2 spheres. In the case of heating the sample there is some damage to the shells, which consist of mainly anatase crystals.

Preformed metal oxide nanoparticles have been successfully coated on polymer spheres by the use of the layer-by-layer method. This involves the coating of the template spheres with polyelectrolyte layers, which are oppositely charged to the metal oxide nanoparticles to be deposited. Alternating the polyelectrolyte and nanoparticle deposition has led to the successful formation of silica [67, 68] and titania [69] coated PS spheres. Using this approach preformed crystalline nanoparticles can be deposited on the organic spheres and crystalline hollow spheres can be obtained without the need of calcination. On removal of the template and the polymer interlayers by heating, hollow spheres of the inorganic material can be obtained [68–70]. This procedure is described in detail in the chapter by Dr Frank Caruso.

4.1.2
Porous Spheres

Porous polymer spheres with a range of diameters, pore sizes, and composition (or surface functionality) can be purchased. Antonietti et al. have shown that porous polymer spheres could be cast to form monolithic materials [33]. Water-swollen polystyrene sulfonate spheres were cast in an aqueous TMOS/HCl solution. The other latices that had been used in this procedure gave spherical pores on removal of the organic material as casting occurred around the particles (see Sect. 2.2.1); however, these polystyrene sulfonate spheres were infiltrated by the precursor solution and subsequent polyreaction, followed by removal of the organic spheres, gave a mesoporous silica material without spherical pores.

Instead of casting the complete solution to form a monolith, individual porous spheres can be templated to form porous inorganic beads, which maintain porosity from the template. As with any of the discrete organic templates, experimental conditions need to be optimized to ensure that aggregation of the beads does not occur and that there is not excess inorganic material forming on the template surface.

PS cross-linked with divinylbenzene beads, with diameters of either 15 or 30 μm that could also be functionalized with hydroxy or amine groups, were impregnated with precursor, for the fabrication of titania spheres, before being placed in excess water for hydrolysis and condensation reactions to occur [71]. To form silica porous spheres the silica precursor was added to an acidic aqueous solution containing the beads, and the system was heated at 70°C. On calcination the polymeric material was removed giving the inorganic spheres with inner pores. The surface functionality of the beads played an important role in the success of templating with the non-functionalized beads being most suitable for the titania, and the hydroxy or amine groups being required for silica sphere formation. Images of the titania spheres, obtained from SEM and TEM are shown in Fig. 6.

4.2 Elongated Materials

Biological tubes of high aspect ratio, including phospholipid tubes and tubular structures formed by the tobacco mosaic virus, and the inorganic materials obtained by templating them, are described first before moving to a smaller scale, the templating of carbon nanotubes. Organic crystals are also included in this section as they have been used successfully to give hollow elongated inorganic structures.

4.2.1 *Tubes and Nanotubes*

Chiral phospholipid molecules aggregate spontaneously to form tubes with diameters of 500 nm and lengths of ~50–100 μm. Diacetylenic phosphatidylcholine structures were first coated in 1993 by Baral and Schoen [72] with silica nanoparticles. The tubule dispersion was mixed with Ludox (a silica sol with a particle diameter of 10–15 nm and negative surface charge at pH 8.2) and allowed to stand for up to 9 days, during which time a white precipitate formed. TEM analysis of the collected precipitate showed a film with a thickness of about 50 nm, composed of silica particles, on the hollow cylindrical templates. The adsorption of the nanoparticles to the headgroups of the phospholipid is believed

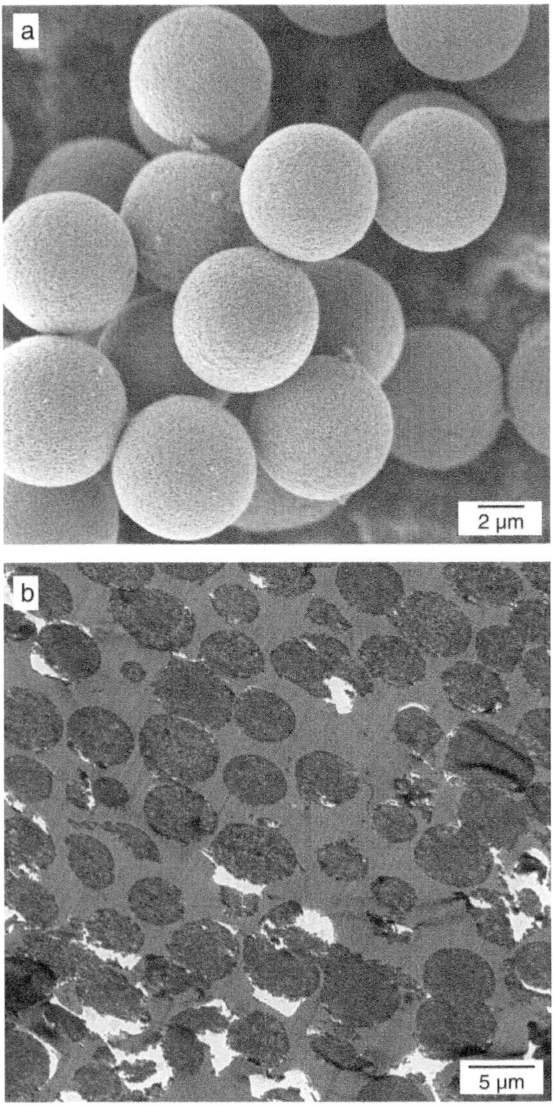

Fig. 6. a SEM image and b TEM image of the porous titania spheres obtained on templating of PS cross linked with divinyl benzene porous beads. Adapted with permission from [71]. Copyright 2002 Wiley-VCH

to be due to electrostatic interactions. Heating can be used to remove the tubules, giving hollow cylindrical silica materials with lower aspect ratios (length decreases to 15–25 μm). This structure is different to that obtained by simultaneous phospholipid crystallization and silica polymerization, where helical silica-lipid multilamellar microstructures were obtained [1].

The tobacco mosaic virus is a hollow protein nanotube with inner and outer diameters of 4 and 18 nm, respectively, and lengths of 300 nm [73]. The virus is stable over a range of pH values and has been used as a template for the formation of inorganic-organic hybrid nanotubes with reaction pHs ranging from 2.5 to 9. These tubular structures were templated using sol-gel chemistry. A silica coating of 3 nm formed on the outer surface of the tobacco mosaic virus tube when a suspension of the hollow tubes was mixed with an acidified TEOS/ethanol solution [73]. TEM analysis showed that the white precipitate obtained is composed of longer silica coated nanotubes due to end-to-end self assembly of the virus.

Carbon nanotubes with inner diameters of 2–8 nm and lengths of up to 1 μm have been used for the fabrication of both metal oxide nanotubes and nanorods, as well as having application in the coated form. Satishkumar et al. [74] used multiwalled carbon nanotube templates for the formation of silica, alumina, and vanadium oxide nanotubes. After stirring the carbon nanotubes in the precursor solutions (TEOS, aluminum isopropoxide/H_2O, or a vanadium pentoxide gel) the solids obtained were heated to give inorganic coatings on the organic nanotubes. Further heating under air leads to the removal of carbon, giving well preserved SiO_2 and V_2O_5 tubes. The tubular structure was not well maintained for Al_2O_3 on removal of the template.

Zirconia nanotubes were also obtained using a similar method with a zirconium propoxide precursor [75]. After oxidizing the carbon, zirconia tubes with a diameter of ~40 nm, 6 nm wall thickness, and several micrometers long were obtained. The ZrO_2 was composed of mixed crystal phases (monoclinic and tetragonal). Increased temperature treatment led to collapse of the nanotubes. The addition of yttria in a slightly modified procedure gave a more stable nanotube structure with similar wall thicknesses. The yttria-stabilized zirconia had a cubic structure.

Extending the work further to other oxides gave V_2O_5, WO_3, MoO_3, RuO_2, and IrO_2 coated carbon nanotubes [76]. TEM images of some examples are shown in Fig. 7. However, on removal of the carbon, nanorods of V_2O_5, WO_3, MoO_3 (and with further treatment MoO_2), Sb_2O_5, RuO_2, and IrO_2 were formed. Nanotubes of MoO_3 and RuO_2 were also obtained. The nanorods had a range of diameters (from 10 to 200 nm) and lengths up to a few micrometers.

Nanoparticles have also been used for coating carbon nanotubes. The multiwalled carbon nanotubes (average length 50 μm) were combined with a silica sol that was prepared by mixing the silane precursor (either methyl, ethyl, or propyl-trimethoxysilane) with ethanol and hydrochloric acid [77]. After thorough mixing and then drying, a porous film exhibiting fibrous structures is obtained. Varying the amount of sol added to the carbon nanotubes resulted in different loading amounts of silica to carbon. The electrochemical properties of the composite film also altered with changes in silica to carbon nanotube composi-

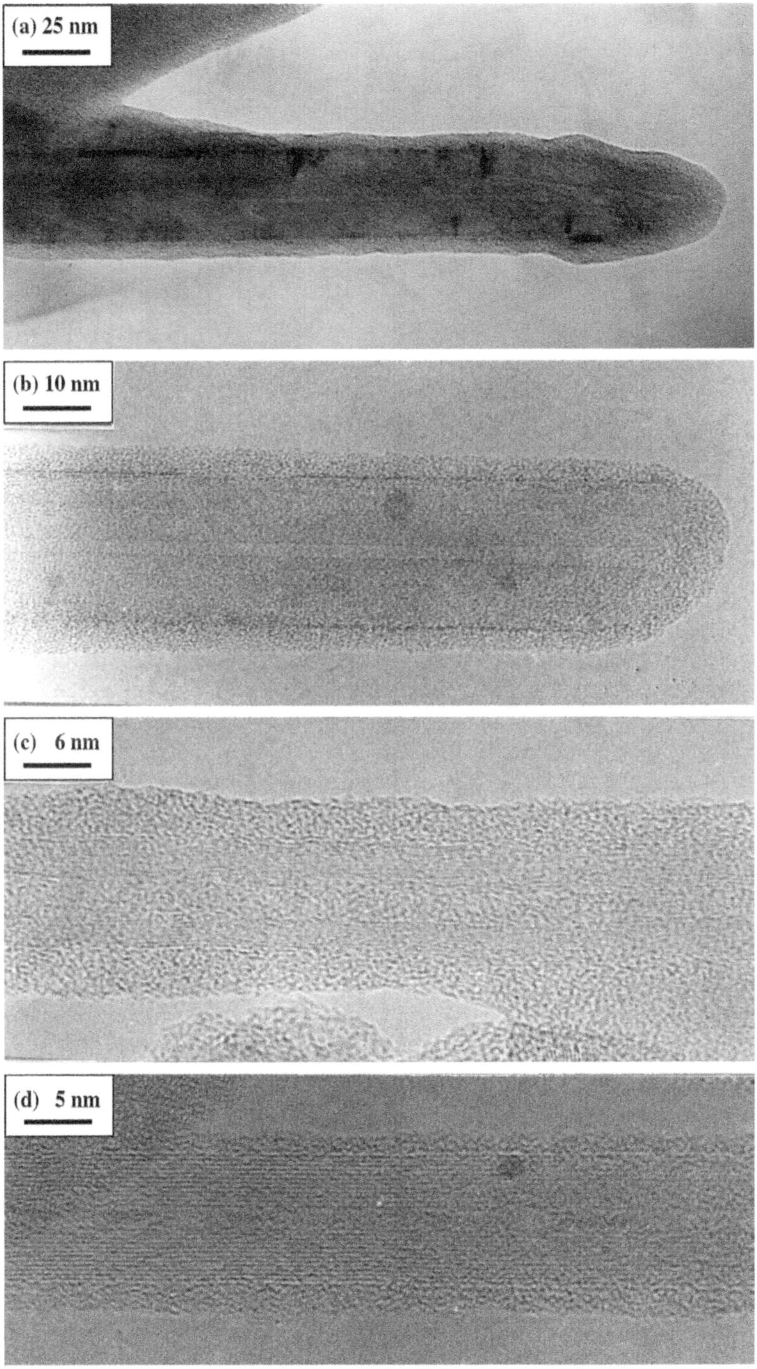

Fig. 7a–d. Metal oxide coated carbon nanotubes: **a** SiO_2; **b** V_2O_5; **c** WO_3; **d** MoO_3. Adapted with permission from [76]. Copyright 2000 The Royal Society of Chemistry

tion with reduction in double layer capacitance with either increasing silica concentration or increasing chain length of the alkoxide.

4.2.2
Organic Crystals

Ammonium *dl*-tartrate crystallizes in ethanol to give needle-like crystals, which can be used as templates for surface specific sol-gel polymerization [78]. The addition of NH_4OH to an ethanol solution containing TEOS and *dl*-tartaric acid gave a precipitate. As ammonium *dl*-tartrate crystals are soluble in water, the water formed on silica condensation and the water washing of the precipitate removes the template to yield smooth, open-ended silica tubes (length 200–300 μm) with wall thicknesses ranging from 30 to 300 nm [79, 80]. The size of the tubes could be varied by the addition of base to solution before the silica precursor. Triangular- or rectangular-shaped (0.7–6.5 μm wide), elongated silica structures (length 50–100 μm) could also be obtained by the crystallization of ammonium oxalate hydrate [78].

5
Summary

The examples given in this chapter demonstrate how organic materials of various structure can be utilized for the formation of either novel hybrid systems or structured metal oxide materials by nanocasting or nanocoating techniques. The two procedures discussed – (i) the use of sol-gel polymerization within or around the organic mold and (ii) nanoparticle infiltration and adhesion to the organic mold – have been used to form structured porous monoliths, films, and individual entities composed of metal oxide or hybrid materials. Composite structures can have significantly different properties to both the organic and inorganic starting materials. On removal of the organic template, novel inorganic structures are produced with structural properties that can lead to enhanced performance or new applications. For example, lower densities, larger surface areas, ordered pore systems, film or monolith formation, and higher flow throughput are beneficial in numerous applications. In general, porous metal oxide materials with controlled structure formation can be used in catalytic applications, chromatography, as containers for encapsulation and delivery, fillers, in quantum optics, as sensors, separators, in solar energy applications, or as pigments for paper or paints.

Fabrication of new organic templates with interesting structures coupled with further control of the ordering or dispersity of current templates will be a prime task for future application of nanocasting and nanocoating techniques.

Acknowledgement. CNR Rao is thanked for providing original photos for Fig. 7. Jan H. Schattka is especially appreciated for assistance with figure preparation. Markus Antonietti, Frank Caruso, Atul Deshpande, Michael Giersig, Andreas Greiner, Hans-Peter Hentze, Anders Larsson, Ulrika Meyer, Helmuth Möhwald, Jan H. Schattka, Dmitry G. Shchukin, and Andrei Susha are thanked for fruitful discussions and collaborations. The Max Planck Society is acknowledged for funding.

References

1. Mann S, Burkett SL, Davis SA, Fowler CE, Mendelson NH, Sims SD, Walsh D, Whilton NT (1997) Chem Mater 9:2300
2. Brinker CJ, Scherer GW (1990) Sol-gel science: the physics and chemistry of sol-gel processing. Academic Press, California
3. Yoldas BE (1986) J Mater Sci 21:1086
4. Rouquérol J, Avnir D, Fairbridge CW, Everett DH, Haynes JH, Pericone N, Ramsay JDF, Sing KSW, Unger KK (1994) Pure Appl Chem 66:1739
5. Antonietti M, Göltner C, Hentze H-P (1998) Langmuir 14:2670
6. Antonietti M, Caruso RA, Göltner CG, Weissenberger MC (1999) Macromolecules 32:1383
7. Caruso RA, Giersig M, Willig F, Antonietti M (1998) Langmuir 14:6333
8. Caruso RA, Antonietti M, Giersig M, Hentze H-P, Jia J (2000) Chem Mater 13:1114
9. Schattka JH, Shchukin DG, Antonietti M, Caruso RA (2002) Chem Mater 14:5103
10. Breulmann M, Davis SA, Mann S, Hentze H-P, Antonietti M (2000) Adv Mater 12:502
11. Caruso RA, Schattka JH (2000) Adv Mater 12:1921
12. Caruso RA, Antonietti M (2002) Adv Funct Mater 12:307
13. Stein A (2001) Microporous Mesoporous Mater 44/45:227
14. Blanford CF, Yan H, Schroden RC, Al-Daous M, Stein A (2001) Adv Mater 13:401
15. Velev OD, Jede TA, Lobo RF, Lenhoff AM (1997) Nature 389:447
16. Velev OD, Jede TA, Lobo RF, Lenhoff AM (1998) Chem Mater 10:3597
17. Holland BT, Blanford CF, Stein A (1998) Science 281:538
18. Wijnjoven JEGJ, Vos WL (1998) Science 281:802
19. Yang P, Deng T, Zhao D, Feng P, Pine D, Chmelka BF, Whitesides GM, Stucky GD (1998) Science 282:2244
20. Park SH, Qin D, Xia Y (1998) Adv Mater 10:1028
21. Holland BT, Blanford CF, Do T, Stein A (1999) Chem Mater 11:795
22. Gundiah G, Rao CNR (2000) Solid State Phys 2:877
23. Yi G-R, Moon JH, Yang S-M (2001) Chem Mater 13:2613
24. Subramania G, Constant K, Biswas R, Sigalas MM, Ho K-M (1999) Appl Phys Lett 74:3933
25. Subramania G, Manoharan VN, Thorne JD, Pine DJ (1999) Adv Mater 11:1261
26. Subramania G, Constant K, Biswas R, Sigalas MM, Ho K-M (1999) J Lightwave Technol 17:1970
27. Meng Q-B, Gu Z-Z, Sato O, Fujishima A (2000) Appl Phys Lett 77:4313
28. Meng Q-B, Fu C-H, Einaga Y, Gu Z-Z, Fujishima A, Sato O (2002) Chem Mater 14:83
29. Gates B, Yin Y, Xia Y (1999) Chem Mater 11:2827
30. Park SH, Xia Y (1998) Chem Mater 10:1745
31. Park SH, Xia Y (1998) Adv Mater 10:1045
32. Akin FA, Zreiqat H, Jordan S, Wijesundara MBJ, Hanley L (2001) J Biomed Mater Res 57:588
33. Antonietti M, Berton B, Göltner C, Hentze H-P (1998) Adv Mater 10:154
34. Jiang P, Bertone JF, Colvin VL (2001) Science 291:453
35. Martin CR (1996) Chem Mater 8:1739
36. Lakshmi BB, Patrissi CJ, Martin CR (1997) Chem Mater 9:2544
37. Patrissi CJ, Martin CR (1999) J Electrochem Soc 146:3176
38. Patrissi CJ, Martin CR (2001) J Electrochem Soc 148:A1247
39. Chia S, Urano J, Tamanoi F, Dunn B, Zink JI (2000) J Am Chem Soc 122:6488
40. Bognitzki M, Hou H, Ishaque M, Frese T, Hellwig M, Schwarte C, Schaper A, Wendorff JH, Greiner A (2000) Adv Mater 12:637
41. Bognitzki M, Czado W, Frese T, Schaper A, Hellwig M, Steinhart M, Greiner A, Wendorff JH (2001) Adv Mater 13:70
42. Caruso RA, Schattka JH, Greiner A (2001) Adv Mater 13:1577
43. Caruso RA, Greiner A (unpublished results)
44. Davis SA, Burkett SL, Mendelson NH, Mann S (1997) Nature 385:420
45. Terech P, Weiss RG (1997) Chem Rev 97:3133

46. Ono Y, Nakashima K, Sano M, Kanekiyo Y, Inoue K, Hojo J, Shinkai S (1998) Chem Commun 1477
47. Ono Y, Kanekiyo Y, Inoue K, Hojo J, Shinkai S (1999) Chem Lett 23
48. Jung JH, Ono Y, Shinkai S (1999) J Chem Soc Perkin Trans 2:1289
49. Jung JH, Ono Y, Shinkai S (2000) Langmuir 16:1643
50. Jung JH, Ono Y, Shinkai S (2000) Chem Lett 636
51. Kobayashi S, Hanabusa K, Hamasaki N, Kimura M, Shirai H, Shinkai S (2000) Chem Mater 12:1523
52. Kobayashi S, Hanabusa K, Suzuki M, Kimura M, Shirai H (1999) Chem Lett 1077
53. Jung JH, Ono Y, Shinkai S (2000) Angew Chem Int Ed 39:1862
54. Jung JH, Ono Y, Hanabusa K, Shinkai S (2000) J Am Chem Soc 122:5008
55. Jung JH, Kobayashi H, von Bommel KJC, Shinkai S, Shimizu T (2002) Chem Mater 14:1445
56. Jung JH, Amaike M, Nakashima K, Shinkai S (2001) J Chem Soc Perkin Tran 2 10:1938
57. Tamaru S-I, Takeuchi M, Sano M, Shinkai S (2002) Angew Chem Int Ed 41:853
58. Jung JH, Shinkai S, Shimizu T (2002) Nano Lett 2:17
59. Jung JH, Ono Y, Sakurai K, Sano M, Shinkai S (2000) J Am Chem Soc 122:8648
60. von Bommel KJC, Jung JH, Shinkai S (2001) Adv Mater 13:1472
61. Clavier GM, Pozzo JL, Bouas-Laurent H, Liere C, Roux C, Sanchez C (2000) J Mater Chem 10:1725
62. Bamnolker H, Nitzan B, Gura S, Margel S (1997) J Mater Sci Lett 16:1412
63. Shiho H, Kawahashi N (2000) Colloid Polym Sci 278:270
64. Zhong Z, Yin Y, Gates B, Xia Y (2000) Adv Mater 12:206
65. Lu Y, Yin Y, Xia Y (2001) Adv Mater 13:271
66. Imhof A (2001) Chem Mater 13:3579
67. Caruso F, Lichtenfelder H, Giersig M, Möhwald H (1998) J Am Chem Soc 33:8523
68. Caruso F, Caruso RA, Möhwald H (1998) Science 282:1111
69. Caruso RA, Susha A, Caruso F (2001) Chem Mater 13:400
70. Caruso F, Caruso RA, Möhwald H (1999) Chem Mater 11:3309
71. Meyer U, Larsson A, Hentze H-P, Caruso RA (2002) Adv Mater 14:1768
72. Baral S, Schoen P (1993) Chem Mater 5:145
73. Shenton W, Douglas T, Young M, Stubbs G, Mann S (1999) Adv Mater 11:253
74. Satishkumar BC, Govindaraj A, Vogl EM, Basumallick L, Rao CNR (1997) J Mater Res 12:604
75. Rao CNR, Satishkumar BC, Govindaraj A (1997) Chem Commun 1581
76. Satishkumar BC, Govindaraj A, Nath M, Rao CNR (2000) J Mater Chem 10:2115
77. Gavalas VG, Andrews R, Bhattacharyya D, Bachas LG (2001) Nano Lett 1:719
78. Miyaji F, Davis SA, Charmant JPH, Mann S (1999) Chem Mater 11:3021
79. Nakamura H, Matsui Y (1995) J Am Chem Soc117:2651
80. Nakamura H, Matsui Y (1995) Adv Mater 7:871

Top Curr Chem (2003) 226: 119–172
DOI 10.1007/b10827

Mineral Liquid Crystals from Self-Assembly of Anisotropic Nanosystems

Jean-Christophe P. Gabriel[1], Patrick Davidson[2]

[1] Nanomix, Inc., 5980 Horton Street, Suite 600, Emeryville, CA 94608, USA
E-mail: jcgabriel@nano.com
[2] Laboratoire de Physique des Solides, Bât. 510, UMR 8502 CNRS, Université d'Orsay, 91405 Orsay, France, *E-mail: davidson@lps.u-psud.fr*

In this article we review the mesogenic properties of the mineral liquid crystals (MLCs) based on molecular nanowires: $Li_2Mo_6Se_6$; nanotubes: Imogolite and $NaNb_2PS_{10}$; molecular ribbons: V_2O_5; exfoliated single sheets: smectic clays and $H_3Sb_3P_2O_{14}$; nanorods: Boehmite (γ-AlOOH), Akaganeite (β-FeOOH), Goethite (α-FeOOH); platelets: Gibbsite ($Al(OH)_3$); disks: $Ni(OH)_2$; bio-mineral hybrids. We then propose numerous phases that could lead to the discovery of new MLCs. We finally review how the properties of these mesophases and their collective behavior have been used for making mesoporous composites with anisotropic properties, for making nanodevices and solar cells, as well as how they could be used to allow the measurement of residual dipolar couplings in NMR studies of biomolecules.

Keywords. Liquid crystals, Complex Fluids, Self-assembly, Mineral, Inorganic, Nanosystem, Nanoparticle, Nanotube, Nanowire, Nanorod, Anisotropy, NMR

List of Abbreviations

AAO	Anodic Aluminum oxide
AOT	(2-Ethylhexyl)sulfosuccinate
CNT	Carbon nanotube
CVD	Chemical vapor deposition
DLVO	Deryagin Landau Verwey Overbeek
DMF	Dimethylformamide
DMSO	Dimethylsulfoxide
FFEM	Freeze fracture electron microscopy
FWHM	Full width at half maximum
HFCVD	Hot filament chemical vapor deposition
HPC	Hydroxypropylcellulose
HT	High temperature
MLC	Mineral liquid crystals
NMF	*N*-Methylformamide
NR	Nanorod
NT	Nanotube
NW	Nanowire
PBLG	Poly(γ-benzyl-L-glutamate)
PDMS	Poly(dimethylsiloxane)
PVA	Poly(vinylalcohol)
SANS	Small Angle Neutron Scattering
SAXS	Small angle X-ray scattering
TEM	Transmission electron microscopy

TMV Tobacco Mosaic Virus
USAXS Ultra small angle X-ray scattering

1 Introduction

Liquid crystalline phases, which were discovered more than a century ago, are phases intermediate between crystalline and liquid ones [1]. They combine both the fluidity of the liquids and the anisotropy of crystals and they are usually based on anisotropic building blocks (molecules, polymers, micelles, aggregates etc...). Liquid crystalline materials are extremely diverse since they range from DNA to Kevlar (a high strength synthetic polymer) and from small organic molecules used in displays to self assembling amphiphilic soap molecules. A common feature of all these materials is that they are organic [2] or organometallic – ca. inorganic – and currently there are more than 80,000 examples of organic liquid crystalline compounds [3]. In contrast, there are only a dozen lyotropic mineral liquid crystals known and characterized to date [4]. One reason is that there are only a few low-dimensional (one-dimensional or two-dimensional) purely-inorganic objects known to dissolve or to be dispersed in water or in other organic solvents. The solubility criteria of mineral moieties, their interactions with solvents, and the structures and physical properties of such fluids are currently a subject of intense research activity [4, 5]. In addition, organic synthesis techniques allow numerous structural variations of liquid crystalline molecules to be made whereas the solution chemistry of molecular mineral objects, i.e., prepared by high temperatures solid state reactions, is still being developed [6].

Concerning the use of the term "mineral" to name this family of liquid crystals, one can argue that the term inorganic would be more appropriate. However, inorganic liquid crystals have long been used for organometallic liquid crystals [7]. Therefore in order to avoid any confusion between these fairly chemically different families, and taking into account that a large number of these liquid crystals occur naturally in nature, we think that the use of the old fashioned but adequate "mineral" adjective taken sensus largo is more specific that an alternative such as "purely inorganic", to name this subclass of the inorganic liquid crystals family.

In spite of their scarcity, lyotropic mineral liquid crystals have actually been known for many decades: Zocher, a German scientist, reported as early as 1925 the observation of the unusual behavior of colloidal aqueous suspensions of vanadium pentoxide (V_2O_5) [8]. He compared these suspensions with the newly discovered liquid crystalline phases and inferred that these colloids must be constituted from needle-like objects mutually oriented in the same direction. We know now that this behavior in fact arises from V_2O_5 ribbons forming a nematic phase. Later, in 1938, Langmuir examined suspensions of bentonite clay platelets and recognized their anisotropic and fluid-like properties [9]. However, the key theoretical development came with the theory that Onsager devised to describe the nematic ordering of suspensions of rod-like particles such as the Tobacco Mosaic Virus (TMV) [10]. This model provides an extremely useful frame to understand and predict this kind of phase transition in mineral and organic systems. This model which is based on simple assumptions describes in quantitative terms the phase diagram for any suspension of anisotropic rod-like (or disk-like) particles interacting solely through a hard-core potential and predicts that it should undergo a nematic/isotropic first-order transition (i.e., with phase coexistence) and that temperature has no effect on this transition. Numerical simulations have also proved to be very powerful techniques to gain insight into such phenomena [11].

From the experimental point of view, a simple way to identify a liquid crystalline phase is to observe its texture using polarized light microscopy. Unlike usual isotropic liquids, liquid crystalline phases are birefringent, interact with linearly polarized light, and can be observed between crossed polarizers. The detection of birefringence when the fluid phase is at rest proves that it is liquid crystalline. The textures observed under polarized light are due to the existence of topological defects and their analysis gives information on the symmetry of the phase. For instance, the most common defects of the nematic phase are called disclination lines. These are one-dimensional singularities in the local average direction (director) of the anisotropic moieties. Thus, unless special caution is taken to remove these defects, nematic samples are usually "powders" (i.e., random distributions) of small nematic domains (a "nematic domain" is defined as a region of the sample which shows ideal nematic orientational correlation).

Nematic phases can often be aligned by the action of a magnetic field. The alignment is due to the anisotropy of the mesophase that reorients to minimize its free energy in the field. Therefore, the director is aligned along or perpendicular to the field direction, depending on the sign of the magnetic anisotropy of the building blocks. The alignment of the phase in an external field suppresses the topological defects and produces a single domain that is suitable for more detailed physical studies. Electric fields produce similar effects: this is the basis of most applications of nematic phases in display technology.

Small angle X-ray scattering (SAXS) is a powerful technique [12] to study the molecular organization of lyotropic nematic phases on a length scale ranging from 3 to 300 nm. In this range, the detailed atomic structure of the particles is irrelevant and the scattering is sensitive to only the particle shape and its electronic contrast with the solvent. In this respect mineral liquid crystals are particularly convenient because their contrast is usually much larger than that of their organic counterparts. SAXS also allows a clear-cut demonstration of the symmetry and structure of a complex fluid, avoiding possible and usual pitfalls that can come when one restricts itself to only a texture analysis.

In contrast to organic liquid crystals, mineral liquid crystals can differ largely in size from one system to another. They actually range from molecular nanowires or nanotubes to anisotropic crystallites such as nanorods. Molecular nanowires can be considered as one-dimensional objects where the diameter is precisely defined by the molecular unit that makes up the chain (typically 5–25 Å). In contrast, molecular disks (nanosheets) are two-dimensional objects where the thickness of the disks is defined at the molecular scale but the diameter is polydisperse (100–10,000 Å). Finally, liquid crystalline suspensions may be formed from anisotropic crystallites whose dimensions are all in the colloidal range (100–10,000 Å) so that these moieties are two to three orders of magnitude larger than usual organic liquid crystals. This classification defines the outline of this review article.

2 Molecular Nanowires

2.1 Dispersion of a Conducting Nanowire: $Li_2Mo_6Se_6$

Among all the phases that have been discovered after the report of high critical field superconductivity within the Chevrel-Sergent phases (MMo_6X_8–M=Cu, Pb, Sn, etc.; X=S, Se, Te), $Li_2Mo_6Se_6$ is singular since it contains $^1_\infty[Mo_6Se_6]^{2-}$ infinite conducting molecular nanowires [13, 14] and, it has been demonstrated to be soluble in highly polar solvents such as *N*-methylformamide (NMF) or dimethylsulfoxide (DMSO) [15][1]. It is this last property that prompted us to investigate such solutions and therefore led us to study mineral liquid crystals. The infinite mineral polymer $^1_\infty[Mo_6Se_6]^{2-}$ (Fig. 1d) can be regarded as a stack of dimerized $Mo_3Se_3^-$ units (Fig. 1a), and therefore is the end member (n=∞) of the $(Mo_3Y_3^-)_nY_2$ series (Y=chalcogen) (Fig. 1b, c). The study of static solutions of $Li_2Mo_6Se_6$ in NMF ($c>10^{-2}$ mol l^{-1}) with an optical microscope under polarized light (crossed polarizers) has demonstrated their birefringent and thus anisotropic nature. Both threaded and Schlieren textures were revealed [16], which are textures usually observed in the case of organic nematic phases and prove the mesogenic nature of these solutions.

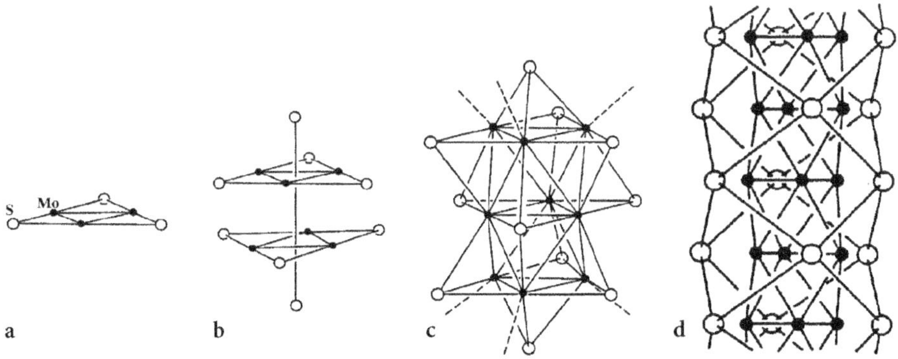

Fig. 1a–d. $Li_2Mo_6Se_6$ can be seen as the end member of the family $(Mo_3Se_3^-)_nSe_2$ made from the alternating stacking of $Mo_3Se_3^-$ motifs with an extra chalcogen capping each extremity: **a** n=1; **b** n=2; **c** n=4; **d** n=∞

[1] Note that $^1_\infty[Mo_6Se_6]^{2-}$ solutions are not stable for more than a few hours when in contact with dioxygen because of the decomposition of the chains. However, they can be kept for months and even years if kept or sealed under an inert atmosphere.

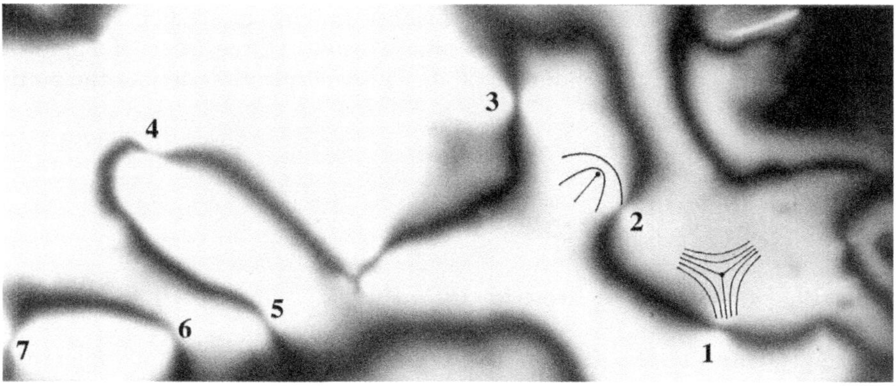

Fig. 2. Photograph of a typical Schlieren texture obtained from a 10^{-1} mol l^{-1} solution of $Li_2Mo_6Se_6$ in *N*-methylformamide in the nematic phase taken using an optical microscopy with crossed polarizers with a magnification of 1000x. (Reprinted from [4b], copyright (2000) from John Wiley and Sons)

For example, Fig. 2 shows a Schlieren texture, which is typical of nematic distortions of the director field (visible as dark branches usually named *brushes*). These distortions are induced by the perpendicular anchoring on the microscope slide of topological defects, called disclination lines, which are numbered from 1 to 7 in Fig. 2. These disclination lines can be classified into two groups according to their local topology as the dark brushes seen around these defects can rotate either clockwise or counterclockwise as the crossed polarizers are rotated simultaneously while the sample is kept fixed. This is due to the two different possible defect topologies illustrated in Fig. 2 for cases 1 and 2 [1].

The structure of these fluids was investigated by X-ray diffraction. When the chains are fully exfoliated the full width at half maximum (FWHM) of the (002) peak characteristic of the distance separating two $Mo_3Se_3^-$ moieties along the chain is of the same order as the experimental resolution; hence the coherence length of the chains is at least 1000 Å [17].

The anisotropic X-ray scattering pattern shows that there is no long-range three-dimensional positional order, demonstrating that this fluid is of a nematic liquid-crystalline nature (Fig. 3a). Moreover, the observation of an interference peak in Fig. 3b proves the existence of short-range positional correlations at ~70 Å, due to the separation of the chains, and probably arising from their repulsive electrostatic interactions. Indeed, charged particles confined in a finite volume at thermodynamic equilibrium repel each other and therefore stay at a well defined distance. Note that the diffuse ring of Fig. 3b appears anisotropic, indicating partial orientation of the chains along the capillary axis.

In addition to nematic ordering observed for the system $Li_2Mo_6Se_6$/solvent, the monodispersity in diameter of the wires should allow one to observe a nematic-hexagonal transition at high concentration. Note that, when the proper solvent is used (in order to avoid formation of bundles), one can expect to be able to tune the distance between the wires just by changing the concentration.

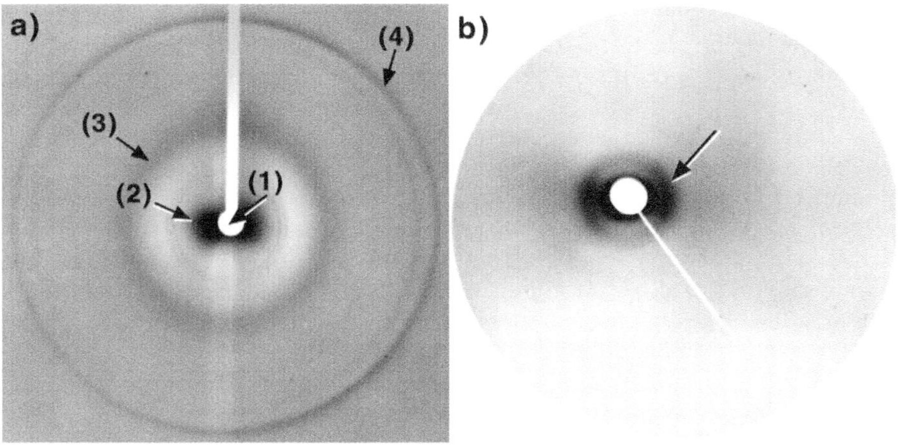

Fig. 3a,b. X-ray diffraction patterns obtained from a capillary filled with 10^{-1} mol l^{-1} solution of $Li_2Mo_6Se_6$/NMF: **a** using wide angle scattering. (**1** is the beam stop shadow, **2** the small angle scattering, **3** the diffuse scattering ring due to the solvent, **4** the 002 diffraction peak); **b** small angle scattering showing the presence of a diffuse ring at $q=9 \cdot 10^{-2}=2\pi/70$ Å^{-1} (*arrow*)

This is indeed a typical behavior for dispersion of 1-D nanostructures (2-D swelling).

More recently, the molecular nanowire $Mo_6Se_6^{2-}$ has become a fairly popular subject of investigation due to the rise of nanotechnology [18] and the scarcity of well-defined molecular wires available that can be dispersed in solution. In one such study, $Mo_6Se_6^{2-}$ nanowires were dispersed and recondensed using a surfactant. Dense phases were obtained with lamellar or hexagonal ordering of the wires, depending upon the surfactant used. Furthermore, the distance between the $Mo_6Se_6^{2-}$ wires in these condensates could be adjusted by varying the number of carbon atoms in the aliphatic chain of the surfactant [19]. In a similar dispersion-recondensation study, P. Yang and co-workers used the self-assembly properties of $Li_2Mo_6Se_6$-solvent to perform a cation exchange of the Li with more bulky cations such as Na^+, K^+, Rb^+, Cs^+, NMe_4^+. This was performed by retaining the lithium cation in solution using 12-crown-4 ether [20].

Other studies focused on the electrical properties of discrete nanowires in a similar way that has been done with nanotubes. However the lack of stability of this phase when in contact with oxygen makes these studies experimentally difficult [21].

The example of $Li_2Mo_6Se_6$ illustrates one synthetic method towards low dimensional mineral compounds, commonly referred to as the "shake and bake" method. It usually implies the mixing of the starting materials followed by a high temperature treatment. Then, in order to obtain a liquid crystal, one needs to bring these charged low-dimensional mineral anions into solution. Thus, how can one control, within a given ternary or quaternary system, the dimensionality of the synthesized polymeric compounds? This problem has been addressed in the literature and an answer, based upon the 8-N electron rule, can be found when using alkali metals (or other strong electropositive metals) as cations [22, 23]. Transferring electrons from the alkali metal to a 3-D polymeric covalent framework usually results in some of the covalent bonds being replaced by negative charges, hence the dimensionality of the framework is reduced, producing covalent negatively charged fragments, separated by alkali cations. By increasing the alkali metal content, further dismantling of the framework structure can be induced to the point where the residual framework is of molecular dimensions. This method is most efficient when large alkali metals such as cesium are used, because increasing the separation between the low dimension anionic moieties minimizes their electrostatic repulsion. A consequence of this is that a synthesis successful with cesium may not necessarily work with lithium (for example, the synthesis of $Li_2Mo_6Se_6$ differs from the one of $In_2Mo_6Se_6$ in the fact that it cannot be achieved directly at high temperature from the native element but required a two step synthesis, including an ion exchange procedure). This can be a problem in order to get soluble phases since it is empirically observed that in an alkali series of a given anion the highest solubility is obtained with the smallest cation [24]. Thus, a delicate balance needs to be found in order to synthesize new soluble low dimensional phases and, when the targeted phase cannot be directly synthesized with the smallest cation, a cation exchange additional step should be applied. This can either be tried in the solid state by a clever use of the different sublimation temperature to induce the cation exchange, or indirectly in solution by using high concentration of the replacing cation. In the latter case, it should be noted that the smallest cation that can be used is in fact H^+, for phases that can handle acidic conditions.

2.2 Imogolite: A Natural Nanotube, Aqueous Synthesis, and Composite Materials

Imogolite is a natural hydrated aluminum silicate first discovered in 1962 in the clay fraction of Japanese volcanic ash and it is found widely distributed in recent volcanic deposits. Since 1977, it is possible to synthesize it in the laboratory from dilute aqueous solutions of hydroxyaluminum cations and orthosilicic acid [25] or, by reaction of tetraethylorthosilicate with an aluminum source (aluminum chloride or aluminum *sec*-butoxide) [26, 27], at temperatures ranging from 25 to 100 °C.

This compound is made up of hollow rigid molecular tubes of diameter 25 Å and of variable length (Fig. 4). These can be dispersed in acidified water solutions [28]. A Japanese group has demonstrated that, in a specific range of concentrations, these suspensions phase separate into an isotropic liquid phase and a birefringent one [29]. Temperature has no influence on this phase transition. Fingerprint textures have been observed under polarized light, which, though highly reminiscent of cholesteric (i.e., chiral nematic) textures, have finally been attributed to the existence of a pleated structure on the basis of electron microscopy investigations of freeze-dried samples [30]. These studies show that the pleated sheets consist of a nematic organization of imogolite tubes. This prompts a comparison with the precholesteric states of DNA or collagen [31]. The phase transition could be explained on the basis of the Onsager

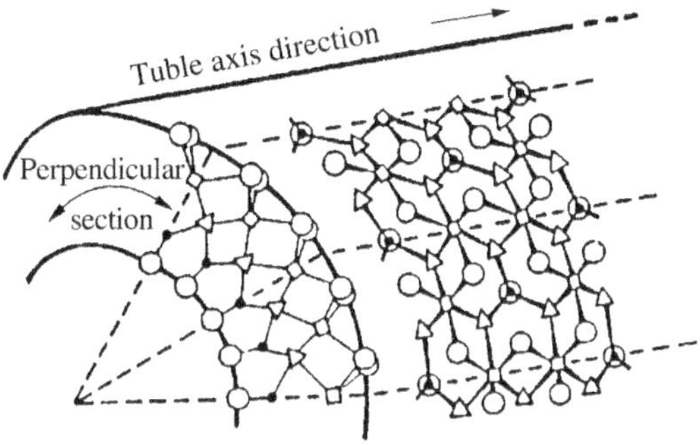

Fig. 4. Schematic view of part of an imogolite cylinder formed from twelve gibbsite structural type units, each corresponding to ($Al_2O_3 \cdot SiO_2 \cdot 2H_2O_2)_2$) and indicated by *dotted lines* (Reprinted from [29a], copyright (1986) from John Wiley and Sons)

model and the specific effects of polydispersity of the tube length were studied in detail.

The same group has also obtained composite materials based on imogolite and water-soluble polymers such as hydroxypropylcellulose (HPC) and poly (vinylalcohol) (PVA) [32]. HPC is a rigid rod polymer that has a cholesteric phase, whereas PVA is a random coil amorphous polymer. Significantly enhanced mechanical properties were only obtained in the case of the HPC composite materials.

2.3 Solvent Dependent Coiling of a Chain into a Nanotube: $NaNb_2PS_{10}$

In a recent search for other soluble mineral polymers, we discovered that KNb_2PS_{10} was soluble in hot DMSO. Our attention was drawn to this structure because we suspected the $^1_8[Nb_2PS_{10}]^-$ chain to be of intermediate rigidity between the rigid $Mo_6Se_6^{2-}$ molecular nanowire and the highly flexible $KPdPS_4$. The latter was shown to be highly flexible and behaving very similarly to organic flexible polymers [6]. In order to increase the solubility of KNb_2PS_{10} we synthesized its sodium equivalent $NaNb_2PS_{10}$ which, as expected, proved to be much more soluble in polar solvents. When observed under polarized light between crossed polarizers, concentrated solutions in NMF show strong flow birefringence. Such flow birefringence can arise either from a preferred orientation of anisotropic rigid molecules induced by the flow (like logs aligned with the flow in a river), or from stretching of entangled, folded, semi-flexible polymers (like drawn spaghetti).

SAXS investigation of the structure of these fluids showed a complex structure with two distinct features. The first signal demonstrated that upon dilution a 2-D swelling occurred, typical of suspensions of one-dimensional moieties. The second signal however was independent of the dilution and could not be explained by the chain structure; it is therefore the signature of a superstructure.

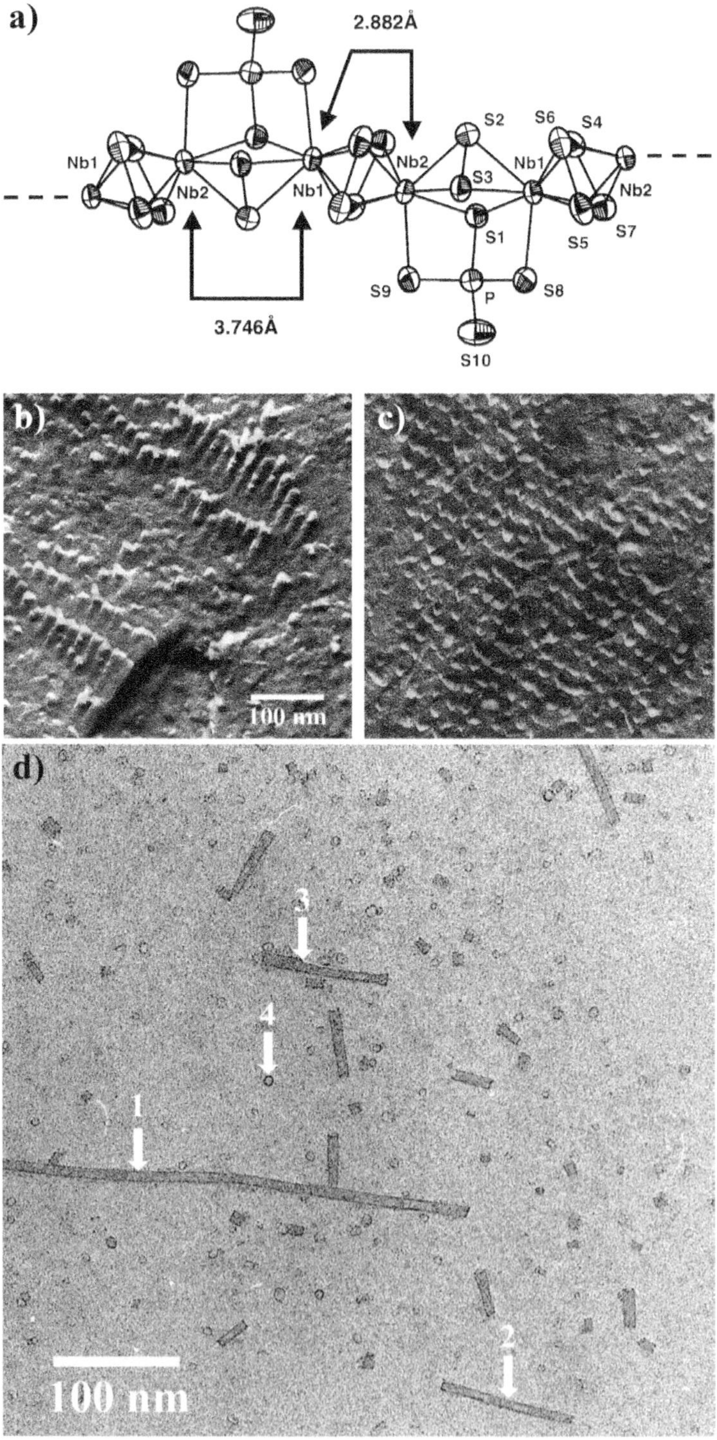
a)
2.882Å
3.746Å
Nb1
Nb2
Nb1
Nb2
Nb1
Nb2
S1
S2
S3
S4
S5
S6
S7
S8
S9
S10
P
b)
100 nm
c)
d)
3
4
1
2
100 nm

Furthermore, at higher concentration, thin diffraction lines were superimposed on the previously described signals, indicating that a first order transition toward a columnar mesophase (H) of hexagonal symmetry occurred. Partial alignment and even the melting (above 240 s^{-1}) of this hexagonal MLC could be obtained using a Couette shear cell. Such a melting had only been reported once prior to this study in the case of a hexagonal mesophase of a surfactant [33].

These results put together provide strong support for the presence in solution of cylinders having a well-defined diameter. However, a difference of behavior depending upon the solvent used was observed. If now DMF was used as the solvent, only the signature of 0-D moieties was detected. In addition no transition towards a hexagonal phase could be found.

To understand further and confirm our interpretation of the SAXS data, we complemented these results with a study using freeze fracture electron microscopy (FFEM) as well as low dose TEM (Fig. 5). This allowed us to show that in NMF $NaNb_2PS_{10}$ folds into single walled nanotubules having wall thickness of 1.6 nm (Fig. 5), whereas only globular coils can be observed in DMF. Furthermore, at high concentration in NMF these nanotubules tend to aggregate into bundles packed in a well-defined hexagonal arrangement [34].

A question still remains about the structure of the polymer within the walls of these nanotubules. The very low stability of theses tubes under TEM irradiation, even when applying low doses, precluded us so far from observing high-resolution images. An electron diffraction pattern highly reminiscent of a helicoidal organization of the polymer was observed but this still needs to be confirmed and studied in more detail.

3
Molecular Ribbons: V_2O_5, a MLC that Can be Readily Aligned in Both Magnetic and Electric Fields

Another way of obtaining suspensions of anisotropic mineral moieties is by the spontaneous condensation of dissolved molecular species. A typical example of this process is the synthesis of V_2O_5 ribbons by using "chimie douce" (soft chemistry) techniques (Fig. 6) [35, 36]. At pH≈2, V(V) species exist in an octahedral coordination with a V=O bond pointing along the Oz axis, three V-OH bonds in the xOy plane, and two bonded H_2O molecules to fill the coordination sphere. Beyond a concentration threshold of 10^{-3} mol l^{-1}, these vanadate species spontaneously condense in the xOy plane by two different reactions respectively called olation and oxolation reactions (olation: $V\text{-}OH+V\text{-}OH_2 \rightarrow V\text{-}OH\text{-}V+H_2O$; oxolation: $V\text{-}OH+V\text{-}OH \rightarrow V\text{-}O\text{-}V+H_2O$) to form ribbons 1 nm thick, about 25 nm

Fig. 5. **a** View of the $^1_\infty[Nb_2PS_{10}]^-$ chain showing the atom labeling and selected Nb-Nb bond lengths. **b,c** Arrays of nanotubules as observed by FFEM ($NaNb_2PS_{10}$ in NMF, 0.140 mol l^{-1}) with the cryofracture plane being arbitrary or perpendicular to the nanotubule axes for b and c, respectively. **d** Image obtained by low dose TEM of very dilute solutions of $NaNb_2PS_{10}$ in NMF showing single-wall nanotubules with: 10 nm ∅ (1), 7 nm ∅ (2), a 7–10 nm ∅ defect (3), and small rings corresponding to tubes seen along their principal axis (4). (Adapted from [34], copyright (2002) from the American Chemical Society)

a

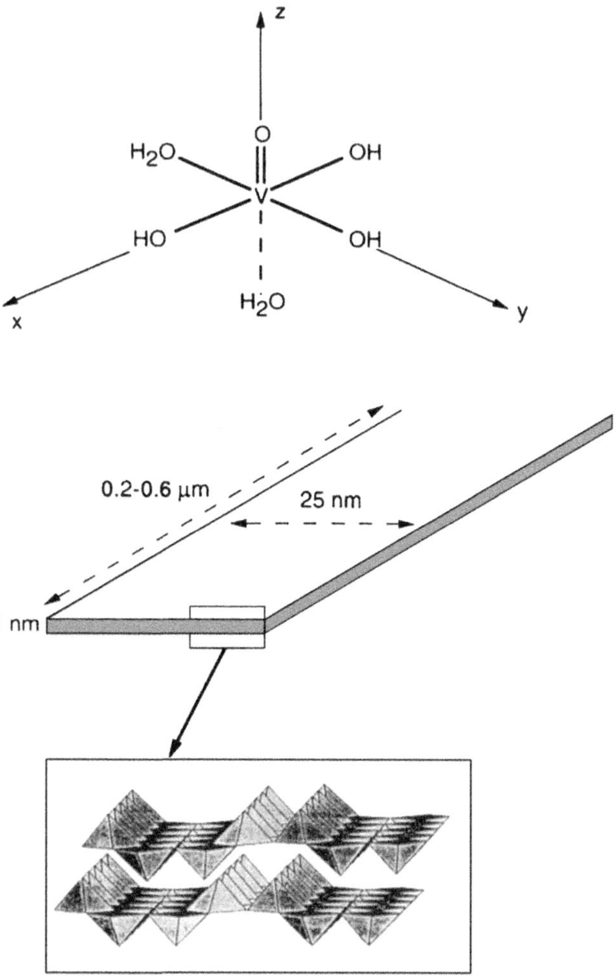

b

c

Fig. 6. **a** The molecular species that condense through olation and oxolation reactions into V_2O_5 ribbons **b** whose molecular structure is shown in **c**. (Reprinted from [4b], copyright (2000) from John Wiley and Sons)

wide, and about 200–600 nm long. The thickness is precisely defined by the ribbon molecular structure whereas both width and length is subject to some polydispersity, which is, however, usually rather small. The structure of these oxide ribbons is close to that of orthorhombic V_2O_5. The condensation reaction can be performed in several ways leading to similar ribbon structures. One of the most popular methods, as it avoids the introduction of foreign ions, consists of the acidification of an aqueous $NaVO_3$ solution by passing it through an ion-exchange resin column. The synthesis of V_2O_5 ribbons is a complex non-equilibrium process in which electrical charges play an important role. During the olation and oxolation reactions, V(V) charged molecular species are consumed and uncharged condensed moieties are produced. These condensed moieties do not experience electrostatic repulsion and hence flocculation of the suspension occurs under the influence of attractive Van der Waals interactions. However, this flocculation is only transient because subsequent adsorption and dissociation of water molecules produce V-OH groups, the ionization of which creates negative electrical charges on the surface of the ribbons and therefore repulsive electrostatic interactions become important again and ensure the stability of the colloid. Dark red gels and sols comprised of entangled ribbons dispersed in water are thus obtained.

The colloidal suspensions produced in this way are stable for years and, a few days after synthesis, their properties do not vary with time any more. In particular, they form a nematic liquid crystalline phase at volume fractions Φ larger than 0.7% [37]. Moreover, these suspensions undergo a sol/gel transition at $\Phi=1.2\%$ which divides the nematic domain into a nematic sol and a nematic gel. The optical textures of samples sealed in flat glass capillaries and observed under polarized light microscopy are indeed typical of a nematic phase. The nematic sol phase readily displays Schlieren textures whereas the nematic gel often shows the so-called "banded texture" of sheared nematic polymers caused by filling the flat glass capillaries with gel samples. Moreover, at volume fractions larger than 5%, SAXS experiments with samples aligned by shear, proved that V_2O_5 suspensions are in a biaxial nematic gel state [38].

An outstanding feature of the V_2O_5 nematic sols is that they can be aligned in quite moderate magnetic fields [39]. For example, for samples 0.1 mm thick, at a volume fraction of $\Phi\approx0.8\%$, a magnetic field larger than 0.3 Tesla will align the V_2O_5 ribbons parallel to the field. The defects observed by microscopy then vanish and the texture becomes completely homogeneous, as expected for a nematic single domain. This means that mineral non-ferromagnetic objects can thus be aligned by a purely external non-invasive action. Once the magnetic field has been removed, the samples still keep their nematic single domain properties, which makes them suitable for other physical studies. The alignment of a $\Phi=0.7\%$ sample takes about 2 h at 0.3 T and about 5 min at 1 T. These times increase rapidly when Φ gets closer to the gelation threshold. Thus, V_2O_5 sols align as fast or even faster than usual lyotropic rigid rod polymers such as poly(*p*-benzamide) and poly(γ-benzyl-L-glutamate) (PBLG) which respectively take hours or fractions of hours to align under similar conditions [40]. The magnetization of the phase is very small however and has not even been determined by a Squid magnetometer or by torque measurements. This means that the magnetization of the phase is actually comparable to that of pure water, a result which is understandable if one remembers the very small V_2O_5 volume fraction ($\Phi=0.7\%$) involved. Consequently, the microscopic origin of the magnetic susceptibility and its anisotropy is still unknown. It has been proposed that the magnetization could arise either from a diamagnetism possibly related to the anisotropic shape of the ribbons or from the paramagnetism of V(IV) impurities always present at the level of a few percent. Nevertheless, this phenomenon illustrates a characteristic feature of "soft matter", namely that a very small perturbation can have a huge effect on the mesophase due to its collective nature. In other words, the mineral building blocks of a nematic phase need not have strong magnetic properties for the phase to align in a magnetic field. In contrast, it was very difficult to induce any appreciable alignment in the isotropic phase of V_2O_5 ribbons.

An interesting application of the magnetic alignment is the following (Fig. 7). A single domain of the nematic phase can first be obtained by application of a magnetic field. Then, if this single domain is suddenly rotated in the field, it will undergo a transient hydrodynamic instability. This instability appears as a periodic pattern of microdomains in which the average direction of the ribbons is modulated. The pattern looks different depending upon the final magnetic field direction. This instability is classical in the field of liquid crystals and was thoroughly analyzed by Srajer et al. [41]. It can be used to micropattern a mineral material with periods in the range 10–100 μm in a very simple and cheap way. The lifetime of this instability varies from minutes to hours depending on the magnetic field intensity.

The production of nematic single domains also allows more detailed structural studies to be performed. For instance, the SAXS pattern (Fig. 8) of an aligned sample of V_2O_5 sol shows two diffuse spots located along a direction perpendicular to the director. These spots arise from the lateral interference between ribbons and their positions give the typical distance (~50 nm) between ribbons in the plane perpendicular to the director. This pattern is very similar to those of the nematic suspensions of the Tobacco Mosaic Virus (TMV), a typical example of rod-like organic polymer. Using a well-documented procedure, one can derive the nematic order parameter S from the distribution of scattered intensity [42]. This parameter takes values between 0, for a completely isotropic phase, and 1 for perfectly aligned moieties. The value obtained for V_2O_5, S=0.75, is in very good agreement with the predictions from the Onsager model and also, as expected, the nematic/isotropic phase transition is first order (with phase coexistence) and temperature has no influence on the nematic ordering (see also Sect. 5.1.). However, an increase of ionic strength, through salt (NaCl) addition had unexpected consequences. The nematic phase was actually stabilized with respect to the isotropic phase, in contrast to the behavior usually observed with this kind of lyotropic liquid crystals. This was attributed to Van der Waals interactions among ribbons that become increasingly important as electrostatic interactions are screened [43].

The anisotropy of the nematic phase also affects the properties of the solvent, in this case water. A recent ^{2}H NMR study of single domains of V_2O_5 sols has

Fig. 7a–d. Texture photographs in polarized light of transient hydrodynamic instabilities induced by a sudden change of the magnetic field orientation: **a** starting from a powder sample; **b** the magnetic field is applied along the main capillary axis; **c** the magnetic field direction is now perpendicular to the main capillary axis, but still in the plane of the figure. In consecutive stripes, the ribbons point in two different but symmetrical directions with respect to the polarizer-analyzer directions, giving rise to a zigzag pattern; **d** the magnetic field direction is now perpendicular to the plane of the figure. The dark regions are areas where the ribbons are perpendicular to the plane of the figure whereas the bright regions are areas where the ribbons still lie in the plane of the figure. (Reprinted from [39], copyright (1997) from John Wiley and Sons)

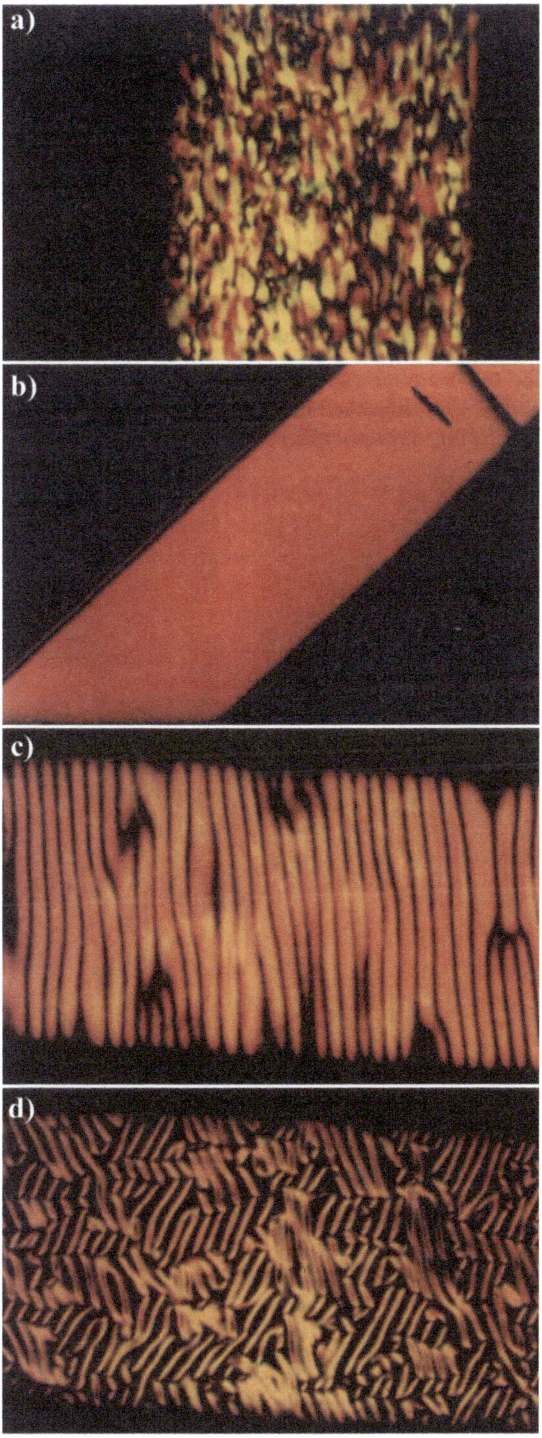
a)
b)
c)
d)

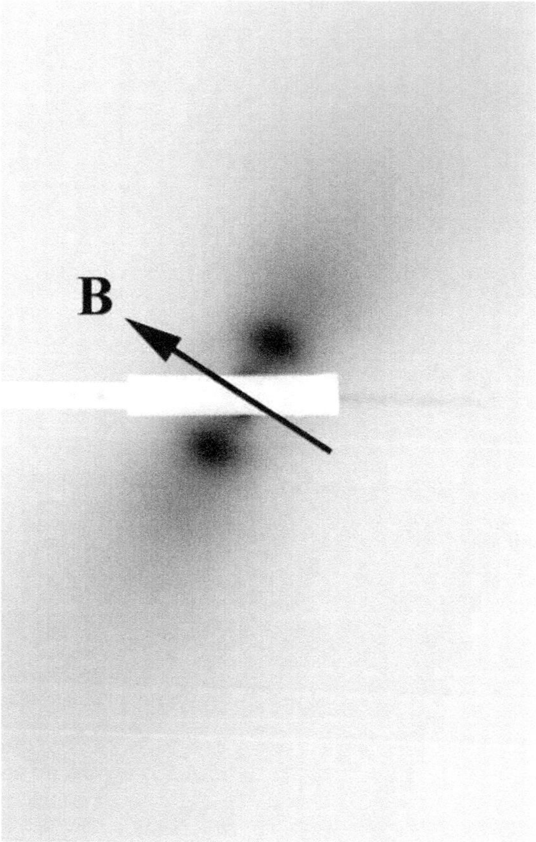

Fig. 8. Small angle X-ray scattering pattern of a sample aligned in a 1 T magnetic field. B is the field direction, which is set at an arbitrary angle of ≈40° to the main capillary axis. The *white rectangle* at the center of the figure is due to the beam stop

shown that, even at volume fractions as low as 0.7%, water molecules still experience a residual anisotropic interaction due to a situation of fast exchange between the bulk and sites located on the surface of the ribbons [44]. This is surprising and indicates that water molecules adsorbed on the surface of the ribbons still undergo fast motions. Thus, the behavior of the solvent faithfully reflects the anisotropy of the nematic phase, a feature that has subsequently been used to study the reorientation dynamics of the mesophase.

The magnetic alignment of these materials may be useful for applications when highly oriented samples are needed. V_2O_5 gels are classical host materials for the intercalation of guest species. For instance, the oxidative intercalation-polymerization of organic molecules, such as aniline or thiophene, has led to the synthesis of hybrid organic-inorganic materials, in which organic electrically conducting polymers are inserted between V_2O_5 ribbons giving rise to novel sandwich compounds. This process, carried out in a magnetic field, may yield highly oriented conducting polymers of better quality.

Since one of the most important applications of liquid crystals lies in the electro-optic display industry, it was natural to examine the influence of an electric field on the nematic phase of V_2O_5 suspensions [45]. A major problem, due to the lyotropic nature of this liquid crystal, is the screening of the field by mobile ions. This was overcome by the use of a high frequency (300 kHz) a.c. field and of blocking electrodes. Home-made cells (thickness 10–75 μm) were prepared with glass plates coated with indium tin oxide electrodes covered with SiO_2 and polyimide layers. The cells were filled by capillarity. Applying fields of about 1 V μm^{-1} was enough to align the phase from an initially planar texture to a homeotropic one. This reorientation is reversible so that switching cycles could be produced, using effective voltages of about 10 V. The response times are in the region of 1 s. In addition, the isotropic phase shows a very strong birefringence induced by the electric field. These properties formally demonstrate that mineral liquid crystals would have the potential for simple slow display applications.

4 Molecular Disks

4.1 Clays, Nematic Ordering vs Gelation

Clays of the montmorillonite family are lamellar aluminosilicates [46] used in many industrial processes and in products such as paints, softeners, and composite materials [47]. They swell when brought into contact with water, which is due to the insertion of water molecules between the sheets. Complete exfoliation can be induced leading to dispersions of disk-like particles of 10 Å thickness and 300–3000 Å in diameter, depending on the variety of clay used. These clay platelets bear a rather large surface electrical charge so that electrostatic interactions between them must be considered and are actually responsible for the colloidal stability of these dispersions. These suspensions have been widely studied as model colloids and also because they form physical thixotropic gels.

As early as 1938, Langmuir observed the phase separation of clay suspensions into an isotropic phase and a birefringent gel at the macroscopic level in test-tubes [9]. However, in the same report, he noted that this property of phase separation was gradually lost with time, which he tentatively explained by the incorporation of impurities diffusing from the glass tubes. He also compared this system to normal liquid crystals. Later, in 1956, Emerson observed a banded texture similar to that displayed by the Tobacco Mosaic Virus [48]. The investigation of clay suspensions from the structural point of view has been recently resumed. However, the study of the nematic order of suspensions of montmorillonite clays is in fact complicated by their gel properties. In spite of sustained efforts to understand its nature, the gelation mechanism has not yet been fully elucidated [49].

The role of the surface electrical charge is certainly crucial because the properties of the suspensions vary considerably with their ionic strength.

Quite recently two systems of the montmorillonite family, namely bentonite $[Na_x(Al_{2-x}Mg_x)(Si_4O_{10})(OH)_2,zH_2O]$ and laponite B $[(Na_2Ca)_{x/2}(Li_xMg_{3-x})(Si_4O_{10})(OH)_2, z\ H_2O]$ (an industrial form of hectorite) have been studied under polarized light using both the naked eye and optical microscopy [50]. The microscopic textures of concentrated gels (Fig. 9) are typical nematic ones. However, the nematic-isotropic phase separation has only been achieved very exceptionally, which seems to be a typical feature of these systems. In contrast with V_2O_5 ribbons, as the clay concentration increases, the suspensions, initially isotropic-liquid, become isotropic-gel and finally birefringent-gel (Fig. 10a,b). Temperature seems to have no influence on the nematic organization; the system is athermal as expected from the Onsager description that also applies to discs [51]. The isotropic gels show very large pretransitional effects such as flow and shock birefringence (Fig. 10c). The question about the thermodynamic nature of the birefringence of these gels naturally arises. In other words, is this birefringence due only to the freezing of some accidental flow birefringence or is it really intrinsic? Some simple experiments point to the thermodynamic nature of the nematic ordering. First, for one system, Langmuir has reported the phase separation of these suspensions. Second, the optical textures of these birefringent gels still slowly evolve over a few weeks after their preparation because topological defects anneal. However, the birefringence of these gels does not de-

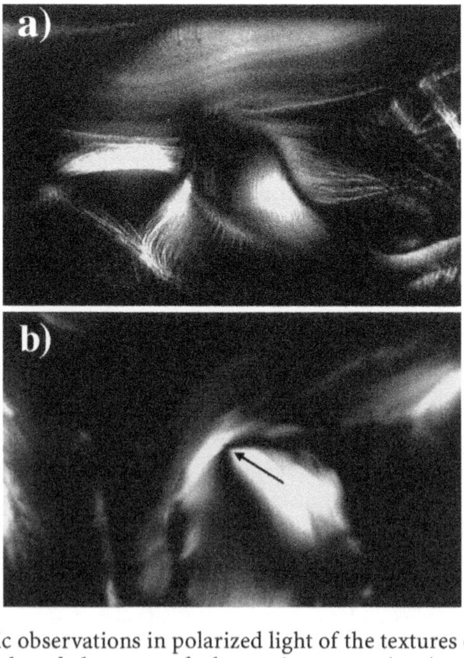

Fig. 9a,b. Microscopic observations in polarized light of the textures of the aqueous clay suspensions: **a** nematic threaded texture of a bentonite suspension (concentration 0.044 g/cm^3, magnification 50×); **b** detail of a 1/2 disclination line (*arrow*) in a laponite suspension (concentration 0.034 g/cm^3, magnification 100×) (Reprinted from [50], copyright (1996) from the American Chemical Society)

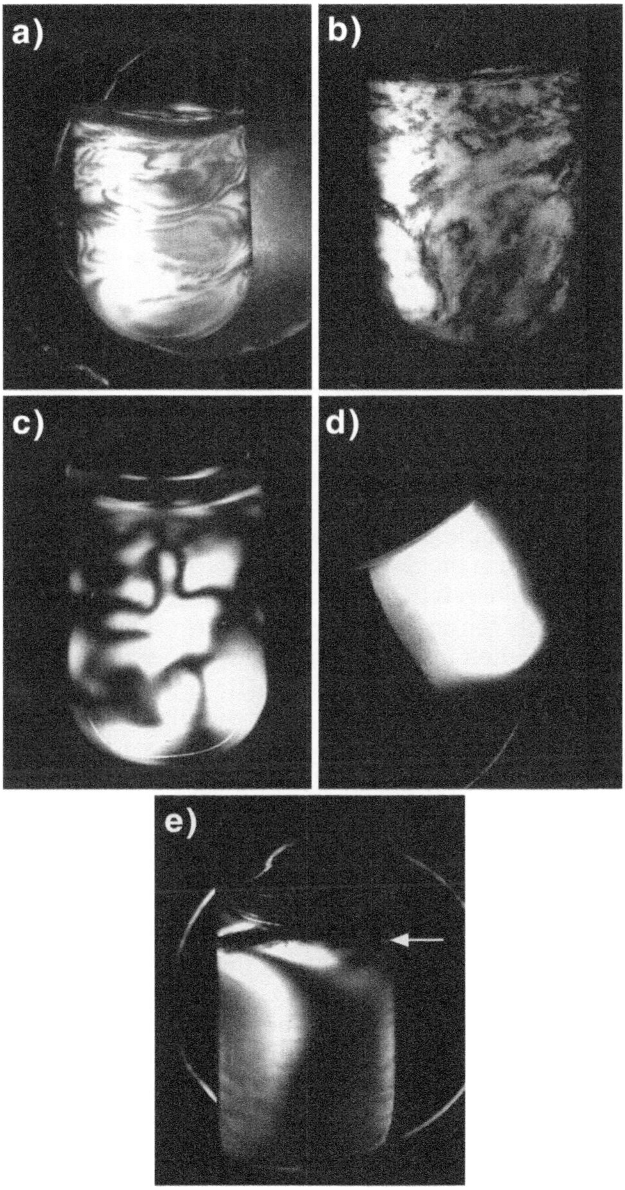

Fig. 10a–e. Tubes of aqueous clay suspensions observed between crossed polarizers: **a** bentonite suspensions of concentration 0.043 g/cm^3; **b** laponite suspensions of concentration 0.043 g/cm^3; **c** flow birefringence in a bentonite suspension of concentration 0.019 g/cm^3 – the test tube was previously vigorously shaken and was photographed during relaxation; **d** example of an oriented laponite suspension over a centimeter length scale – the sample looks bright because its principal axes are close to 45° with those of the polarizers; **e** an initially isotropic sample of bentonite gel (c=0.020 g/cm^3, bottom) in contact with brine (5 mol/l) turns birefringent. The *arrow* points to the interface between the gel and the brine. (Reprinted from [4b], copyright (2000) from John Wiley and Sons)

crease over many years. This means that, at least during the first weeks, the particles are not completely frozen and that any accidental flow birefringence could in fact relax. Third, birefringent gels may be prepared in test tubes by slowly concentrating isotropic sols over weeks. This process does not involve any flow. In this way, nematic single domains as large as a cubic centimeter were reproducibly prepared (Fig. 10d). Moreover, birefringence appears at the same concentration as that determined on samples prepared in the usual way. Then, the mechanical history of the sample seems irrelevant in this respect. Therefore, these studies show that a nematic order prevails in these suspensions over a distance that can be as large as 1 cm. Nevertheless, this discussion not only emphasizes that one should not confuse plain flow birefringence with true nematic ordering but also illustrates the difficulties of working with gels.

Very recently, well-aligned samples of laponite gels ($c=0.02$ g cm^{-3}) were prepared by slowly concentrating isotropic sols directly in Lindemann glass capillary tubes suitable for SAXS experiments, instead of test-tubes as described in the previous paragraph. The SAXS patterns observed are clearly anisotropic (Fig. 11a), which demonstrates the nematic orientational order at the particle scale over the whole sample [52]. Moreover, the SAXS patterns also show that the positional correlations of the clay particles are very weak, which is a well-known feature of these laponite gels [53]. This allows one to model easily the SAXS pattern (Fig. 11b) by only considering the particle form factor and the Maier-Saupe orientational distribution function that classically describes the nematic order [1]. By fitting the SAXS pattern, the nematic order parameter is obtained, $S=0.55\pm0.05$. This value compares quite well with those of usual liquid crystals (0.4–0.8) and shows that the nematic order of these materials should always been kept in mind when elaborating experimental procedures.

The phase diagram of these suspensions has been established versus clay volume fraction and ionic strength (Fig. 12) allowing the behavior of the sol-gel as well as the isotropic-nematic phase transitions to be followed. At high salt concentration (>0.3 mol l^{-1} for bentonite and $>9\times10^{-3}$ mol l^{-1} for laponite), the repulsive electrostatic interactions are screened sufficiently so that short-range attractive interactions can take over inducing the flocculation of the suspension. Very strikingly, in the case of bentonite (Fig. 12a), the nematic phase is stabilized upon increasing the ionic strength: an increase of the ionic strength induces both gelation and the isotropic-anisotropic phase transitions to occur at lower clay concentration. If one follows a vertical line on this diagram, and for example starts with a liquid suspension of bentonite (1.5 wt%) in pure water, at a concentration of [NaCl]=0.02 mol l^{-1} the suspension is an isotropic gel and at [NaCl]=0.1 mol l^{-1} it is a birefringent gel. Therefore, adding some brine on an isotropic gel of bentonite of concentration slightly under the isotropic/nematic transition allows one to induce the phase transition without any slightest flow, another hint at the thermodynamic origin of this transition (Fig. 10e). This effect, though unexpected, was still predicted by a statistical physics model incorporating electrostatic interactions [54]. Furthermore, a similar phase diagram was recently reported by Mourchid and Levitz on a closely related form of laponite, in a thorough and detailed study [55]. The same authors have fully discussed the relation between gelation and nematic ordering in this system. They

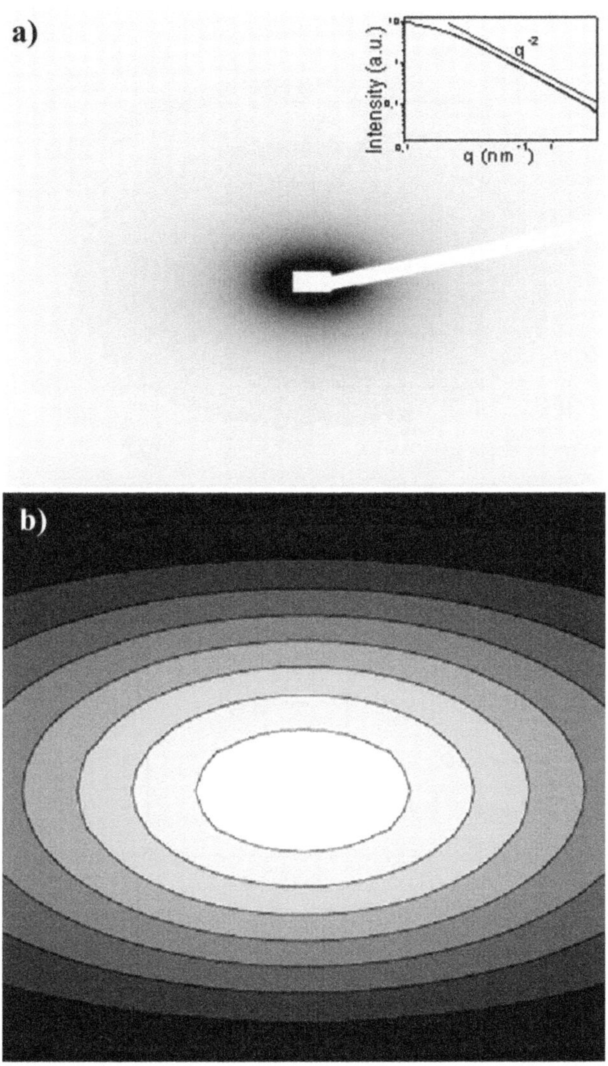

Fig. 11. a Anisotropic SAXS pattern of an aligned sample of clay gel in a horizontal capillary. Insert: SAXS intensity vs scattering vector modulus of an unoriented sample of the same concentration. (The *straight line* shows the q^{-2} dependence typical of the "intermediate regime" of plate-like particles). **b** Calculation of the SAXS pattern, with L=1 nm, R=15 nm, and S=0.55, that gives good agreement with the experimental one. (Reprinted from [52], copyright (2002) from EDP Sciences)

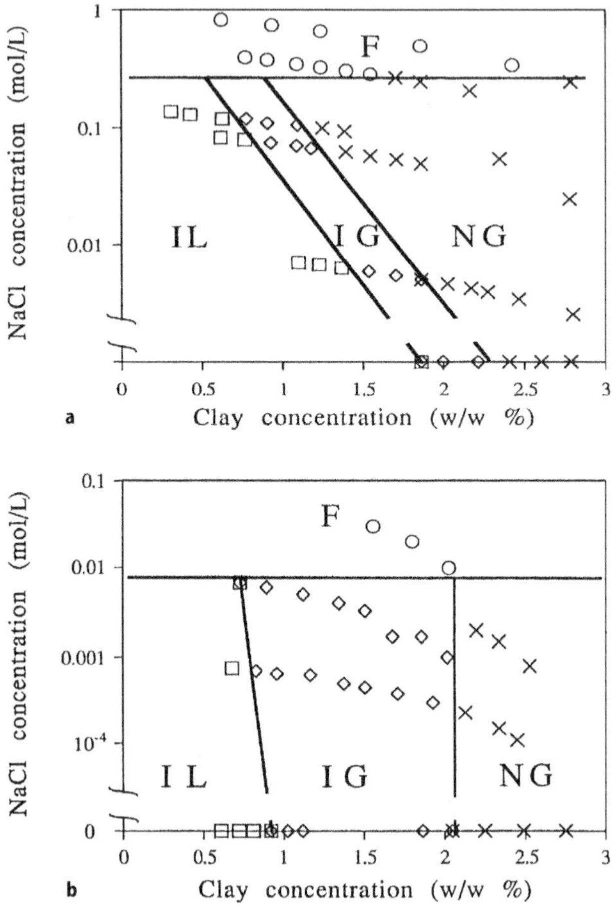

Fig. 12a,b. Phase diagram of clay suspensions vs clay and NaCl concentrations. (*circles*, F) Flocculated samples; (*squares*, IL) isotropic liquid samples (*lozenges*, IG) isotropic gel samples; (*crosses*, NG) nematic gel for: **a** bentonite; **b** laponite. (Reprinted from [4b], copyright (2000) from John Wiley and Sons)

have also explored the phase diagram at very low ionic strength and found the existence of a soft solid stabilized by long-range electrostatic repulsions [56].

In addition, it should be noted that there is now quite a few experimental studies that directly or indirectly document the existence of nematic order in clay gels [57]. Among them, let us particularly mention the direct visualization of aligned domains by X-ray fluorescence microscopy [58]. These clay-rich domains may arise from the nematic/isotropic phase separation that takes place at the micron length scale but does not extend to the macroscopic scale due to gelation, as discussed above. Another recent study leads to very similar conclusions [59]: Magnetic spherical colloidal particles of maghemite were incorporated as probes into laponite gels. A microscopic phase separation was also observed in the region of the phase diagram where the nematic/isotropic phase separation was expected. Finally, all these various studies raise the same and yet open question: what is the interplay (if any) of the nematic/isotropic phase transition with the gelation of these clay suspensions [60]?

4.2 A Unique Liquid-Crystalline Lamellar Phase Comprised of Mineral Nanosheets: $H_3Sb_3P_2O_{14}$

Until very recently, a liquid-crystalline lamellar phase based on extended, covalent, solid-like sheets had never been reported, even in the huge field of organic lyotropic phases. However, the field of mineral compounds provides a wealth of original moieties that are not usually obtained from organic synthesis techniques. Any general chemistry textbook will show a wealth of low-dimensionality mineral compounds and among them, many are two-dimensional ones. The problem then is to try to disperse them in a suitable solvent without alteration. We have already discussed in a previous section the principles that lead to the successful dispersion of mineral moieties and this approach will be further illustrated with the following example [61]. The members of the series of solid acids $H_nM_nZ_2O_{3n+5}$ (with, for example, M=Sb, Nb, Ta; Z=P, As; n=1, 3) are known to display a crystalline lamellar structure [62] [2]. Among them, $H_3Sb_3P_2O_{14}$ was found to disperse fully in water to yield homogeneous, transparent, colorless suspensions of extended covalent sheets of 1 nm thickness and at least 300 nm in diameter. The observation (Fig. 13) in polarized light of a series of test-tubes

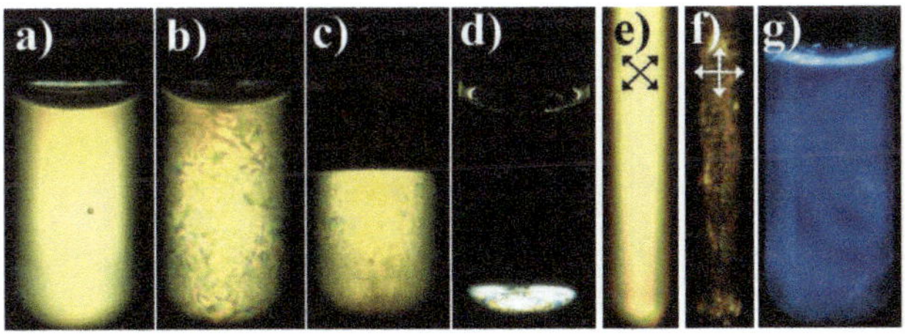

Fig. 13a–g. Naked-eye observation of samples. Test-tubes filled with aqueous suspensions of $H_3Sb_3P_2O_{14}$ single-layers, observed between crossed polarizers (a–e) (the isotropic phase appears dark): **a** 2 ml of birefringent gel phase (ϕ=1.98%) (topological defects are so dense that the texture appears homogenous at the scale of this photograph); **b** 2 ml of birefringent fluid phase (ϕ=0.93%); **c** 2 ml of a biphasic sample with an overall volume fraction ϕ_o=0.65%; **d** 2 ml of a biphasic sample with an overall volume fraction ϕ=0.03%; **e, f** magnetically aligned sample observed in a 5-mm NMR tube that has been immersed 10 min in a 18.7 T field at 50 °C, in two different orientations compared to the polarizer/analyzer system represented by *arrows* (ϕ=0.75%); **g** sample iridescence (ϕ=0.75%) observed in natural light and due to light scattering by the $H_3Sb_3P_2O_{14}$ layers stacked with a period of 225 nm. (Reprinted by permission from Nature [61], copyright (2001) Macmillan Publishers Ltd)

[2] In some cases, test tubes have shown temporarily a rainbow set of colors (see www graphical abstract) (induced by strong centrifugation for example), prior to reaching a single color equilibrium state.

filled with suspensions of decreasing volume fractions shows that: i) samples of high volume fractions (ϕ>1.78%) form birefringent gels; ii) for 1.78%>ϕ > 0.75%, the samples form a fluid birefringent phase; iii) for 0.75%>ϕ, the suspensions demix into a bottom birefringent fluid phase and a top isotropic one.

SAXS experiments (Fig. 14) have proved that the birefringent phase (sol and gel) has a lamellar structure and that its period can be continuously tuned from 1 to 225 nm according to dilution. In fact, very dilute samples are iridescent because the lamellar structure then has a period comparable to the wavelengths of visible light [62]. A large number (up to 10) of lamellar reflections were observed by SAXS, even at high dilutions, which shows the strength of the interactions between the mineral sheets. Further dilution leads to phase separation as excess water is expelled from the lamellar phase. Adding salt destabilizes the liquid-crystalline phase, eventually leading to flocculation of the colloid, indicating that these interactions are most probably of an electrostatic nature; this is hardly surprising as the sheets bear a rather large electrical charge. In addition, wide-angle X-ray scattering experiments have revealed the existence of sharp diffrac-

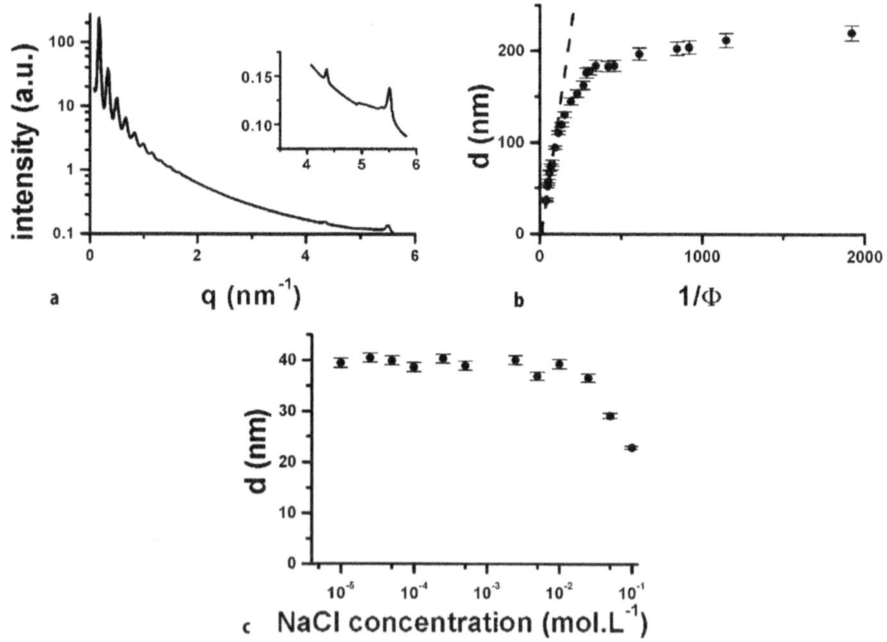

Fig. 14a–c. USAXS study of "powder" samples: **a** an example of scattered intensity vs scattering vector modulus q, showing ten orders of reflection due to the lamellar period as well as, in the inset, the thin diffraction lines due to the two-dimensional positional order within the covalent layers (ϕ=2.0%); **b** variation of the lamellar period, d, with inverse volume fraction (including two data points at very small d extracted from [62]) – the *dashed straight line* represents the one-dimensional swelling behavior of the lamellar phase d=δ/ϕ with δ=1.05 nm. A crossover from this law to a plateau is observed when entering the biphasic regime; **c** variation of the lamellar period d with salt molarity at constant volume fraction (ϕ=1.9%). (Reprinted from [61], copyright (2001) from Nature Publishing Group)

tion lines due to the crystalline order within the mineral sheets that do not fold or crumple but are fully preserved in suspension.

The mineral lamellar phase is readily aligned by shear in a Couette cell (Fig. 15) and the covalent sheets align parallel to the shearing surfaces. Dilute samples could be aligned at rather low shear rates (30 s^{-1}) and showed the best orientations. Moreover, the mesophase also aligns in high magnetic fields with the sheets perpendicular to the field, which provides a simple way to produce large (centimetric) oriented lamellar samples. This property was exploited by using the lamellar phase as a medium to induce a weak alignment of biomole-

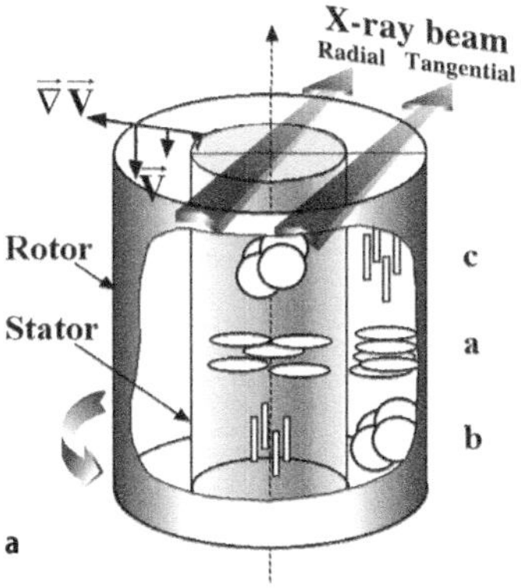

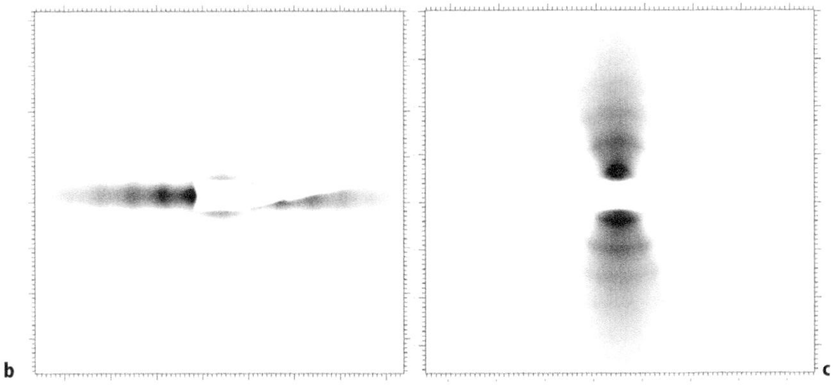

Fig. 15a–c. USAXS studies of aligned samples: **a** scheme of the Couette shear cell illustrating the radial and tangential geometries with respect to the X-ray beam (*flat arrows*). We also depict within the cell the three different types of lamellar orientations (a, b, and c) as seen by the X-rays in both geometries (the layers are sketched as *circular disks*); **b** USAXS 2-D scattering pattern obtained with a sample aligned in a Couette shear cell, in the tangential geometry, showing that the layers (d=175 nm) mostly belong to the c orientation; **c** USAXS 2-D scattering pattern of a vertical capillary filled with a suspension of $H_3Sb_3P_2O_{14}$ (ϕ=0.75%), obtained 20 min after its alignment using an 18.7-T magnetic field applied along its long axis. The vertical localization of the interference peaks (d=215 nm) indicates that the layers are oriented perpendicular to the magnetic field (in this case, since no mechanical shear is applied, the sample relaxes back to a powder within 1–2 h once removed from the magnetic field). (Reprinted from [61], copyright (2001) from Nature Publishing Group)

cules, which proves helpful when trying to determine their structure by NMR (see Sect. 6, Emerging Fields section).

5 Anisotropic Nanoparticles

5.1 Boehmite, a Model System to Test the Onsager Theory

The boehmite system (γ-AlOOH), originally studied by Zocher and Török [63] and Bugosh [64] was further developed by Lekkerkerker and coworkers [65]. They extended the hydrothermal preparation pioneered by Bugosh [64] by starting from an aqueous aluminum alkoxide mixture acidified with hydrochloric acid [65a]. They studied the phase behavior of both charge stabilized aqueous dispersions of colloidal boehmite rods [65b,c] as well as sterically stabilized colloidal boehmite rods in an organic solvent (cyclohexane) [65d–f].

Using the aqueous suspensions as obtained from the hydrothermal treatment, the I–N transition was observed, but took a very long time (months) [65b]. On the other hand, by treating the boehmite dispersions with hydrolyzed aluminum polycations, the phase transition is complete in a few days [65c]. In retrospect, it appears that the boehmite suspensions prepared by Bugosh also contain aluminum polycations. The sterically stabilized boehmite dispersions undergo the I–N phase transition. The biphasic gap is much wider than predicted by the Onsager theory for a monodisperse sample [65e], but can be explained by taking into account the polydispersity [65f]. The rate of the I–N transition strongly depends on the concentration. Upon increasing concentration from c_i to c_n, polarization microscopy indicates a crossover from nucleation and growth to spinodal decomposition [66].

At this point, let us detail the conditions of validity and the main predictions of the Onsager theory. This model applies to very anisotropic rod-like (or disk-like) particles. An important quantity is the particle aspect ratio, L/D, defined as the ratio of its length L over its diameter D. The particles should only interact through a purely hard-core potential, which means that particles do not interact at all if they do not touch, but they also cannot interpenetrate each other. This statistical physics model is based on a balance of two kinds of entropies. As concentration of rod-like moieties increases, the system undergoes nematic ordering because its loss of orientational entropy is more than compensated by a gain in translational entropy. The predictions of the model are the following. The nematic/isotropic transition should be first-order, i.e., with phase coexistence. The concentrations of the coexisting nematic and isotropic phases are given by c_n=4.2D/L and c_i=3.3D/L respectively. The nematic order parameter should jump from 0 to 0.8 at the transition. Temperature has no effect on this transition and the system is called athermal. Thus, the predictions of the Onsager model and of more recent theories based on it were confirmed by the detailed and precise investigations using the suspensions of grafted boehmite rods dispersed in cyclohexane. Figure 16 illustrates the first-order nematic/isotropic phase transition of these grafted boehmite rods. The effect of controlled polydispersity was

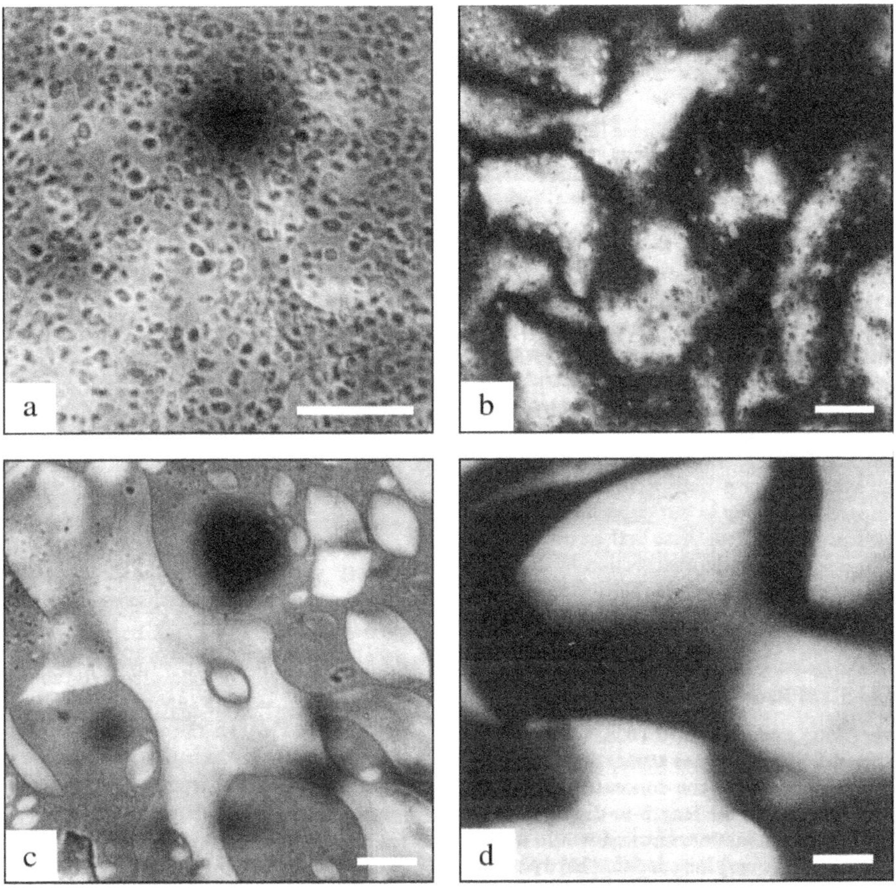

Fig. 16. **a** Optical micrograph of the nucleation of very small nematic droplets. **b** Nematic droplets and underneath a continuous nematic lower phase with Schlieren. **c** Grown-out tactoids. **d** Schlieren texture after completion of phase separation. All observations with crossed polarizers. The length of the *bar* represents 20 μm (Reprinted from [65e], copyright (1993) from the American Chemical Society)

also examined and the complex coexistence of two different nematic phases with the isotropic liquid phase, which is a consequence of polydispersity predicted by theory, was proved experimentally.

Non-grafted boehmite rods experience a much more complicated interaction potential as positive surface electrical charges now play an important role. In this case, the phase diagram has to be discussed in the frame of both the Onsager model of nematic ordering and the DLVO (named after: B.V. Deryagin, L. Landau, E.J.W. Verwey, and J.T.G. Overbeek) theory of colloidal stability, which describes colloidal stability as a balance between repulsive electrostatic and attractive van der Waals interactions [67, 68]. At low ionic strength, electrostatic repulsion dominates so that the phase stability is essentially described by the

Onsager model. However, at large ionic strength, these electrostatic repulsions are screened and van der Waals attractions come into play. Indeed, the strength of these latter interactions depends through the Hamaker constant on the electronic contrast between the mineral moieties and the solvent. Mineral moieties being electron rich are likely to show a large Hamaker contrast with the solvent. In the case of boehmite rods, these strong van der Waals attractions lead to the formation of dense space-filling isotropic gels. It should be noted at this point that the percolation threshold of very anisotropic rods is inversely proportional to their aspect ratio so that gels can be formed at rather low volume fractions. Note that a rather similar situation also prevails for the aqueous dispersions of V_2O_5 ribbons [69].

The same authors also investigated the dynamic properties of these suspensions by light scattering and rheology both in the isotropic liquid phase and in the nematic one. Even though this constitutes one of the most comprehensive studies of its kind for suspensions of rod-like moieties, we shall not further discuss their findings, as they are well described [68, 70].

5.2 Gibbsite and the Onsager Transition for Disks

In his seminal work of 1948, Onsager had already predicted the existence of a nematic phase for suspensions of disk-like particles. However, it was only fifty years later that Van der Kooij and Lekkerkerker clearly proved this conjecture by carefully investigating suspensions of gibbsite, $Al(OH)_3$, hexagonal platelets [71]. These particles were sterically stabilized by grafting them with a layer of polyisobutene before being dispersed in toluene. The aspect ratio of these moieties, defined as the ratio of their diameter ($\approx$170 nm) to their thickness ($\approx$15 nm), is approximately 11. In a given range of volume fraction (0.16–0.17), these suspensions spontaneously phase separate (Fig. 17) into a nematic and an isotropic phase as predicted by Onsager, and also more recently confirmed by computer simulations [72]. The birefringent nematic phase, being denser than the isotropic one, sediments at the bottom of the tube. Moreover, the relative proportion of the nematic phase increases linearly with the gibbsite volume fraction, as expected (Fig. 17a–d). The isotropic upper phase shows streaming birefringence when the tube is slightly shaken, indicating the presence of free alignable platelets. The lower phase is permanently birefringent due to nematic ordering and flows like a moderately viscous liquid when the tube is tilted, both properties proving its liquid crystalline nature. This phenomenon proves that the nematic order of gibbsite suspensions of volume fraction larger than 0.17 is really of a thermodynamic nature. In addition, the range of volume fraction of the biphasic gap is in relatively good agreement with the theory. This experimental study also proves that grafting particles with a polymer layer ensures that they strictly interact through a hard-core potential. Indeed, the thickness of the polymer layers prevents the particles from approaching each other at short range where electrostatic and van der Waals interactions come into play.

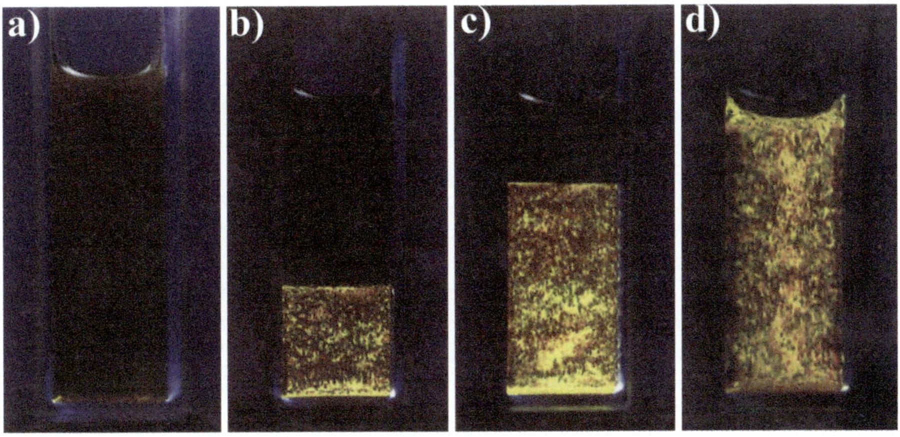

Fig. 17a–d. Isotropic-nematic phase separation as observed between crossed polarizers, at volume fractions ranging from; **a** 15.9; **b** 16.5; **c** 16.9; **d** 17.4 vol.%. (Reprinted from [71], copyright (1998) from the American Chemical Society)

5.3
Liquid Crystalline Hexagonal Phase of Disks: $Ni(OH)_2$

Liquid crystalline phases can show not only long-range orientational order as nematic phases do but also long-range positional order. When this positional order is one-dimensional, the mesophase is called lamellar or smectic; when it is two-dimensional, it is called columnar. The latter case is often found with thermotropic liquid-crystal disk-like molecules. Such molecules stack in columns that assemble on a 2-D lattice of hexagonal, rectangular, or oblique symmetry. The molecules in a given column only show 1-D liquid-like order and the uncorrelated columns are free to slide past each other, which ensures the mesophase fluidity [73].

Very recently, a hexagonal columnar mesophase has been reported in two different mineral systems. In the first example, nickel hydroxide ($Ni(OH)_2$) hexagonal plate-like particles have been grown by controlled precipitation [74]. They were then coated with a polyacrylate layer in order to avoid flocculation by steric repulsion. The bare particles are about 90 nm in diameter and about 10 nm thick, with an aspect ratio of 9:1. The polymer layer being about 4 nm thick, the particles, when coated, have an aspect ratio of about 5:1. The nickel hydroxide particles are charged and the strength of the electrostatic interactions can be tuned by adding salt. Observations of macroscopic samples with the naked eye, in polarized light, show, in a given range of concentration, the coexistence of two phases – a top isotropic phase and a bottom birefringent one. Small Angle Neutron Scattering (SANS) experiments on "powder" (i.e., unoriented) samples displayed five diffraction peaks. The first two, located at small scattering angles, in a ratio $1:3^{1/2}$, suggest the existence of a two-dimensional hexagonal lattice whereas the last three, located at wider angles, in a ratio 1:2:3, correspond to the stacking of the plates in columns. Very interestingly, this mineral phase looks therefore quite similar to the hexagonal columnar mesophase of usual disk-like organic molecules. This interpretation was further confirmed by detailed SAXS experiments on aligned samples [75] and the experimental phase diagram was compared to that predicted by computer simulations for plate-like particles [76]. This system also allows one to balance delicately electrostatic and steric interactions by the control of the ionic strength, a possibility that was explored to some extent by the authors. The flow properties of this hexagonal phase, upon shear in a Couette cell, were also investigated [77]. It was found that, at low shear rates, the phase is aligned with columns in the flow direction but the particles have their normals tilted by 20° with respect to the flow direction. As the shear rate is increased, a kind of dynamical transition is observed to a layered state in which the layers and the particles are oriented with normals parallel to the shear gradient. The kinetics of alignment and relaxation were also investigated in detail.

The second example is found in the system of gibbsite disks that we already described in a previous section. In fact, this system shows not only an isotropic/nematic phase transition at a volume fraction of about 20% but also a nematic/columnar phase transition at around 40% [78]. The hexagonal phase is similar in nature to that of the $Ni(OH)_2$ particles but this system provides the additional opportunity to investigate the influence of polydispersity, a parameter that often plays a major role in determining whether a given phase will actually be observed. Two gibbsite systems were prepared that differ by their polydispersities. It was first found in the two systems that the coexisting nematic and columnar phases had different polydispersities. The columnar phase has a markedly smaller polydispersity than the nematic one so that the phase transition could be used to improve the size distribution of a suspension. It was then observed that, in the system of higher polydispersity, the hexagonal phase seemed to be replaced by a lamellar one at higher volume fraction. This nicely illustrates the disruptive effect of the diameter polydispersity, an effect that was indeed awaited. In fact, the relative stability of the nematic, columnar, and lamellar phases are not only controlled by the volume fraction but also by a subtle balance between plate diameter and thickness polydispersities.

Suspensions of colloidal gibbsite platelets of very large thickness polydispersity but of rather monodisperse diameter display a very unusual phenomenon [79]. In a limited range of volume fraction, these suspensions demix into a nematic *upper* phase in coexistence with an isotropic *bottom* phase. This is in stark contrast with the usual case where the nematic phase is denser than the isotropic one. This behavior was interpreted as due to a fractionation effect related to the strong thickness polydispersity as the thicker platelets are largely expelled from the nematic phase and preferentially occupy the isotropic phase. Note that a theoretical model was especially developed in order to account for these quite peculiar observations [80]. Moreover, this model predicts that the nematic phase may demix into two separate nematic phases of differing densities and compositions.

5.4 Mixtures of Rods and Plates

The liquid crystalline behavior of mixtures of rod-like and plate-like particles is intrinsically much richer than those of rods and plates considered separately.

This is nicely illustrated by mixing the suspensions of goethite rods and gibbsite plates already described in this review. One of the motivations to undertake this kind of study was to find a new type of nematic phase, called biaxial [1]. Even though this biaxial nematic was not found here, the mixtures do present a remarkably rich phase diagram with four different liquid crystalline phases [81]. Up to five phases coexist in a region of the phase diagram. Two of these mesophases are nematic but are rod-rich and plate-rich respectively, another one is the columnar phase already described above, the nature of the fourth mesophase is still unknown, and the last phase is the isotropic one. The observation of a five-phase equilibrium region implies that the phase diagram cannot be explained without explicitly taking into account the polydispersity of both rods and plates. Actually, the two nematic phases differ not only in rod/plate proportion but also in particle dimensions. Strong fractionation effects also occurred in the columnar phase, resulting in a more pronounced columnar ordering than observed with suspensions of plate-like particles only. A theory adapted from the Onsager model was successfully developed to account for the corner of the phase diagram that shows the isotropic phase, the rod-rich and plate-rich nematic phases, and all the possible equilibria between them [82]. Finally, it was observed that the formation of the rod-rich nematic phase proceeds by the mechanism of spinodal decomposition whereas that of the plate-rich nematic phase proceeds by a nucleation-growth process [83].

5.5
β-FeOOH, a Colloidal Smectic Phase

In his search for ordering phenomena in suspensions of mineral crystallites, Zocher and Jacobsohn observed, in 1929, iridescent layers at the bottom of a flask filled with a colloidal solution of β-FeOOH (akaganeite) [84]. He called these layers “Schiller layers”, which means iridescent layers in German. β-FeOOH forms fairly well monodisperse nanorods 500 nm long and 100 nm in diameter. The iridescence is due to the sedimentation of these rod-like particles, which form layers at the bottom of the flask [85]. The thickness of each layer, roughly the particle length, is comparable to the wavelength of light, explaining the iridescence. A lamellar structure is thus formed which is similar to the molecular organization of thermotropic smectic phases, but here the periodicity is about 100 times larger. This colloidal liquid crystalline phase coexists with the supernatant phase, which is a dispersion of individual rods. The thermodynamics of this phase transition could also be understood in the frame of the DLVO theory. Again, temperature has very little influence on the phase stability, pointing to a purely steric potential. Moreover, detailed studies [86] of dried Schiller layers by atomic force microscopy has shown that the layers can have two types of internal structure: the rods can pack on a distorted square lattice and be tilted with respect to the layer normal or they can assume a liquid-like ordering within the layers. Similar phenomena seem to exist in suspensions of tungstic acid, but they are less well documented. Finally, this example proves that very monodisperse mineral particles can form a colloidal smectic phase.

5.6
Suspensions of Goethite Nanorods

Goethite (α-FeOOH), which should not be confused with akaganeite (β-FeOOH), is one of the most widespread iron oxides [87]. It is widely used as a pigment in industry. Aqueous suspensions of goethite nanorods are readily obtained by raising the pH of an aqueous $Fe(NO_3)_3$ solution. The resulting precipitate is aged at room temperature, centrifuged, and rinsed several times, and finally dispersed at pH 3. At this pH, which represents a compromise between chemical and colloidal stabilities, due to the large surface charge density (0.2 C m^{-2}), electrostatic interactions prevent flocculation. The nanorods are strongly polydisperse with average length of 150 nm, width of 25 nm, and thickness of 10 nm. Goethite suspensions display a liquid crystalline phase at volume fractions of approximately 10%.

Despite the fact that bulk goethite is a typical antiferromagnet, the nematic suspensions of goethite nanorods display quite unexpected magnetic properties [88]. For instance, they align in very weak magnetic fields as the alignment (Frederiks) transition takes place at 20 mT for samples 20 μm thick. This field intensity is at least an order of magnitude smaller than observed in usual thermotropic and lyotropic liquid crystals. This mineral liquid crystal is therefore extremely sensitive to a magnetic field. Moreover, nematic suspensions of

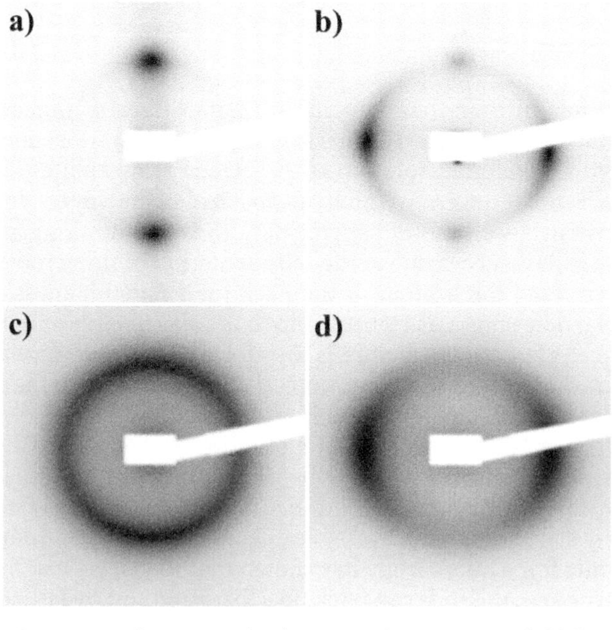

Fig. 18a–d. SAXS patterns of: **a** a nematic phase sample in a 30-mT field; **b** a nematic phase sample in a 625-mT field; **c** an isotropic phase sample in a 270-mT field; **d** an isotropic phase sample in a 900-mT field. (Reprinted from [88], copyright (2002) from the American Physical Society)

goethite align parallel to the field at low field intensities (≤350 mT) but, oddly enough, they reorient perpendicular to the field at large field intensities (≥350 mT) (Fig. 18). This behavior is again in sharp contrast with those of usual liquid crystals that align, at all field intensities, either parallel or perpendicular to the field, depending on their nature. It should also be noted that this property is also observed with initially isotropic goethite suspensions that show huge magnetic-field induced anisotropies and orientation reversal around 350 mT as well. Then, this phenomenon simply reflects the individual properties of the nanorods, even though they are probably enhanced by collective effects in the nematic phase. These quite peculiar magnetic properties probably arise from a competition between a nanorod remanent magnetic moment due to uncompensated surface spins and a negative anisotropy of its magnetic susceptibility.

6
Emerging Fields

6.1
Self Assembly of Nanorods and Nanowires

In the recent past, and following the discovery of nanotubes by Iijima [89], there has been an enormous increase in the number of groups and reports concerning the synthesis of nanorods, nanowires, as well as nanotubes. Most of the elements of the periodic table have now been used (N.B. the difference between nanorods and nanowires is that nanorods usually have a small aspect ratio when nanowires have a high one). These particles, which present a wide range of properties such as semiconducting or metallic behavior, could lead to numerous new liquid crystalline phases and by clever exploitation of their self-assembly properties, numerous new materials with a variety of applications could be synthesized. Since the number of such phases is increasing so rapidly, we have summarized the most studied systems and what we consider to be the most original in Table 1, although we must add that this is not an exhaustive list.

6.1.1
Synthesis

Dozens of methods to synthesize nanotubes, nanowires, and nanorods have been reported that can be found in the references included in Table 1. In addition to the most well known ones, such as hot plasmas, laser ablation, chemical vapor deposition, high temperature solid state and hydrothermal synthesis, filling/coating of carbon nanotubes and similar types of materials, three methods have been developed that enable the synthesis of a wealth of new anisotropic nanoparticles.

The first one makes use of the pores of membranes as templates for the electrolytic growth of metal nanorods. This technique consists of coating one side of a membrane with a metal electrode (such as gold or platinum) and then electrodepositing a metal or conducting polymer within the pores of the membrane [90]. This method controls more closely the filling of the pores compared to other

Table 1. A selection of the most common or original nanotubes, nanorods, and nanowires and their method of synthesis [131]

Composition	Synthesis method(s)	Type	Reference
C	Arc, laser ablation, CVD, templating etc.	NT	[98a, 130]
Ag	AAO templating	NW, NR	[131]
Au	AAO templating	NW, NR	[132, 133]
Bi	AAO templating	NW	[134]
C (diamond)	AAO templating	NW	[135]
Co	AAO templating	NW	[136, 137]
Cu	AAO templating	NW, NR	[95, 138]
Fe	AAO templating		[139]
Ni	AAO templating		[137]
Ge	Laser ablation		[140]
Si	Laser-assisted catalytic growth, laser ablation, catalytic growth, HT synthesis, thermal evaporation SiO, CVD		[141, 142]
Vanadium oxides(V_2O_5, VOx, $Na_2V_3O_7$)	Solution, hydrothermal, NT AAO templating, others	NT, NR	[143–145]
TiO_2	AAO templating, self assembly, sol-gel, wet chemistry	NT, NW	[144, 146, 147]
SnO_2	AAO templating	NW	[10, 148]
Al_2O_3	Anodisation	NT, NW	[141, 149]
In_2O_3	AAO templating	NW	[150]
ZnO	AAO templating, PVD	NW	[144, 147, 151, 152]
WO_3	AAO templating, NT templating	NW, NR	[144, 146, 147]
SiO_2	AAO templating (mesoporous), templated coprecipitation, wet chemistry, laser ablation, carbothermal reduction, sol-gel	NT, NW	[144, 153]
Co_3O_4	AAO templating,	NW	[144]
MnO_2	AAO templating,	NW	[144]
MoO_3	NT templating	NT, NR	[145]
Sb_2O_5	NT templating	NT, NR	[145]
MoO_2	NT templating	NT, NR	[145]
RuO_2	NT templating	NT, NR	[145]
IrO_2	NT templating	NT, NW	[145]
Ga_2O_3	High T, arc discharge, carbothermal reduction	NW	[152, 154]
CuO	Solid state (grinding)	NR	[155]
α-Fe_2O_3	Controlled precipitation	NR	[156]
BaCrO4	Reverse micelle templating	NW	[96]

Table 1 (continued)

Composition	Synthesis method(s)	Type	Reference
$BaWO_4$	Reverse micelle templating	NR	[157]
$YBa_2Cu_3O_7$	Laser ablation	NR	[158]
MoS_2	Catalyzed transport reaction, H_2 reduction, C60 and CNT templating, other	NT	[159–161]
WS_2	H_2 reduction of WS_3, WO_x+H_2S, other	NT	[160–162]
WSe_2	HT reaction	NT	[163]
$W_xNb_yS_2$	HT reaction	NT	[164]
$W_xMo_yC_zS_2$	$W_xNb_yO_z$+H_2S	NT	[165]
NbS_2	H_2 reduction of NbS_3	NT	[166]
TaS_2	H_2 reduction of TaS_3	NT	[166]
$FeTe_2$	Hydrothermal	NR	[167]
$SiSe_2$	Theoretical study	NW	[168]
Bi_2Te_3	AAO templating	NW	[169]
$LiMo_3Se_3$	HT synthesis+cation exchange	NW	[13, 170]
$NaNb_2PS_{10}$	HT synthesis	Nanotubules	[34]
Cu_2S	Cu (surfactant treated)+H_2S	NW	[171]
ZnS	HT conversion of ZnO	NR	[172]
CdS	Thermal decomposition	NR, NW	[173]
CdSe	AAO templating	NR, NW	[97, 174]
BN	Arc, laser ablation, CVD, templating etc.	NT	[175–180]
GaN	Use of growth promoter, CNT templating	NW	[152, 181]
AlN	HT synthesis	NW, NR	[152]
SiN	Si-SiO_2+N_2+2C (CNT templating)	NR	[182]
InP	Laser-assisted catalytic growth	NW	[183]
GaAs	Laser-assisted catalytic growth, laser ablation	NW	[184, 185]
SiC	Arc, HT reaction via CNT templating, carbothermal reduction, HFCVD	NW, NR	[186]
TiC	HT reaction	NW	[187]
MC (M=transition metals, Ge, Ta, Gd, Hf, La)	Various	NR	[188]

nanometer sized materials that have been used, such as the pores of membranes, zeolites, or mesoporous materials [91]. It even allows one, by changing the voltage during the electrodeposition, to synthesize nanowires made of periodically alternating metals [92]. Note that these "striped" metal nanoparticles, in which the stripes are comprised of different metals such as gold, silver, platinum, and nickel, can be visualized using bright-field optical microscopy, which allows them to be used as nano-barcodes [93]. Indeed, in an analogous fashion to conventional barcoding, this technology enables the creation of an extremely large numbers of unique and identifiable particles by varying the width and composition of the stripes. Individual particles can be identified by reading the striped pattern using an optical microscope. Nanobarcoded particles can be used to carry out simultaneously large numbers of biological assays in a small volume.

The second approach takes advantage of tools developed in the field of "*Chimie douce*" typically applied for synthesizing metal nanoparticles, semiconductors or oxides. By analogy to biomineralization, in which minerals are induced to form within an isolated space (a cell or a cell compartment), reverse micelles have been used [94]. In these complex fluids, small compartments of supersaturated aqueous solutions are stabilized within an oil using a surfactant (for example (2-ethylhexyl)sulfosuccinate, AOT) A precipitation is then induced inside each of these nanoreactors (either by chemical reaction of the precursors or by changing the temperature), which leads to the synthesis of nanoparticles whose size and shape is determined by the size and shape of the micelle. Anisotropic particles were grown by the addition of salts and/or co-surfactants that changed the micelle from spherical into roughly rod-like shapes [95]. A further development involved controlling the interfacial activity of the reverse micelles so that the nanoparticle synthesis and their self-assembly was coupled to give superlattices of shape controlled nanoparticles [96]. Face growth kinetics could even be used to tune the shape of the particles when the structure of the material targeted was highly anisotropic [97].

The third and most recent approach is an extension of the thoroughly studied carbon nanotube growth mechanism, namely the use of a growth promoter that helps in directing the growth of a material in a highly anisotropic manner [98]. The idea is to use a growth promoter nanoparticle (the diameter of which will drive the diameter of the nanowire) that can form a liquid alloy with the nanowire material of interest (Fig. 19). A careful study of the phase diagram allows the specific composition to be determined and the synthesis temperature necessary to achieve simultaneously the coexistence of the liquid alloy and the solid nanowire material. The liquid alloy nanoparticles act as preferred sites for the adsorption/dissolution of new reactant. When this alloy becomes supersaturated, it also acts as the nucleation site for recrystallization, leading to the growth of a nanowire.

6.1.2
Properties

However, apart from the examples developed above, the study of the self-assembly, in relation to the mesogenic properties for these suspensions of nanotubes,

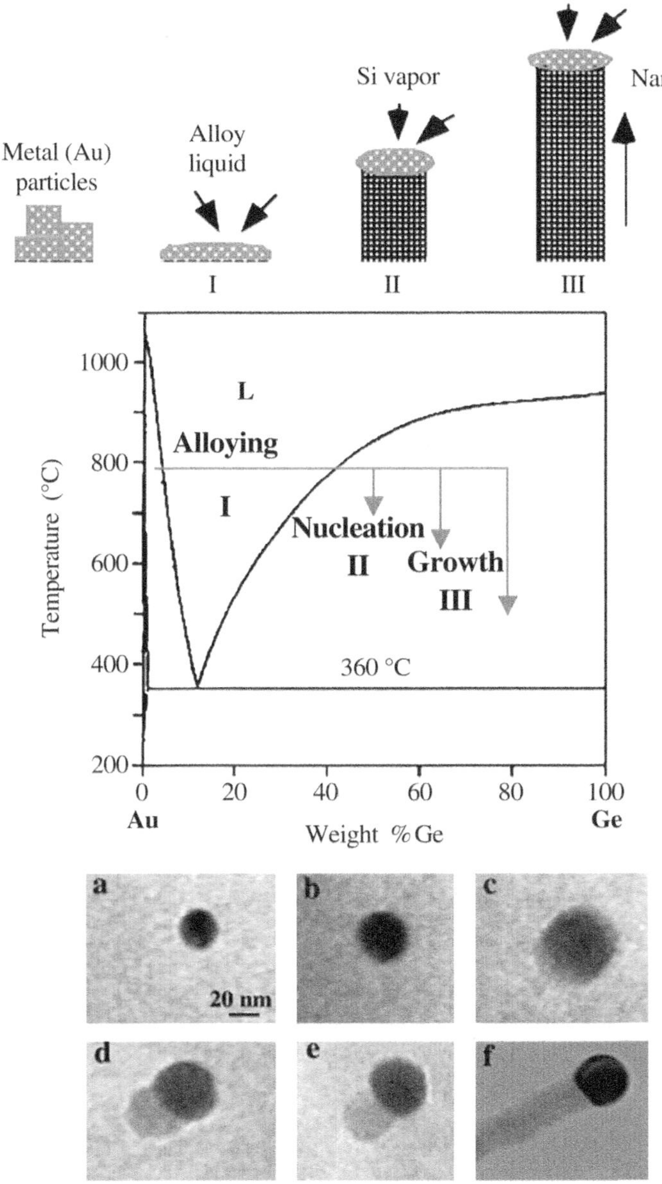

Fig. 19. Schematic illustration of vapor – liquid – solid nanowire growth mechanism including three stages: I) alloying, II) nucleation, and III) axial growth, as exemplified with the Au-Ge binary phase diagram (Au is a good solvent for the Ge nanowire growth because of the existence of the Au-Ge eutectic). This mechanism has been observed by in situ TEM that recorded images during the process of nanowire growth with: **a** Au nanoclusters in solid state at 500 °C; **b** alloying is initiated at 800 °C (Au now exists mostly in solid state); **c** liquid Au/Ge alloy; **d** nucleation of a Ge nanocrystal on the alloy surface; **e** Ge nanocrystal elongates with further Ge condensation; **f** eventually forms a wire. (Reprinted from [98b], copyright (2002) from John Wiley and Sons)

nanowires, and nanorods, is still in its infancy. A first challenge is to be able to reach a high enough concentration so as to obtain the Onsager nematic transition. We have seen that in the case of highly anisotropic charged particles, concentrations in the range of few percent by weight are usually necessary in order to observe such a transition. With much lower aspect ratio particles, such as nanorods or with uncharged moieties such as nanotubes, much higher concentrations are required. Also, many of these colloids are often too bulky and therefore do not form stable concentrated colloidal suspensions on their own. In most cases studied up to now, this problem has been overcome by stabilization of the suspension using either surfactants or by wrapping the colloids within a polymer therefore allowing highly concentrated suspensions to be obtained.

Another difficulty is that often only very small amounts of these anisotropic moieties can be synthesized at a time, limiting the study of the self-assembly properties to transmission electron microscopy analysis [135]. This makes it difficult to study in detail the phase diagram in concentration, usually a fairly sample consuming process. However, different approaches are currently being devised that give a taste of the huge potential that the self-assembly of these nanowires and nanorods can lead to.

A first way to circumvent the problem is to study the self-assembly of these anisotropic moieties within a 2-D Langmuir-Blodgett film. This was recently illustrated by P. Yang and co-workers using suspensions of $BaCrO_4$ nanorods synthesized by the reverse micelle approach [157]. Langmuir-Blodgett films obtained on the water surface and collected on a TEM grid display typical 2-D nematic and smectic organizations, and their topological defects. In Yang's report, a 2-D smectic to 3-D nematic transition is briefly mentioned that gives an insight into the bulk properties of such nanorods. However, as always in the case of TEM studies, it is difficult to determine exactly the role of drying. Note that in a related set of experiments, although the film was deposited by dip coating (directly from the bulk of the solution), the self-assembly properties of the carbon nanotubes bundles [99] allow the formation of fairly well oriented and therefore birefringent films of bundles. Such self-assembly of carbon nanotubes into a structure highly reminiscent of a nematic mesophase has been long awaited and discussed in meetings. Some alignment successes have even been reported when one combines nanotubes with some other liquid crystals, either lyotropic [100] or thermotropic [101]. Based upon our calculation using the Onsager model and our trials, in the case of bulk suspensions of neutral carbon nanotubes, we think that what has precluded the observation of a transition similar to what has been reported in the case of $Li_2Mo_6Se_6$ nanowire, imogolite or $NaNb_2PS_{10}$ nanotubes has been that the isotropic-nematic critical concentration is still higher than the highest concentration yet obtained. In the Zhou and co-workers' report, nanotube bundles have been shortened using a strong etching solution. The consequence is that the nanotubes are probably much more charged than the neutral pristine ones (due to the addition of carboxylic groups), which is known to decrease the critical transition concentration. We therefore expect for these suspensions that the pH of the solution should have a dramatic effect on the self-assembly properties reported.

Another method that has been developed for rapid screening of organic complex fluid phase diagrams consists of studying the drying of a suspension within a capillary tube, directly on the beamline of a synchrotron. It is not an easy method for the quantitative determination of phase diagrams, but one may determine some phase transitions. Unfortunately, thermodynamic equilibrium is not usually reached and therefore this technique cannot be recommended.

Finally, a study of fairly monodisperse CdSe semiconducting nanorods functionalized with amphiphilic molecules in a suspension of cyclohexane has recently been published that shows preliminary results indicative of a typical nematic behavior [102]. Note that these nanorods have been synthesized by the pyrolysis of organometallic precursors of Cd and Se in hot surfactant mixture allowing a good control of the aspect ratio. This state of the art particle size and shape control method should allow one to customize accurately their semiconducting properties.

6.2
Use of MLC for the Structure Determination of Biomolecules by NMR

The first reported use of the mesogenic properties of MLC did not arise where one could have first expected it, i.e., liquid crystal displays, but in NMR. The classic NMR strategy for determining the conformation of a biomolecule [103] involves exploiting the combination of the scalar coupling, ^3J, to obtain dihedral angle information and the ^1H-^1H dipolar cross-relaxation rate, that has a $(1/r^6)$ dependence, where r is the internuclear distance. The limitation of this approach is that only short-range structural information is obtained (a few bonds for the scalar coupling and distances inferior to 5 Å for the relaxation). As a consequence, if only few constraints per nucleus can be detected (e.g., multi-domain proteins, oligosaccharides, or oligonucleotides), the accumulation of errors can lead to an imprecise overall structure. In order to circumvent this problem, a different approach has recently been proposed using, in addition, structural data that are defined relative to an absolute molecular frame [104, 105]. Particularly interesting are methods that exploit residual dipolar couplings [105, 106]. However, as dipolar couplings are almost averaged to zero (tenths of Hz) because of the almost isotropic overall tumbling of the solute molecule, it is therefore necessary to make this tumbling motion more anisotropic. This can be achieved by replacement of the aqueous phase by an anisotropic medium such as a magnetically oriented mesophase [106]. As a consequence new couplings, D, add to the usual scalar couplings, J, to split the signals by a few Hz. By using pairs of spins separated by well known distances (any C-H or N-H bond) one can finally extract more precise and longer range structural information.

A few organic/biological liquid crystals (LC) have been used successfully for this purpose: surfactant-based bicelles, purple membranes, phages, and cellulose microcrystals [107]. We recently proposed using MLC to generate an anisotropic medium. Their inherent advantages are numerous: (i) some MLCs can be easily aligned when a magnetic field is applied, using only very small amounts of mineral materials (1–3 wt%, to be compared with 5–30% for bicelles, 10% for purple membranes, 5% for phages, and 8% for cellulose); (ii) these MLCs, which are sometimes formed using polar solvents other than water, present nematic phases that are stable on a very long time scale (few years) and over a wide temperature range (the liquid state temperature range of the solvent), in stark contrast with the organic LC systems whose stability is often an experimental problem for this type of NMR method [108]; (iii) the dissolved biomolecule can simply be recovered after the experiment by flocculation of the mineral colloid; (iv) finally, the absence of ^1H and ^{13}C nuclei renders isotope labeling of the studied molecule unnecessary, whereas, in the case of organic liquid crystals, the NMR signal is strongly dominated by the liquid crystal resonances due to the high ratio of LC molecules to biomolecules. This is obviously a major advantage especially for systems such as oligosaccharides or oligonucleotides for which labeling is chemically challenging and particularly expensive.

We therefore demonstrated the potential of such MLC for NMR structural characterization by reporting residual C-H dipolar couplings observed for non-labeled saccharide containing the Lewis^X motif of various flexibility dissolved in aqueous suspension of V_2O_5 [109], as well as $H_3Sb_3P_3O_{14}$ [61, 110, 111].

6.3 Use of Flow Alignment and Pretransitional Effect Toward Applications

We have seen that it can be difficult to reach the critical concentration required to observe an isotropic-anisotropic transition because concentrated suspensions of colloids are not always stable. However, orientation of flexible polymers as well as of anisotropic particles in suspension can be induced by flow, a phenomenon that has long been observed, reported, and studied. This phenomenon is especially strong when a pretransitional effect exists, which can be easily observed by the naked eye on a sample that is shaken between crossed polarizers (see for example the section on clays). In these systems, birefringence is induced via mechanical forces, like the shear stresses in a laminar flow ("Maxwell-dynamo-optic effect").

One can therefore take advantage of this alignment effect when anisotropic materials are desired. In a first example, the mesomorphic properties of $Li_2Mo_6Se_6$ that we had previously demonstrated were used to synthesize: (i) highly oriented conducting thin films by spin coating the pure material (after solvent evaporation); (ii) oriented composites by extrusion [112].

Another example comes from the fact that up to recently one could not store beer in plastic bottles because gas diffusion (in and out) through the bottle walls was too large. Composites have now been developed by dispersing clay particles within a polymer. The particles are aligned by flow during extrusion, parallel to the bottle walls. Because of this orientation, the clay particles can play a barrier role and drastically reduce gas diffusion. The charge of the clay also allowed improvements of the mechanical properties of these composites materials [113]. Highly oriented extruded polymer-carbon nanotube composites have been obtained using flow orientation [114] that show very anisotropic properties, including electrical conductivity (with high conductivity only along the axis of extrusion). Such anisotropic conductivity can be useful for some applications but not for the conducting composites used in the new generation of reversible fuses that are widely employed nowadays by the semiconducting and electronic industry [115].

More recently, the flow alignment of suspensions of nanowires has been combined with microfluidic technology [116]. In this development, suspensions of nanowires/nanotubes are flowed within microchannels formed in a PDMS mold and deposited on a silicon wafer substrate (Fig. 20a,b). This allows parallel arrays of NW aligned along the flow direction when the mold is removed to be obtained. Multiply crossed NW arrays can then be built by changing the flow direction sequentially by iterating this process using a succession of molds with channels oriented along different axes (Fig. 20b).

6.4 Composites Materials

We have seen that only in a few cases have some composites materials based on MLC been made. Out of those only in the case of $Li_2Mo_6Se_6$ was the mesogenic

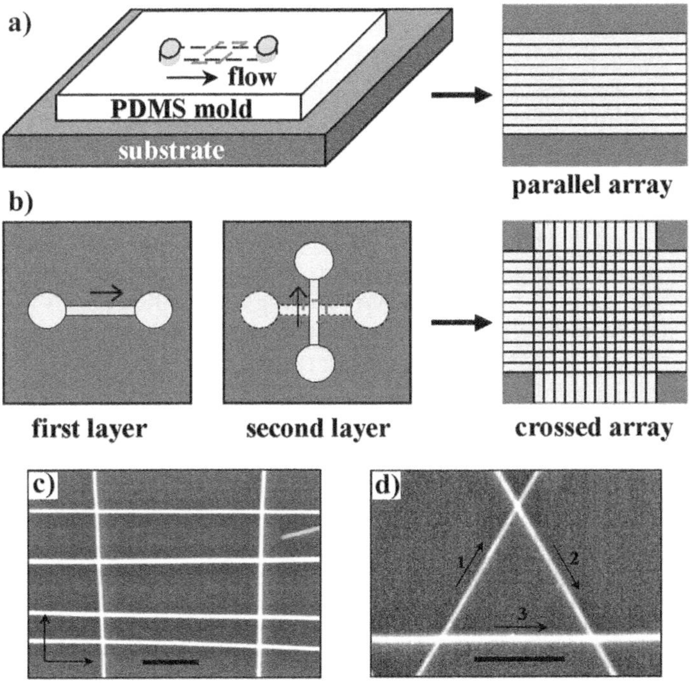

Fig. 20a–d. Schematic of fluidic channel structures for flow assembly and layer-by-layer assembly of crossed NW arrays: **a** NW assembly is carried out by controlling the flow of a NW suspension inside the channel (made from the mold-substrate assembly) – arrays of NWs are oriented parallel to the flow direction on the substrate (after removal of PDMS mold); **b** multiple crossed NW arrays are created by changing the flow direction sequentially in a layer-by-layer assembly process – typical SEM image of crossed arrays of NWs obtained in an assembly process (flow directions are highlighted by *arrows* in the images) comprising: **c** two sequential steps with orthogonal flow directions (InP NWs); **d** three sequential steps with 60° angles between flow directions (GaP NWs). The *scale bars* correspond to 500 nm in c and d. (Reprinted from [116], copyright 2001 American Association for the Advancement of Science)

property purposely used to organize the final material. We develop below two other recent examples.

6.4.1
Mesoporous Materials Based on MLC Templates

We recently showed that it is possible to synthesize a mesoporous composite material by direct assembly of anisotropic hollow objects [117]. In this study we used stable colloidal suspensions of well defined tubular structure made from the scrolling of the sheets themselves obtained by the exfoliation of the acid-exchange $K_4Nb_6O_{17}$ with tetra-(n-butyl)-ammonium hydroxide in water [118]. Detailed physicochemical studies of these colloidal suspensions at high concentration are still in progress [119]; however we thought they would be good candidates to make a silicate-nanotubule composite. The synthesis of our composite material (CMI-1) was adapted from the procedure of Goltsov et al. [120], but now using the organized nanotubules instead of an organic surfactant. Nitrogen physisorption experiments performed on this composite exhibited type IV behavior (hysteresis), typical of materials with large pores with restricted openings (ink bottle). This material has an N_2 Brunauer-Emmett-Teller (BET) surface area of 102 m^2/g, a mean pore size of 14.6 nm, and a single point (p/p0=0.965) pore volume of 0.32 cm^3/g. This mean pore size is in agreement with the mean inner diameter determined by TEM (15.5 Å). This use of mineral liquid crystals represents a new approach for the synthesis of composite porous materials and thus diversifies the types of objects that can be assembled in this way, offering the possibility of synthesizing new materials with a variety of properties.

This work has been further developed using V_2O_5 nanoribbons instead of the $[Nb_6O_{17}^{4-}]_n$ nanotubules [121]. The use of a magnetic field during the condensation of the silicate framework allowed us to obtained centimeter sized single-domain monoliths in which the nematic mineral liquid crystals are well aligned. Total removal of the MLC can also easily be achieved leaving transparent mesoporous birefringent silica with complete retention of the magnetically aligned channels director, even for centimeter sized samples.

6.4.2
Hybrid Organic-Inorganic Solar Cells

In a recent publication, Alivisatos and co-workers reported the making of hybrid nanorods-polymer solar cells and their properties [122]. These solar cells were made by spin casting of a solution of both poly(3-hexylthiophene) (hole acceptor) and CdSe nanorods (electron acceptor) onto indium tin oxide glass substrates coated with poly(ethylene dioxythiophene) doped with polystyrene sulfonic acid and aluminum as a top contact. Nanorods have been used in composites so as to improve the carrier mobility. Indeed, the latter can be high for some inorganic semiconductors, but it is typically extremely low for conjugated polymers [123]. The use of the nanorods supplies an interface for the charge transfer as well as a direct path for electrical transport. Also, because of their anisotropy, self-assembly of these nanorods is observed by electron microscopy. It shows that the nanorods are partially aligned perpendicular to the substrate plane. The orientation of the nanorods as a function of their aspect ratio is predicted to play a crucial role in the properties of these new devices, as it should have a strong impact on the formation of percolation pathways for electron transport.

6.5 Hybrid Bio-Mineral Mesogens

A superb liquid crystal system was recently created by Belcher and co-workers that allowed them to produced a highly ordered composite material from genetically engineered M13 bacteriophage and zinc sulfide (ZnS) nanocrystals [124]. The mesogenic properties of the bacteriophage [125], which was selected for its specific recognition for ZnS crystal surfaces [126] (Fig. 21), were used to trigger the self-assembly. Indeed, bacteriophages were coupled with ZnS solution precursors and spontaneously evolved into a self-supporting hybrid film material that was ordered both at the nanoscale and at the micrometer scale and which was continuous over a centimeter length scale.

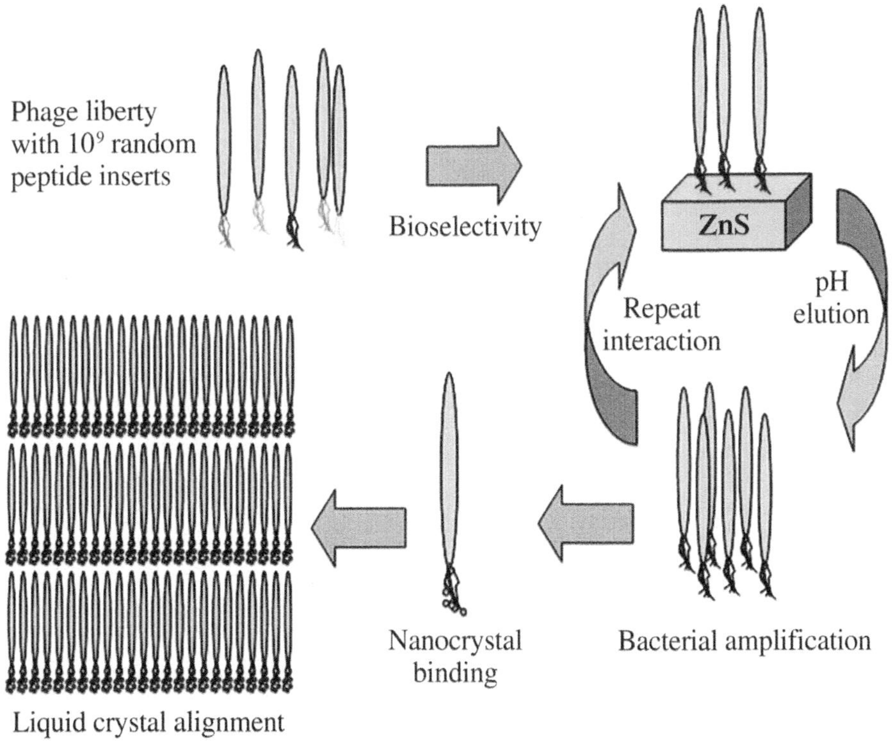

Fig. 21. Schematic diagram of the process used to generate nanocrystal alignment by the phage display method. (Reprinted with permission from [124], copyright 2000, American Association for the Advancement of Science)

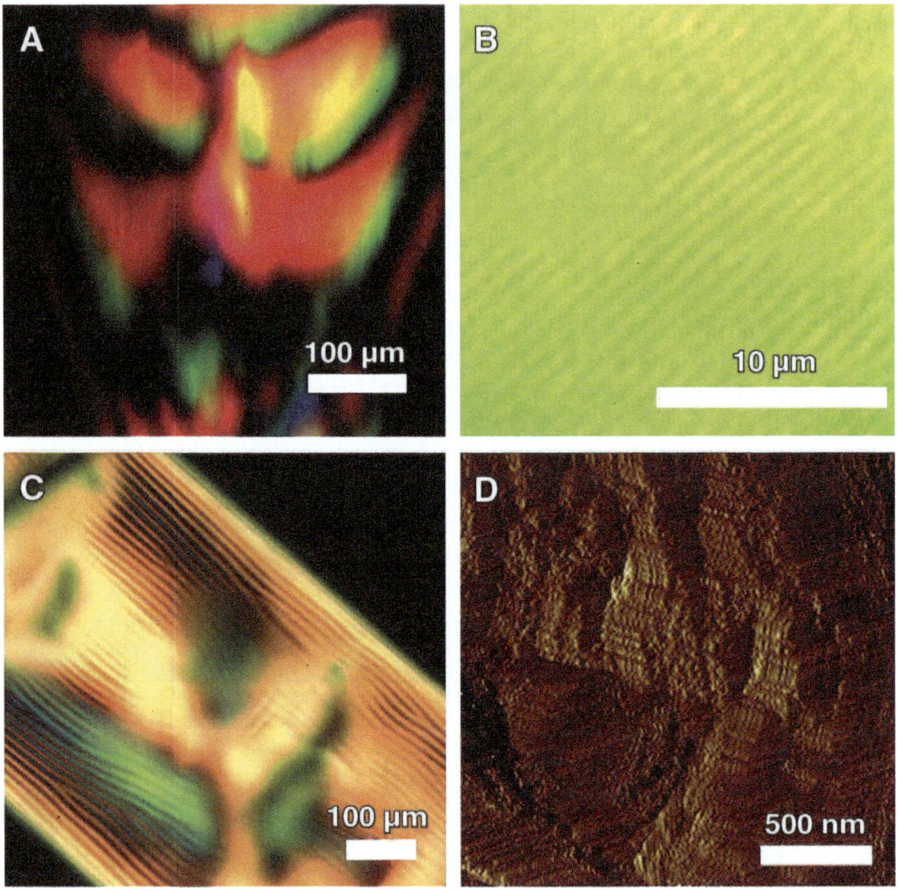

Fig. 22A–D. Characterization of the liquid crystalline suspensions of A7 phage-ZnS nanocrystals (A7-ZnS) and cast film: **A** POM texture of a suspension of A7-ZnS (127 mg/ml); **B, C** their constructive and destructive interference patterns generated from parallel aligned smectic layers when observed using a differential interference contrast filter; **D** texture of the cholesteric phase of an A7-ZnS suspension (76 mg/ml). (Reprinted with permission from [124], copyright 2000, American Association for the Advancement of Science)

In addition, the liquid crystalline phase behavior of these suspensions could be controlled by solvent concentration as well as by the use of a magnetic field. These suspensions and films observed by optical microscopy showed textures indicative of the smectic and cholesteric structure of the hybrid systems (Fig. 22).

7 Perspectives

Let us here briefly discuss a few guidelines useful for the search of new mineral liquid crystalline suspensions. This discussion tries to summarize the subtle balance between excluded volume interactions, electrostatic repulsions, and Van der Waals attractions that is needed to stabilize the nematic phase with respect to the isotropic one and flocculation or recrystallization.

An obvious criterion is that the mineral moieties of interest have to be able to be dispersed in a suitable solvent in large enough concentrations. We have seen several examples of how this can be achieved by exfoliation or by direct self-assembly from molecular precursors. This dispersion step is probably one of the most delicate points to fulfill.

Another related criterion is that the colloidal moieties should be as anisotropic as possible. This can be quantitatively predicted in the frame of the Onsager model. The volume fraction at which nematic ordering occurs is $\Phi \approx 4D/L$, which is 40% in the case of rods of aspect ratio 10 interacting through a purely hard-core potential. This is still quite a large volume fraction. In contrast, rods of aspect ratio 100 will order at a much more reasonable value of 4%. Moreover, both theory and numerical simulations show that particles of aspect ratio smaller than 3–4 never order in a nematic phase whatever their volume fraction. These orders of magnitude illustrate the need for fairly anisotropic moieties.

A parameter that cannot be over looked is the dispersity in size of the particles. Indeed, polydispersity usually prevents long range positional ordering. For example, it is crucial in order to obtain (i) smectic phases to have nanoparticles of fairly homogeneous length and (ii) hexagonal phases to have nanoparticles with diameters as monodisperse as possible.

The charge of the building blocks within mineral liquid crystalline suspensions should also be considered in detail since electrostatic repulsions play two important roles. First, these interactions often ensure colloidal stability against Van der Waals attractions. Second, the existence of clouds of counter-ions increase the effective volume fraction of anisotropic moieties in suspension, which helps to attain the phase transition threshold. Moreover, there is a large entropy gain associated with the release of the counter-ions in solution, which stabilizes the nematic suspension. Therefore, charged moieties are better candidates. However, in the case of neutral moieties, grafting with a polymer layer seems to be an efficient way of preventing particle aggregation.

Flexibility of the moieties is another important parameter because it was recently shown that flexible objects require larger volume fractions to undergo nematic ordering. Flexibility also reduces the nematic order parameter at the transition. Intuitively, very flexible mineral polymers should not show any orientational order at rest, but may display a strong flow birefringence. Thus, any soluble system where the structural unit in the solid state is anisotropic may not necessarily be a lyotropic liquid crystal. For example, in solution a polymer is much less constrained than in the solid state, and hence one must consider the elastic properties of the polymer chain and whether the anisotropic units still exist in solution. As shown recently for the case of the complex fluid with a mineral core based upon the 1-D polyelectrolyte $^{1}_{\infty}[MPS_4]^-$ (M=Ni, Pd) dissolved in DMF [6], the polymer chain is highly flexible and may even, depending upon the metal M and the temperature, fragment and lose its complex fluid properties. High flexibility does not however preclude self-assembly when folding of the polymer leading to unexpected superstructures [34].

We have only described here the mineral mesophases that have been fully and unambiguously characterized. This review article is therefore not exhaustive. Other less well-documented examples can be found in the literature, in particular in another review article [127, 128]. Nevertheless, the diversity of examples described here should draw the attention of the reader to the fact that suspensions based on anisotropic structural units may be liquid crystalline, a feature that could be used to obtain a better understanding/optimization of materials synthesis and processing conditions. Thus, the existence of a nematic domain in the phase diagram together with the possibility of growing reproducibly oriented single domains on the centimeter scale could be used, for instance, in the field of pillared materials to produce better structured materials.

The main interest of mesophases based on mineral moieties is, in contrast to their organic counterparts, that they can be electron rich and so have enhanced electrical, optical, and magnetic properties. In addition, mineral moieties are thermally very stable. Thus, combined with their typically very wide range of temperature stability of mesomorphic organization, it can be envisaged that they can be used for applications under extreme conditions. Another important and common asset of these MLC is their very low cost since some of them can even be found in a natural form. This will allow their use for massive material production.

Finally, since MLC are based on anisotropic nanosystems, the understanding of their collective properties, either pre- or post-transitional, is of prime importance to allow their manipulation and therefore for the development of applications in the field of nanotechnology.

Acknowledgements. We are deeply indebted to many colleagues for instructive discussions and to our coworkers whose listing would be too long. However, we particularly wish to express our gratitude to Patrick Batail and Jacques Livage who have persistently supported us in this new avenue of research, to our PhD students, Olivier Pelletier, Franck Camerel, and Bruno Lemaire as well as to H.N.W. Lekkerkerker and L.M. Bull for their critical review of the manuscript. Finally, we would like to thank Wiley-VCH Verlag GmbH & Co. who kindly granted us permission to use in this article parts of our previous review article [4b], copyright 2000

References

1. De Gennes PG, Prost J (1995) The physics of liquid crystals. Clarendon Press, Oxford
2. Tschierske C (1998) J Mater Chem 8:1485
3. Demus D, Goodby JW, Gray GW, Spiess HW, Vill V (eds) (1998) Handbook of liquid crystals. Wiley, Weinheim
4. a) Davidson P, Batail P, Gabriel JCP, Livage J, Sanchez C, Bourgaux C (1997) Prog Polym Sc 22:913; b) Gabriel JCP, Davidson P (2000) Adv Mater 12:9
5. Gabriel JCP, Uriel S, Boubekeur K, Batail P (2001) Chem Rev 101:2037

6. Sayettat J, Bull LM, Gabriel JCP, Jobic S, Camerel F, Marie AM, Fourmigué M, Batail P, Brec R, Inglebert RL (1998) Angew Chem Int Ed 37:1711
7. Oriol L, Serrano JL (1995) Adv Mater 7:348
8. Zocher H (1925) Z Anorg Allg Chem 147:91
9. Langmuir I (1938) J Chem Phys 6:873
10. Onsager L (1949) Ann NY Acad Sci 51:627
11. See, for instance, Samborski A, Evans GT, Mason CP, Allen MP (1994) Mol Phys 81:263
12. Guinier A, Fournet G (1955) Small angle scattering of X-rays. Wiley, New York
13. Synthesis of $M_2Mo_6Se_6$: a) M=Tl. In: Potel M, Chevrel R, Sergent M (1980) Acta Cryst B 36:1545; b) (M=Li, Na, K, Rb, Cs) Tarascon JM, Hull GW, DiSalvo FJ (1984) Mater Res Bull 19:915
14. Most members of the family, $M_2Mo_6Y_6$ (M=Na, In, K, Rb, Cs, Tl, Y=chalcogen), are pseudo-one-dimensional metals with even a superconductor at 3 K with $Tl_2Mo_6Se_6$: a) Armici JC, Decroux M, Fisher Ø, Potel M, Chevrel R, Sergent M (1980) Solid State Comm 33:607; b) Potel M, Chevrel R, Sergent M, Armici JC, Decroux M, Fisher Ø (1980) J Solid State Chem 35:286
15. Tarascon JM, DiSalvo FJ, Chen CH, Carrol PJ, Walsh M, Rupp L (1985) J Solid State Chem 58:290
16. Gabriel JCP (1993) PhD Thesis, University Paris XI, Orsay
17. a) Davidson P, Gabriel JC, Levelut AM, Batail P (1993) Europhys Lett 21:317; b) Davidson P, Gabriel JC, Levelut AM, Batail P (1993) Adv Mater 5:665
18. Fuhrer MS, Nygard J, Shih L, Forero M, Yoon Y, Mazzoni MSC, Choi H, Ihm J, Louie SG, Zettl A, McEuen PL (2000) Science 288:494
19. Messer B, Song JH, Huang M, Wu YY, Kim F, Yang PD (2000) Adv Mater 12:1526
20. Song JH, Messer B, Wu YY, Kind H, Yang PD (2001) J Am Chem Soc 123:9714–9715
21. Venkataraman L, Lieber CM (1999) Phys Rev Lett 83:5334
22. Bronger W, Müller P (1984) J Less Comm Met 100:241
23. Lu YJ, Ibers JA (1993) Comments Inorg Chem 14:229
24. For example, this was observed in the case of the series $Na_{2-x}Li_xMo_6Se_6$ (x=0–2) [15]
25. a) Farmer C, Fraser AR, Tait JM (1977) Chem Comm 462; b) Wada S-I, Eto A, Wada K (1979) J Soil Sci 30:347; c) Wada S-I (1987) Clays Clay Miner 35:379
26. Barett SM, Budd PM, Price C (1991) Eur Polym J 27:609; d) US Pats 4,252,779 and 4,241,035
27. Koenderink GH, Kluijtmans SGJM, Philise AP (1999) J Colloid Interface Sci 216:429
28. In the case of synthetic Imogolite, it seems that the diameter of the cylinder is 28 Å, slightly more that the natural one
29. a) Kajiwara K, Donkai N, Hiragi Y, Inagaki H (1986) Makromol Chem 187:2883; b) Kajiwara K, Donkai N, Fujiyoshi Y, Inagaki H (1986) Makromol Chem 187:2895; c) Donkai N, Kajiwara K, Schmidt M, Miyamoto T (1993) Makromol Chem Rapid Comm 14:611
30. Donkai N, Hoshino H, Kajiwara K, Miyamoto T (1993) Makromol Chem 194:559
31. a) Livolant F, Leforestier A (1996) Prog Polym Sc 21:1115; b) Giraud-Guille MM (1998) Curr Opin Solid State Mater Sci 3:221
32. a) Hoshino H, Yamana M, Donkai N, Sinigerski V, Kajiwara K, Miyamoto T, Inagaki H (1992) Polym Bull 28:607; b) Hoshino H, Ito T, Donkai N, Urakawa H, Kajiwara K (1992) Polym Bull 29:453
33. Ramos L, Molino F, Porte G (2000) Langmuir 16:5846
34. Camerel F, Gabriel JCP, Davidson P, Schmutz M, Gulik-Krzywicki T, Lemaire B, Bourgaux C, Batail P (2002) Nanoletters 2:403
35. Livage J (1991) Chem Mater 3:578
36. Bailey JK, Nagase T, Pozarnsky GA, MeCartney ML (1990) Mat Res Soc Symp Proc 180:759; b) Pozarnsky GA, McCormick AV (1994) Chem Mater 6:380; c) Livage J (1998) Coord Chem Rev 178/180:999; d) Pelletier O, Davidson P, Bourgaux C, Coulon C, Regnault S, Livage J (2000) Langmuir 16:5295

37. a) Davidson P, Garreau A, Livage J (1994) Liq Cryst 16:905; b) Davidson P, Bourgaux C, Schoutteten L, Sergot P, Williams C, Livage J (1995) Phys II (France) 5:1577; c) Pelletier O, Bourgaux C, Diat O, Davidson P, Livage J (2000) Eur Phys J E 2:191
38. Pelletier O, Bourgaux C, Diat O, Davidson P, Livage J (1999) Eur Phys J B 12:541
39. Commeinhes X, Davidson P, Bourgaux C, Livage J (1997) Adv Mater 9:900
40. Panar M, Beste LF (1977) Macromolecules 10:1401; b) Sridhar CG, Hines WA, Samulski ET (1974) J Chem Phys 61:947
41. Srajer G, Fraden S, Meyer RB (1989) Phys Rev A 39:4828
42. Davidson P, Petermann D, Levelut AM (1995) J Phys II (France) 5:113
43. Pelletier O, Davidson P, Bourgaux C, Livage J (1999) Europhys Lett 48:53
44. Pelletier O, Sotta P, Davidson P (1999) J Phys Chem B 103:5427
45. Lamarque-Forget S, Pelletier O, Dozov I, Davidson P, Martinot-Lagarde P, Livage J (2000) Adv Mater 12:1267
46. Van Olphen H (1977) An introduction to clay colloid chemistry. Wiley, New York
47. a) Burger J, Sourieau P, Combarnous M (1985) Thermal methods of oil recovery. Technip, Paris; b) Pinnavaia TJ (1983) Nature 220:365; c) Kojima Y, Usuki A, Kawasumi M, Okada A, Kurauchi T, Kagimaito O, Kaji K (1995) J Polym Sci B 33:1039
48. EmersonWW (1956) Nature 178:1248
49. a) Mourchid A, Delville A, Lambard J, Lécolier E, Levitz P, (1995) Langmuir 11:1942; b) Mourchid A, Lécolier E, Van Damme H, Levitz P (1998) Langmuir 14:4718; Pignon F, Piau JM, Magnin A (1996) Phys Rev Lett 76:4857; c) Pignon F, Magnin A, Piau JM (1997) Phys Rev Lett 79:4689; d) Pignon F, Magnin A, Piau JM, Cabane B, Lindner P, Diat O (1997) Phys Rev E 56:3281; e) Bonn D, Tanaka H, Wegdam G, Kellay H, Meunier J (1998) Europhys Lett 45:52; f) Bonn D, Kellay H, Tanaka H, Wegdam G, Meunier J (1999) Langmuir 15:7534
50. Gabriel JCP, Sanchez C, Davidson P (1996) J Phys Chem 100:11,139
51. a) Forsyth PA, Marcelja JS, Mitchell DJ, Ninham BW (1978) Adv Colloid Interface Sci 9:37; b) Eppenga R, Frenkel D (1984) Mol Phys 52:1303; c) Frenkel D (1989) Liq Cryst 5:929
52. Lemaire BJ, Panine P, Gabriel J-CP, Davidson P (2002) Europhys Lett (in press)
53. Morvan M, Espinat D, Lambard J, Zemb T (1994) Colloid Surf A 82:193
54. Forsyth PA, Marcelja JS, Mitchell DJ, Ninham BW (1978) Adv Colloid Interface Sci 9:37
55. Mourchid A, Levitz P (1998) Phys Rev E 57:R4887 and references cited therein
56. Levitz P, Lécolier E, Mourchid A, Delville A, Lyonnard S (2000) Europhys Lett 49:672
57. a) Saunders JM, Goodwin JW, Richardson RM, Vincent B (1999) J Phys Chem B, 103:9211; b) Ramsay JDF, Lindner P (1993) J Chem Soc Faraday Trans, 89:4207; c) Ramsay JDF, Swanton SW, Bunce J (1990) J Chem Soc Faraday Trans 86:3919; d) DiMasi E, Fossum JO, Gog T, Venkataraman C (2001) Phys Rev E, 64:061704; e) Porion P, Al Mukhtar M, Faugère AM, van der Maarel JRC, Delville A (2001) J Phys Chem B 105:10505
58. Bihannic I, Michot LJ, Lartiges BS, Vantelon D, Labille J, Thomas F, Susini J, Salomé M, Fyard B (2001) Langmuir 17:4144
59. Cousin F, Cabuil V, Levitz P (2002) Langmuir 18:1466
60. A new study of the salt-induced ordering in lamellar colloids appeared while this paper was in press: Rowan DG, Hansen JP (2002) Langmuir 18:2063–2068
61. Gabriel JCP, Camerel F, Lemaire BJ, Desvaux H, Davidson P, Batail P (2001) Nature 413:504
62. Piffard Y, Verbaere A, Lachgard A, Deniard-Courant S, Tournoux M (1986) Rev Chim Gen 23:766
63. Zocher H, Torök C (1960) Kolloid Zeit170:140; b) Zocher H, Torök C (1960) Kolloid Zeit 173:1; c) Zocher H, Torök C (1962) Kolloid Zeit 180:41
64. Bugosh J (1961) J Phys Chem 65:1789
65. a) Buining PA, Pathmamanoharan C, Jansen JBH, Lekkerkerker HNW (1991) J Ceram Soc 74:1303; b) Buining PA, Philipse AP, Lekkerkerker HNW (1994) Langmuir 10:2106; c) van Bruggen MPB, Donker M, Lekkerkerker HNW, Hugues TL (1999) Colloid Surf 150:115; d) Buining PA, Veldhuizen YSJ, Pathmamanoharan C, Lekkerkerker HNW (1992) Colloid Surf 64:47; e) Buining PA, Lekkerkerker HNW (1993) J Phys Chem 97:11,510; f) Vroege GJ, Lekkerkerker HNW (1993) J Phys Chem 97:3601
66. van Bruggen MPB, Dhont JKG, Lekkerkerker HNW (1999) Macromolecules 32:2256

67. a) Deryagin BV, Landau L (1941) Acta Physicochim URSS 333:55; b) Verwey E, Overbeek JW (1948) Theory of the stability of lyophobic colloids. Elsevier, Amsterdam
68. Israelachvili JN (1991) Intermolecular and surface forces. Academic Press, London
69. Pelletier O, Davidson P, Bourgaux C, Livage J (1999) Progr Colloid Polym Sci 112:121
70. a) van Bruggen MPB, Lekkerkerker HNW, Dhont JKG (1997) Phys Rev E 56:4394; b) Wierenga AM, Philipse AP, Reitsma EM (1997) Langmuir 13:947; c) Wierenga AM, Philipse AP, Lekkerkerker HNW, Boger DV (1998) Langmuir 14:55; d) van Bruggen MPB, Lekkerkerker HNW, Maret G, Dhont JKG (1998) Phys Rev E 58:7668
71. Van der Kooij F, Lekkerkerker HNW (1998) J Phys Chem B 102:7829
72. Bates MA, Frenkel D (1999) J Chem Phys 110:6553
73. Chandrasekhar S, Sadashiva BK, Suresh KA (1977) Pramana 7:471
74. Brown ABD, Clarke SM, Rennie AR (1998) Langmuir 14:3129
75. Brown ABD, Ferrero C, Narayanan T, Rennie AR (1999) Eur Phys J B 11:481
76. Veerman JAC, Frenkel D (1992) Phys Rev A 45:5632
77. Brown ABD, Rennie AR (2000) Phys Rev E 62:851
78. van der Kooij FM, Kassapidou K, Lekkerkerker HNW (2000) Nature 406:868
79. van der Kooij FM, van der Beek D, Lekkerkerker HNW (2001) J Phys Chem B 105:1696
80. Wensink HH, Vroege GJ, Lekkerkerker HNW (2001) J Phys Chem B 105:10,610
81. van der Kooij FM, Lekkerkerker HNW (2000) Phys Rev Lett 84:781
82. Wensink HH, Vroege GJ, Lekkerkerker HNW (2001) J Chem Phys 115:7319
83. van der Kooij FM, Lekkerkerker HNW (2000) Langmuir 16:10,144
84. Zocher H, Jacobsohn K (1929) Kolloid Beih 28:167
85. Heller W (1935) Compt Rend 201:831; b) Heller W (1980) Polymer colloids II. Fitch E (ed) Plenum Press, New York, and references cited therein
86. a) Maeda Y, Hachisu S (1983) Colloids Surf 6:1; b) Maeda H, Maeda Y (1996) Langmuir 12:1446
87. Cornell RM, Schwertmann U (1996) The iron oxides. VCH, Weinheim
88. Lemaire BJ, Davidson P, Ferre J, Jamet JP, Panine P, Dozov I, Jolivet JP (2002) Phys Rev Lett 88:125,507
89. Iijima (1991) Nature 354:56
90. a) Ozin GA (1992) Adv Mater 4:612; b) Martin CR (1994) Science 266:1961; c) Martin CR (1996) Acc Chem Res 28:61
91. Wu CG, Bein T (1994) Science 266:1013
92. a) Piraux L, George JM, Despres JF, Leroy C, Ferain E, Legras R, Ounadjela K, Fert A (1994) Appl Phys Lett 65:2484; b) Fert A, Piraux L (1999) J Magn Magn Mater 200:338
93. a) Martin BR, Dermody DJ, Reiss BD, Fang MM, Lyon LA, Natan MJ, Mallouk TE (1999) Adv Mater 11:1021; b) Nicewarner-Pena SR, Freeman RG, Reiss BD, He L, Pena DJ, Walton ID, Cromer R, Keating CD, Natan MJ (2001) Science 294:137
94. Lisiedki I, Pileni MP (1993) JACS 115:3887
95. a) Tanori J, Pileni MP (1995) Adv Mater 7:862; b) Tanori J, Pileni MP (1997) Langmuir 13:639; c) Filankenbo, Pileni MP (2000) Appl Surf Science 164:260
96. Li M, Schnablegger H, Mann S (1999) Nature 402:393
97. Peng X, Manna L, Yang W, Wickham J, Scher E, Kadavanich A, Alivisatos AP (2000) Nature 404:59
98. a) Hu J, Odom TW, Lieber CM (1999) Acc Chem Res 32:435; b) Wu Y, Yan H, Huang M, Messer B, Song JH, Yang P (2002) Chem Eur J 8:1261
99. Shimoda H, Oh SJ, Geng HZ, Walker RJ, Zhang XB, McNeil LE, Zhou O (2002) Adv Mater 14:899
100. Vigolo B, Penicaud A, Coulon C, Sauder C, Pailler R, Journet C, Bernier P, Poulin P (2000) Science 290:1331
101. Lynch MD, Patrick DL (2002) Nanoletters (in press)
102. Li LS, Walda J, Manna L, Alivisatos AP (2002) Nano Lett 2:557
103. Wüthrich K (1986) NMR of proteins and nucleic acids. Wiley Interscience, New York
104. Tjandra N, Garrett DS, Gronenborn AM, Bax A, Clore GM (1997) Nat Struct Biol 4:443

105. Tjandra N, Omichinski JG, Gronenborn AM, Clore GM, Bax A (1997) Nat Struct Biol 4:732
106. Tjandra N, Bax A (1997) Science 278:1111
107. For a quick review: Prestegard JH, Kishore AI (2001) Curr Opin Chem Biol 5:584 and references therein
108. Losonczi JA, Prestegard JH (1998) J Biomol NMR 12:47
109. Desvaux H, Gabriel JCP, Berthault P, Camerel F (2001) Angew Chem Int Ed 40:373
110. Berthault P, Jeannerat D, Zhang Y, Camerel F, Alvarez-Salgado F, Boulard Y, Sinaÿ P, Gabriel JCP, Desvaux H (submitted)
111. See www graphical abstract that represent biomolecules within $H_3Sb_3P_3O_{14}$ sheets
112. a) Golden JH, DiSalvo FJ, Fréchet JMJ (1995) Chem Mater 7:232; b) Golden JH, DiSalvo FJ, Fréchet JMJ, Silcox J, Thomas M, Ellman J (1996) Science 273:782
113. Here are few examples of patents of composites based on a dispersed clay with enhanced mechanical or gas barrier properties: US4472538, US4739007, US4889885, US4894411, EP0352042, EP0398551, WO9304117, WO9304118, WO9311190, US5385776, US5514734, EP0358415
114. Vigolo B, Penicaud A, Coulon C, Sauder C, Pailler R, Journet C, Bernier P, Poulin P (2000) Science 290:1331
115. Chang J. Personnal communication
116. Huang Y, Duan X, Wei Q, Lieber CM (2001) Science 291:630
117. Camerel F, Gabriel JCP, Batail P (2002) Chem Comm 1926
118. Saupe GB, Waraksa CC, Kim HN, Han YJ, Kaschak DM, Skinner DM, Mallouk TE (2000) Chem Mater 12:1556
119. A nice optical study of this MLC ($K_4Nb_6O_{17}$ sols) appeared while this paper was in Press: Miyamoto N, Nakato T (2002) Adv Mater 14:1267
120. Goltsov YG, Matkovskaya LA, Smelaya ZV, Il'in VG (1999) Mendeleev Commun 241
121. a) Camerel F (2001) PhD Thesis, Nantes University, France; b) Camerel F, Gabriel JCP, Batail P (2002) In Press, Adv Funct Mater
122. Huynh WU, Dimer JJ, Alivisatos AP (2002) Science 295:2425
123. Borazano L, Carte SA, Scott JC, Malliaras GG, Brock PJ (1999) Appl Phys Lett 74:1132
124. Lee SW, Mao C, Flynn CE, Belcher AM (2002) Science 296:892
125. a) Dogic Z, Fraden S (1997) Phys Rev Lett 78:2417; b) Dogic Z, Fraden S (2000) Langmuir 16:7820; c) Lapointe J, Marvin DA (1973) Mol Cryst Liq Cryst 19:269; d) Issaenko A, Harris SA, Lubensky TC (1999) Phys Rev E 60:578
126. It has been shown recently that engineered viruses can recognize specific semiconductor surfaces using the method of selection by combinatorial phage display: Whaley R, English DS, Hu EL, Barbara PF, Belcher AM (2000) Nature 405:665
127. Sonin AS (1998) Colloid J 60:129
128. A report of a possible new MLC appeared while this manuscript was in press: Backov R, Morgan AN, Lane S, Perez-Cordero EE, Williams K, Meisel MW, Sanchez C, Talham DR (2002) Mol Cryst Liq Cryst 376:127
129. A review of oxidic nanotubes and nanorods has been published while this manuscript was in press: Patzke GR, Krumeich F, Nesper R (2002) Angew Chem Int Ed 41:2446
130. Here are few examples of reviews on the subject: a) Sinnott SB, Andrews R (2001) Crit Rev Solid State Mater Sciences 26:145; b) Rao CNR, Satishkumar BC, Govindaraj A, Nath M (2001) Chem Phys Chem 2:78; c) Rakov EG (2000) Uspekhi Khimii 69:41; d) Ajayan PM (1999) Chem Rev 99:1787
131. a) Zhang QM, Li Y, Xu DS, Gu ZN (2001) J Mater Sci Lett 20:925; b) Zhu JJ, Liao XH, Zhao XN, Chen HY (2001) Mater Lett 49:91; c) Lin SW, Yue J, Gedanken A (2001) Adv Mater 13:656; d) Jana NR, Gearheart L, Murphy CJ (2001) Chem Comm 2001:617; e) Bhattacharyya S, Saha SK, Chakravorty D (2000) Appl Phys Lett 77:3770; f) Kyoung M, Lee M (1999) Opt Commun 171:145; g) Link S, El-Sayed MA (1999) J Phys Chem B 103:8410; h) Sloan J, Wright DM, Woo HG, Bailey S, Brown G, York APE, Coleman KS, Hutchison JL, Green MLH (1999) Chem Comm 1999:699; i) Zhou Y, Yu SH, Cui XP, Wang CY, Chen ZY (1999) Chem Mater 11:545; j) Korgel BA, Fitzmaurice D (1998) Self-assembly of

silver nanocrystals into two-dimensional nanowire arrays. Adv Mat 10:661; k) Han YJ, Kim JM, Stucky GD (2000) Chem Mater 12:2068

132. a) Jana NR, Gearheart L, Murphy CJ (2001) J Phys Chem B 105:4065; b) Mbindyo JKN, Reiss BD, Martin BR, Keating CD, Natan MJ, Mallouk TE (2001) Adv Mat 13:249; c) Wang BL, Yin SY, Wang GH, Buldum A, Zhao JJ (2001) Phys Rev Lett 86:2046; d) Ramsperger U, Uchihashi T, Nejoh H (2001) Appl Phys Lett 78:85; e) Yu JS, Kim JY, Lee S, Mbindyo JKN, Martin BR, Mallouk TE (2000) Chem Comm 2000:2445
133. Nikoobakht, Wang ZL, El-Sayed MA (2000) J Phys Chem B 104:8635
134. a) Wang XF, Zhang LD, Zhang J, Shi HZ, Peng XS, Zheng MJ, Fang J, Chen JL, Gao BJ (2001) J Phys D 34:418; b) Choi SH, Wang KL, Leung MS, Stupian GW, Presser N, Morgan BA, Robertson RE, Abraham M, King EE, Tueling MB, Chung SW, Heath JR, Cho SL, Ketterson JB (2000) J Vac Sci Technol A 18:1326
135. Masuda H, Yanagishita T, Yasui K, Nishio K, Yagi I, Rao TN, Fujishima A (2001) Adv Mater 13:247
136. a) Henry Y, Ounadjela K, Piraux L, Dubois S, George JM, Duvail JL (2001) Eur Phys J B 20:35; b) Cao HQ, Xu Z, Sang H, Sheng D, Tie CY (2001) Adv Mater 13:121; c) Yang SG, Zhu H, Ni G, Yu DL, Tang SL, Du YW (2000) J Phys D Appl Phys 33:2388; d) Puntes VF, Krishnan KM, Alivisatos AP (2001) Science 291:2115
137. Ounadjela K, Ferre R, Louail L, George JM, Maurice JL, Piraux L, Dubois S (1997) J Appl Phys 81:5455
138. a) Molares MET, Buschmann V, Dobrev D, Neumann R, Scholz R, Schuchert IU, Vetter J (2001) Adv Mater 13:62; b) Peng LQ, Ju X, Wang SC, Xian DC, Chen H, He YJ (1999) Chin Phys Lett 16:126; c) Setlur AA, Lauerhaas JM, Dai JY, Chang RPH (1996) Appl Phys Lett 69:345; d) Peng LQ, Ju X, Wang SC, Xian DC, Chen H, He YJ (1999) Chin Phys Lett16:126
139. Park SJ, Kim S, Lee S, Khim ZG, Char K, Hyeon TJ (2000) Am Chem Soc 122:8581
140. Zhang YF, Tang YH, Wang N, Lee CS, Bello I, Lee ST (2000) Phys Rev B 61:4518
141. Tang CC, Fan SS, Li P, de la Chapelle ML, Dang HY (2001) J Cryst Growth 224:1171
142. Gu Q, Dang HY, Cao J, Zhao JH, Fan SS (2000) Appl Phys Lett 76:3020; b) Wang N, Tang YH, Zhang YF, Lee CS, Bello I, Lee ST (1999) Chem Phys Lett 299:237; c) Wang N, Tang YH, Zhang YF, Lee CS, Lee ST (1998) Phys Rev B 58(N24):R16024
143. a) Krumeich F, Muhr HJ, Niederberger M, Bieri F, Nesper R (2000) Z Anorg Allg Chem 626:2208; b) Reinoso JM, Muhr HJ, Krumeich F, Bieri F, Nesper R (2000) Helv Chim Acta 83:1724; c) Niederberger M, Muhr HJ, Krumeich F, Bieri F, Gunther D, Nesper R (2000) Chem Mater 12:1995; d) Muhr HJ, Krumeich F, Schonholzer UP, Bieri F, Niederberger M, Gauckler LJ, Nesper R (2000) Adv Mater 12:231; e) Millet P, Henry JY, Mila F, Galy J (1999) J Solid State Chem 147:67678; f) Krumeich F, Muhr HJ, Niederberger M, Bieri F, Schnyder B, Nesper R (1999) J Am Chem Soc 121(N36):8324; g) Spahr ME, Stoschitzki-Bitterli P, Nesper R, Haas O, Novak P (1999) J Electrochem Soc 146:278,083; h) Spahr ME, Bitterli P, Nesper R, Muller M, Krumeich F, Nissen HU (1998) Angew Chem Int Ed 37:1263; i) Pillai KS, Krumeich F, Muhr HJ, Niederberger M, Nesper R (2001) Solid State Ionics 141:185
144. Lakshmi BB, Patrissi CJ, Martin CR (1997) Chem Mater 9:2544
145. Satishku BC, Govindaraj A, Nath M, Rao CNR (2000) J Mater Chem 10:2115–2119
146. a) Zhang M, Bando Y, Wada K (2001) J Mater Res 16:1408; b) Kasuga T, Hiramatsu M, Hoson A, Sekino T, Niihara K (1998) Langmuir 14:3160–3163; c) Hoyer P (1996) Langmuir 12:1411; d) Seo DS, Lee JK, Kim H (2001) J Cryst Growth 229:428; e) Li XH, Zhang XG, Li HL (2001) Chem J Chin Univ 22:130; f) Zhang M, Bando Y, Wada K (2001) J Mater Sci Lett 20:167; g) Zhang SL, Zhou JF, Zhang ZJ, Du ZL, Vorontsov AV, Jin ZS (2000) Chin Sci Bull 45:1533; h) Khitrov G (2000) MRS Bull 25:3; i) Imai H, Takei Y, Shimizu K, Matsuda M, Hirashima H (1999) J Mater Chem 9:2971; j) Lei Y, Zhang LD, Fan JC (2001) Chem Phys Lett 338:231; k) Lei Y, Zhang LD (2001) J Mater Res 16:1138; l) Lei Y, Zhang LD, Meng GW, Li GH, Zhang XY, Liang CH, Chen W, Wang SX (2001) Appl Phys Lett 78:1125; m) Hulteen JC, Martin CR (1997) J Mater Chem 7:1075; n) Kobayashi S, Hanabusa K, Hamasaki N, Kimura M, Shirai H, Shinkai S (2000) Chem Mater 12:1523
147. Lakshmi BB, Dorhout PK, Martin CR (1997) Chem Mater 9:857
148. Zheng MJ, Li GH, Zhang XY, Huang SY, Lei Y, Zhang LD (2001) Chem Mater 13:3859

149. a) Pu L, Bao XM, Zou JP, Feng D (2001) Angew Chem Int Ed 40:1490; b) Zheng MJ, Zhang LD, Zhang XY, Zhang J, Li GH (2001) Chem Phys Lett 334:298
150. Zheng MJ, Zhang LD, Zhang XY, Zhang J, Li GH (2001) Chem Phys Lett 334(N4/6):298
151. a) Li Y, Cheng GS, Zhang LD (2000) J Mater Res 15:2305; b) Huang MH, Wu YY, Feick H, Tran N, Weber E, Yang PD(2001) Adv Mater 13:113; c) Kong YC, Yu DP, Zhang B, Fang W, Feng SQ (2001) Appl Phys Lett 78:407–409; d) Li Y, Cheng GS, Zhang LD (2000) J Mater Res 15:2305; e) Li Y, Meng GW, Zhang LD, Phillipp F (2000) Appl Phys Lett 76:2011
152. Tang CC, Fan SS, de la Chapelle ML, Li P (2001) Chem Phys Lett 333:12
153. a) Zhang M, Bando Y, Wada K, Kurashima K (1999) J Mater Sci Lett 18:1911; b) Chang HJ, Chen YF, Lin HP, Mou CY (2001) Appl Phys Lett 78:3791; c) Wang LZ, Tomura S, Ohashi F, Maeda M, Suzuki M, Inukai K (2001) J Mater Chem 11:1465; d) Wu XC, Song WH, Wang KY, Hu T, Zhao B, Sun YP, Du JJ (2001) Chem Phys Lett 336:53; e) Liu ZQ, Xie SS, Sun LF, Tang DS, Zhou WY, Wang CY, Liu W, Li YB, Zou XP, Wang G (2001) J Mater Res 16:683–686; f) Wang ZL, Gao RPP, Gole JL, Stout JD (2000) Adv Mater 12:1938; g) Harada M, Adachi M (2000) Adv Mater 12:839; h) Lin HP, Mou CY, Liu SB (2000) Adv Mater 12:103; i) Zhang M, Bando Y, Wada K, Kurashima K (1999) J Mater Sci Lett 18:1911; j) Adachi M, Harada T, Harada M (1999) Langmuir 15:7097; k) Zhang M, Bando Y, Wada K (2000) J Mater Res 15:387
154. Park GS, Choi WB, Kim JM, Choi YC, Lee YH, Lim CB (2000) J Cryst Growth 220:494
155. Wang WH, Zhan YJ, Wang GH (2001) Chem Comm 2001:727
156. Vayssieres L, Beermann N, Lindquist SE, Hagfeldt A (2001) Chem Mater 13:233
157. Kwan S, Kim F, Akana J, Yang PD (2001) Chem Comm 2001:447
158. Zhang YF, Tang YH, Duan XF, Zhang Y, Lee CS, Wang N, Bello I, Lee ST (2000) Chem Phys Lett 323:180
159. a) Remskar M, Mrzel A, Skraba Z, Jesih A, Ceh M, Demsar J, Stadelmann P, Levy F, Mihailovic D (2001) Science 292:479; b) Hsu WK, Chang BH, Zhu YQ, Han WQ, Terrones H, Terrones M, Grobert N, Cheetham AK, Kroto HW, Walton DRM (2000) J Am Chem Soc122:10155; c) Seifert G, Terrones H, Terrones M, Jungnickel G, Frauenheim T (2000) Phys Rev Lett 85:146; d) Margulis L, Dluzewski P, Feldman Y, Tenne R (1996) J Microsc 181:68; e) Zhang Q, Huang RB, Liu ZY, Zheng LS (1995) Chem J Chin Univ16:1624; f) Feldman Y, Wasserman E, Srolovitz DJ, Tenne R (1995) Science 267:222; g) Mastai Y, Homyonfer M, Gedanken A, Hodes G (1999) Adv Mater 11:1010
160. Nath M, Govindaraj A, Rao CNR (2001) Adv Mater 13:283; b) Remskar M, Skraba Z, Stadelmann P, Levy F (2000) Adv Mater 12:814; c) Remskar M, Skraba Z, Sanjines R, Levy F (1999) Surf Rev Lett 6:1283; d) Feldman Y, Frey GL, Homyonfer M, Lyakhovitskaya V, Margulis L, Cohen H, Hodes G, Hutchison JL, Tenne R (1996) J Am Chem Soc 118:5362
161. Homyonfer M, Alperson B, Rosenberg Y, Sapir L, Cohen SR, Hodes G, Tenne R (1997) J Am Chem Soc 119:2693
162. a) Zhu YQ, Hsu WK, Terrones H, Grobert N, Chang BH, Terrones M, Wei BQ, Kroto HW, Walton DRM, Boothroyd CB, Kinloch I, Chen GZ, Windle AH, Fray DJ (2000) J Mater Chem 10:2570; b) Rothschild A, Sloan J, Tenne R (2000) J Am Chem Soc 122(N21):5169; c) Zhu YQ, Hsu WK, Grobert N, Chang BH, Terrones M, Terrones H, Kroto HW, Walton DRM, Wei BQ (2000) Chem Mater 12:1190; d) Mackie EB, Galvan DH, Adem E, Talapatra S, Yang GL, Migone AD (2000) Adv Mater 12:495; e) Seifert G, Terrones H, Terrones M, Jungnickel G, Frauenheim T (2000) Solid State Commun 114:245; f) Rothschild A, Frey GL, Homyonfer M, Tenne R, Rappaport M (1999) Synthesis of bulk WS2 nanotube phases. Mater Res Innovations 3:145
163. Tsirlina T, Feldman Y, Homyonfer M, Sloan J, Hutchison JL, Tenne R (1998) Fuller Sci Technol 6:157
164. Zhu YQ, Hsu WK, Terrones M, Firth S, Grobert N, Clark RJH, Kroto HW, Walton DRM (2001) Chem Commun 2001:121
165. a) Hsu WK, Zhu YQ, Firth S, Terrones M, Terrones H, Trasobares S, Clark RJH, Kroto HW, Walton DRM (2001) Carbon 39:1107; b) Hsu WK, Zhu YQ, Boothroyd CB, Kinloch I, Trasobares S, Terrones H, Grobert N, Terrones M, Escudero R, Chen GZ, Colliex C, Windle AH, Fray DJ, Kroto HW, Walton DRM (2000) Chem Mater 12:3541

166. Nath M, Rao CNR (2001) J Am Chem Soc 123:4841
167. Zhang WX, Yang ZH, Zhan JH, Yang L, Yu WC, Zhou GE, Qian YT (2001) Mater Lett 47:367
168. Li W, Kalia RK, Vashishta P (1996) Phys Rev Lett 77:2241
169. a) Sapp SA, Lakshmi BB, Martin CR (1999) Adv Mater 11:402; b) Prieto AL, Sander MS, Martín-González MS, Gronsky R, Sands T, Stacy AM (2001) J Am Chem Soc 123:7160
170. a) Davidson P, Gabriel JC, Levelut AM, Batail P (1993) Nematic liquid crystalline mineral polymers. Adv Mater 5:665; b) Davidson P, Gabriel JC, Levelut AM, Batail P (1993) Europhys Lett 21:317; c) Messer B, Song JH, Yang PD (2000) J Am Chem Soc 122:10,232
171. a) Wang SH, Yang SH (2000) Adv Mater Opt Electron 10:39; b) Wang SH, Yang SH (2000) Chem Phys Lett 322:567
172. Dloczik L, Engelhardt R, Ernst K, Fiechter S, Sieber I, Konenkamp R (2001) App Phys Lett 78:3687
173. Xu DS, Xu YJ, Chen DP, Guo GL, Gui LL, Tang YQ (2000) Chem Phys Lett 325; b) Li Y, Xu DS, Zhang QM, Chen DP, Huang FZ, Xu YJ, Guo GL, Gu ZN (1999) Chem Mater 11:3433
174. Shen CM, Zhang XG, Ki HL (2001) Mater Sci Eng A 303:19
175. a) Rubio A, Corkill JL, Cohen ML (1994) Phys Rev B 49:5081; b) Chopra NG, Luyken RJ, Cherrey K, Crespi VH, Cohen ML, Louie SG, Zettl A (1995) Science 269:966; c) Loiseau A, Willaime F, Demoncy N, Hug G, Pascard H (1996) Phys Rev Lett 76:4737; d) Suenaga K, Colliex C, Demoncy N, Loiseau A, Pascard H, Willaime F (1997) Science 278:653; e) Cumings J, Zettl A (2000) Chem Phys Lett 316:211
176. Shimizu Y, Moriyoshi Y, Komatsu S, Ikegami T, Ishigaki T, Sato T, Bando Y (1998) Thin Solid Films 316:178; b) Shimizu Y, Moriyoshi Y, Tanaka H, Komatsu S (1999) App Phys Lett 75:929
177. a) Han WQ, Bando Y, Kurashima K, Sato T (1998) Appl Phys Lett 73:3085; b) Han WQ, Bando Y, Kurashima K, Sato T (1999) Chem Phys Lett 299:368; c) Golberg D, Bando Y, Han W, Kurashima K, Sato T (1999) Chem Phys Lett 308:337; d) Han WQ, Redlich P, Ernst F, Ruhle M (1999) Chem Mater 11:3620; e) Golberg D, Bando Y, Bourgeois L, Kurashima K, Sato T (2000) Carbon 38:2017; f) Golberg D, Bando Y, Kurashima K, Sato T (2000) Chem Phys Lett 323:185
178. a) Chen Y, Gerald JF, Williams JS, Bulcock S (1999) Chem Phys Lett 299:260; b) Chen Y, Chadderton LT, FitzGerald J, Williams JS (1999) Appl Phys Lett 74:2960
179. Lourie OR, Jones CR, Bartlett BM, Gibbons PC, Ruoff RS, Buhro WE (2000) Chem Mater 12:1808
180. Terauchi M, Tanaka M, Suzuki K, Ogino A, Kimura K (2000) Chem Phys Lett 324:359
181. a) Li JY, Chen XL, Qiao ZY, Cao YG, He M, Xu T (2000) Appl Phys A 71:349; b) Chen CC, Yeh CC (2000) Adv Mater 12:738; c) Han WQ, Fan SS, Li QQ, Hu YD (1997) Science 277:1287; d) Chen et al. (2001) J Am Chem Soc123:2791; e) Li ZJ, Chen XL, Li HJ, Tu QY, Yang Z, Xu YP, Hu BQ (2001) Appl Phys A 72:629; f) He MQ, Minus I, Zhou PZ, Mohammed SN, Halpern JB, Jacobs R, Sarney WL, Salamanca-Riba L, Vispute RD (2000) Appl Phys Lett 77:3731; g) Chen XL, Li JY, Cao YG, Lan YC, Li H, He M, Wang CY, Zhang Z, Qiao ZY (2000) Adv Mater 12:1432; h) Tang CC, Fan SS, Dang HY, Li P, Liu YM (2000) Appl Phys Lett 77:1961; i) Peng HY, Zhou XT, Wang N, Zheng YF, Liao LS, Shi WS, Lee CS, Lee ST (2000) Chem Phys Lett 327:263; j) Cheng GS, Chen SH, Zhu XG, Mao YQ, Zhang LD (2000) Mater Sci Eng A 286:165; k) LilientalWeber Z, Chen Y, Ruvimov S, Washburn J (1997) Phys Rev Lett 79:2835; l) Li JY, Chen XL, Qiao ZY, Cao YG, Lan YC (2000) J Cryst Growth 213:408; m) Cheng GS, Zhang LD, Chen SH, Li Y, Li L, Zhu XG, Zhu Y, Fei GT, Mao YQ (2000) J Mater Res 15:347; n) Duan XF, Lieber CM (2000) J Am Chem Soc 122:188; o) Lee SM, Lee YH, Hwang YG, Elsner J, Porezag D, Frauenheim T (1999) Phys Rev B 60:7788; p) Zhu J, Fan S (1999) J Mater Res 14:1175; q) Lee SM, Lee YH, Hwang YG, Lee CJ (1999) J Korean Phys Soc 34:S253; r) Bedarev DA, Kogitskii SO, Lundin VV (1999) Tech Phys Lett 25:385
182. a) Han WQ, Fan SS, Li QQ, Gu BL, Zhang XB, Yu DP (1997) Appl Phys Lett 71:2271; b) Wu XC, Song WH, Zhao B, Huang WD, Pu MH, Sun YP, Du JJ (2000) Solid State Commun 115:683
183. Duan XF, Huang Y, Cui Y, Wang JF, Lieber CM (2001) Nature 409:66

184. Duan X, Lieber CM (2000) Adv Mater 12:298
185. Shi WS, Zheng YF, Wang N, Lee CS, Lee ST (2001) Appl Phys Lett 78:3304–3306
186. a) Pham-Huu C, Keller N, Ehret G, Ledoux MJ (2001) J Catal 200:400; b) Li YB, Xie SS, Wei BQ, Lian GD, Zhou WY, Tang DS, Zou XP, Liu ZQ, Wang G (2001) Solid State Commun 119:51; c) Gao YH, Bando Y, Kurashima K, Sato T (2001) Scr Mater 44:1941; d) Zhang YJ, Wang NL, He RR, Chen XH, Zhu (2001) J Solid State Commun 118:595; e) Wu XC, Song WH, Huang WD, Pu MH, Zhao B, Sun YP, Du JJ (2001) Mater Res Bull 36:847; f) Li YB, Xie SS, Zou XP, Tang DS, Liu ZQ, Zhou WY, Wang G (2001) J Cryst Growth 223:125; g) Shi WS, Zheng YF, Peng HY, Wang N, Lee CS, Lee ST (2000) J Am Ceram Soc 83:3228; h) Wang ZL, Dai ZR, Gao RP, Bai ZG, Gole JL (2000) Appl Phys Lett 77:3349; i) Liang CH, Meng GW, Zhang LD, Wu YC, Cui Z (2000) Chem Phys Lett 329:323; j) Pan ZW, Lai HL, Au FCK, Duan XF, Zhou WY, Shi WS, Wang N, Lee CS, Wong NB, Lee ST, Xie SS (2000) Adv Mater 12:1186; k) Hu JQ, Lu QK, Tang KB, Deng B, Jiang RR, Qian YT, Yu WC, Zhou GE, Liu XM, Wu JX (2000) J Phys Chem B 104:5251; l) Zhou XT, Wang N, Au FCK, Lai HL, Peng HY, Bello I, Lee CS, Lee ST (2000) Mater Sci Eng A 286:119; m) Tang CC, Fan SS, Dang HY, Zhao JH, Zhang C, Li P, Gu Q (2000) J Cryst Growth 210:595; n) Zhou XT, Lai HL, Peng HY, Au FCK, Liao LS, Wang N, Bello I, Lee CS, Lee ST (2000) Chem Phys Lett 318:58; o) Lai HL, Wong NB, Zhou XT, Peng HY, Au FCK, Wang N, Bello I, Lee CS, Lee ST, Duan XF (2000) Appl Phys Lett 76:294; p) Lu QY, Hu JQ, Tang KB, Qian YT, Zhou G, Liu XM, Zhu JS (1999) Appl Phys Lett 75:507; q) Zhou XT, Wang N, Lai HL, Peng HY, Bello I, Wong NB, Lee CS, Lee ST (1999) Appl Phys Lett 74:3942; r) Meng GW, Zhang LD, Mo CM, Zhang SY, Qin Y, Feng SP, Li HJ (1998) J Mater Res 13:2533; s) Han WQ, Fan SS, Li QQ, Liang WJ, Gu BL, Yu DP (1997) Chem Phys Lett 265:374
187. a) Liang CH, Meng GW, Chen W, Wang YW, Zhang LD (2000) J Cryst Growth 220:296; b) Qi SR, Huang XT, Gan ZW, Ding XX, Cheng Y (2000) J Cryst Growth 219:485; c) Gao Y, Liu J, Shi M, Elder SH, Virden JW (1999) Appl Phys Lett 74:3642
188. a) Fukunaga A, Chu S, McHenry MEJ (1999) Mater Sci Lett 18:431; b) Nesting DC, Kouvetakis J, Smith DJ (1999) Appl Phys Lett 74:958; c) Fukunaga A, Chu SY, McHenry ME (1998) J Mater Res 13:2465; d) Wong EW, Nor BW, Burns LD, Lieber CM (1996) Chem Mater 8:2041; e) Ata M, Hudson AJ, Yamaura K, Kurihara K (1995) Jap J Appl Phys 34:4207; f) Liu MG, Cowley JM (1995) Carbon 33:749; g) Dai HJ, Wong EW, Lu YZ, Fan SS, Lieber CM (1995) Nature 375:769; h) Liu MQ, Cowley JM (1995) Carbon 33:225; i) Ata M, Kijima Y, Hudson AJ, Imoto H, Matsuzawa N, Takahashi N (1994) Adv Mater 6:590; j) Seraphin S, Zhou D, Jiao J, Withers JC, Loutfy R (1993) Nature 362:503

Top Curr Chem (2003) 226: 173–196
DOI 10.1007/b10828

Monodisperse Aligned Emulsions from Demixing in Bulk Liquid Crystals

J. C. Loudet, P. Poulin

Centre de Recherche Paul Pascal – CNRS, avenue A. Schweitzer, 33600 Pessac, France

We present recent findings on the formation of ordered emulsions from bulk demixing in liquid crystalline materials. In contrast to classical phase separations, which lead to randomly distributed macrodomains of various sizes, demixing in liquid crystals leads to remarkably uniform droplets that form ordered arrays. This distinctive behavior arises from the presence of topological defects and elastic distortions around the inclusions formed during the separation. These distortions induce long-range attractions and short-range repulsions. These forces direct the ordering of the microdomains and stabilize them against coalescence; limiting thereby the coarsening mechanism of the separation. We show that the ordering can be controlled on a large-scale by simply controlling the macroscopic alignment of the liquid crystal. We also discuss the influence of an external electric field and demonstrate marked differences with classical emulsions and colloids in isotropic fluids.

Keywords. Liquid crystal, Emulsion, Nematic, Topological defect, Silicone oil

1 Introduction

Phase separations from temperature quenches of binary mixtures start with the formation of small particles that grow and coarsen as time elapses [1]. In solid or liquid media, this coarsening leads to polydisperse dispersions of growing particles that eventually phase separate macroscopically. In the dilute limit of a liquid medium, the domains of dispersed phase consist of droplets that exhibit Brownian motion. When the viscosity of the liquid matrix is not too high, the coarsening is dominated by the coalescence of the droplets, which fuse together when they collide with each other. This classical scenario can be easily observed and is now rather well understood [1]. However, the situation can be completely different if the continuous phase exhibits some liquid crystalline ordering. In that case, the droplets of the dispersed phase can still move due to Brownian motion but the liquid crystal in which they are embedded induces long- and short-range forces that affect the kinetics of the separation and the structures formed over long time periods [2]. It has been indeed shown that colloidal inclusions in liquid crystals experience elastic forces mediated by the distortions of the liquid crystal ordering [3–15]. These forces, which do not exist in isotropic liquids, can lead to anisotropic clusters and stabilize colloids. We show in this paper that they play a predominant role in phase separation mechanisms. Instead of fully phase separating, the mixture self-stabilizes and self-organizes into highly ordered arrays of chains composed of monodisperse colloidal droplets [2]. The main features of this spontaneous ordering can be understood by considering elastic phenomena and topological defects around the droplets. As another distinctive feature in these systems, we present recent findings on the influence of an external electric field [16]. In classical isotropic emulsions and colloidal suspensions, an external field induces the polarization of the particles and their subsequent aggregation and coalescence because of dipole-dipole forces. Polarization of droplets and electrostatic dipole-dipole forces are also present in a liquid crystal. However, there are additional effects because of the modification of the liquid crystal alignment in response to the field. The latter not only induces a direct attraction, as in isotropic liquids, but also an indirect elastic repulsion by imposing different distortions and defects of the liquid crystal. This elastic repulsion can stabilize the droplets against coalescence – thereby leading us to the surprising conclusion that, in contrast to isotropic liquids, the phase separation can be arrested at will by applying an external field.

This paper is divided into six sections. In the next one, we recount some background details about the behavior of colloidal inclusions in liquid crystals. We describe the distortions around inclusions and the induced elastic interactions between them. The basis of an electrostatic analog to explain the anisotropy of colloidal structures in liquid crystals is given. Section 3 is devoted to experimental details related to the systems that allowed the mechanisms discussed in this paper to be observed. In Sect. 4, we present recently reported mechanisms of phase ordering from demixing in a liquid crystal. The structures and the kinetics of the separation are shown to be significantly different from that observed in isotropic binary mixtures. We discuss the experimental observations using the concepts of the second section. We describe in Sect. 5 the effect of an external electrical field and demonstrate marked differences in the behavior of classical electro-rheological fluids [17, 18]. A brief conclusion summarizes the main concepts and experimental results reviewed in this paper.

2 Theoretical Concepts

2.1 Distortions and Defects Around Colloidal Inclusions in Nematic Liquid Crystal

The liquid crystals considered throughout this article are thermotropic nematic liquid crystals composed of rigid, rod-like organic molecules that possess long-range orientational order [19]. The preferred direction of alignment of the molecules is specified by the unit vector field **n**, called the director. The presence of droplets in the nematic host induces long-range distortions of the director field and these distortions mediate strong elastic forces between the particles [3–15]. Another important source of energy in these systems comes from the local interactions between the liquid crystal molecules and the interfaces of both the particles and the boundaries of the sample; this is the anchoring energy.

When the anchoring to a surface is strong, the director makes a well-defined angle with this surface, regardless of the resultant elastic distortions. In this limit, the boundary conditions of **n** are fixed. By contrast, in the limit of weak anchoring, the competition between anchoring energy and elastic distortions may result in different orientations of the director at the interface. The combination of the geometry and the boundary conditions impose global topological constraints that must be satisfied by the director field of the liquid crystal. This can lead to the formation of other topological defects that play a crucial role in determining the elastic droplet-droplet interactions.

A particle with normal (homeotropic) anchoring of the nematic director is equivalent to a topological defect called a radial hedgehog, which creates radial distortions [19]. To satisfy the condition that the liquid crystal must be aligned at long range, a droplet could either nucleate a companion hyperbolic hedgehog (Fig. 1a), or a disclination ring of finite radius encircling the droplet in a "Saturn-ring" configuration (Fig. 1b). The two droplet-defect assemblies satisfy the global boundary conditions imposed on the sample, since the director is homogeneously aligned far from the droplet.

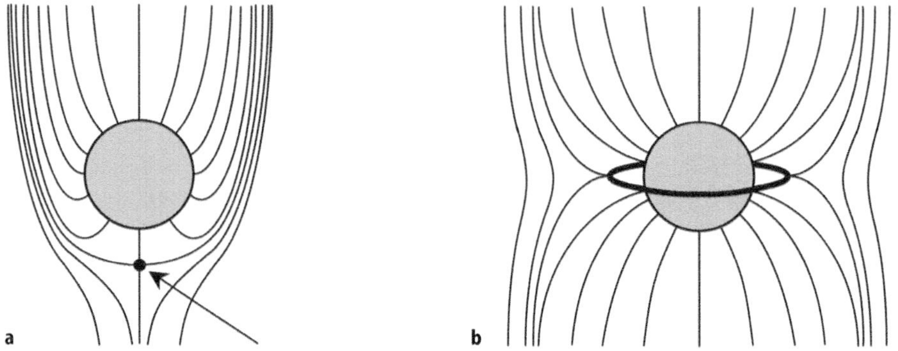

Fig. 1. Schematic diagrams of the director field distortions (*black lines*) around particles in an aligned nematic liquid crystal. For a normal anchoring of the liquid crystal molecules at the surface of the particles, there are two possible configurations. **a** Dipole configuration with a companion point defect (indicated by an *arrow*) located in the immediate vicinity of the particle. **b** Quadrupolar Saturn-ring configuration with a disclination ring surrounding the particle at the equator

When the preferential anchoring on a particle is planar, a tangential configuration is expected [4, 9]. The global boundary conditions are met by the creation of two surface defects, called boojums [20–22], located at the poles of the particle. They are diametrically opposed, and aligned along the axis of the nematic phase. In this paper, we will not consider this particular geometry but instead focus only on systems that exhibit preferential normal anchoring.

The energy scale of an elastic distortion around a particle is of order KR, where K is a typical elastic constant of the nematic liquid crystal [19] and R is the radius of the particle. For a thermotropic liquid crystal, K is approximately 10^{-11} N, and for a colloidal particle, R is approximately one micron; thus the energy scale is a few thousands k_BT, where k_B is the Boltzmann constant and T the temperature. As a result, the entropy of the particles is negligible compared to the elastic interactions. Under these conditions, the structures formed due to attractive interactions remain stable against thermal fluctuations.

A quantitative account of the behavior of the liquid crystal, and the topological defects that result upon addition of particles, can be obtained by considering the free energy of the system, F. In the absence of any external field, it consists only of a bulk and surface term

$$F = F_{el} + F_S = \int d^3r f_{el} + \int dS f_S \,. \tag{1}$$

F_{el} is the elastic Frank free energy which describes the slowly varying spatial distortions of the director; the free energy density f_{el} is a function of the elastic modes of deformation of a nematic liquid crystal and is given by [19, 23]

$$f_{el} = \frac{1}{2}\,[K_1(div\mathbf{n})^2 + K_2(\mathbf{n}\cdot curl\mathbf{n})^2 + K_3(\mathbf{n}\cdot curl\mathbf{n})^2] \tag{2}$$

where K_1, K_2, and K_3 are the splay, twist and bend moduli, respectively.

F_S is the surface free energy which takes into account the interactions of the director with boundaries. A classical expression for the surface free energy density f_S is [24]

$$f_S = \frac{W}{2}\,[1 - (\mathbf{n}\cdot\mathbf{s})^2]\,. \tag{3}$$

The unit vector **s** denotes some preferred orientation of the director at the surface. For homeotropic anchoring it corresponds to the unit vector normal to the surface. W is the coupling constant which merely stands for the strength of the anchoring at the surface of the particle. It varies in the range 10^{-7}–10^{-3} J/m^2 [25].

The director fields adopted in each of the situations described above (see Fig. 1) follows from a minimization of the total free energy F under the constraint that **n** is a unit vector. Different approaches including an analog to electrostatics and computer simulations have been used to study the nematic distortions around particles [3–8, 10–15] and the relative stability of hedgehogs versus Saturn rings. It was shown that the particle-hyperbolic hedgehog pair behaves like a dipole at long range whereas the Saturn ring configuration has quadrupolar symmetry [3–15, 26].

In the experiments reported on inverted nematic emulsions [8, 9], where surfactant-coated water droplets are dispersed in a nematic host, the dipole configuration is almost invariably observed for strong normal anchoring conditions and droplet radii of around one micron. In these conditions, the dipole seems to be more stable than the Saturn ring. The observed separation between the hyperbolic defect core and the droplet is a small fraction of the droplet radius, R. From optical microscopy observations, it was experimentally shown that $r_d/R = 1.2 \pm 0.1$, where r_d is the distance between the defect core and the center of the particle. This is in good agreement with theoretical predictions [5–8, 10–13]. However, for sufficiently small particles or weaker anchoring strength W, a Saturn or surface ring appears to be more stable than the dipole [7, 11, 13]. This allows weaker distortions in the volume of the sample whereas a penalty has to be paid at the surface of the particle because of a weak deviation from the normal. The so called surface ring corresponds to a Saturn ring directly located at the surface of the particle. In the limit of very weak anchoring or very small particles, we can imagine that the nematic accommodates the particle without creating any additional defects and the director field distortions are smooth everywhere.

2.2 Interactions Between Inclusions in Nematic Liquid Crystals

Once the topological configuration around an isolated particle is known, the interactions between the particles can be investigated. Predictions of the long-range two-body interactions can be determined through integration of a free energy density containing f_{el}, plus symmetry-allowed terms which couple the director and its distortions to particle concentration gradients [4, 8, 10, 14].

Because of its dipolar symmetry, the particle-hedgehog pair creates dipole distortions of the director field at long range; leading to elastic dipole–dipole interactions between drops that give rise to their chaining [2, 8–11, 16]. In those chains, the drops do not coalesce because of the in-between hyperbolic hedgehog defect which prevents them from contacting and thereby provides a short-range elastic repulsion. From symmetry, the long-range dipolar two-body interaction can be predicted up to a scaling factor [4, 10]. The lowest order term of the expression for the interaction energy, U_d, between two dipolar droplets located at $\mathbf{r}_\alpha$ and $\mathbf{r}_\beta$ is

$$U_d \propto 4KR^4\left(\frac{1-3\cos^2\theta}{d^3}\right) \tag{4}$$

where $d=|\mathbf{r}_\alpha-\mathbf{r}_\beta|$, θ is the angle between $(\mathbf{r}_\alpha-\mathbf{r}_\beta)$ and the z-axis, the far-field alignment axis of the liquid crystal.

However, the amplitude and the anisotropy of the interactions between drops exhibiting the Saturn ring configuration are completely different. The expression for the quadrupolar interaction energy is [4–7]:

$$U_q \propto K\frac{R^6}{d^5}\,(9-90\cos^2\theta+105\cos^4\theta). \tag{5}$$

This interaction is repulsive when the line joining the centers of two particles makes an angle of 0° or 90° to the z-axis and can be attractive for intermediate angles. Therefore, the structures generated by quadrupolar interactions are significantly different from those generated by dipolar interactions. Instead of chain-like structures, more compact aggregates can be formed [9, 27]. This demonstrates the great importance of the anchoring conditions and of the resultant topological structures around isolated droplets.

Although elegant and efficient at providing qualitative guidelines, the electrostatic analog has its limitations. It assumes a single elastic constant and is only valid at long range, i.e., at length scales where the particles can be viewed as point objects. At short range, more complex phenomena can govern the behavior and the structures of the particles.

3 Experimental Systems

In this section, we present the experimental systems and conditions of references [2] and [16]. These systems allowed some of the main features of monodisperse emulsions obtained from bulk demixing in liquid crystals to be identified.

3.1 Liquid Crystals and Silicone Oils

The system under study is a simple binary mixture composed of a silicone oil and a thermotropic liquid crystal forming the continuous phase. The silicone oils were provided by Aldrich (Aldrich, 37848-8, poly(dimethylsiloxane-*co*-methylphenylsiloxane)) and Fluka (Fluka, 85427-DC 710, poly(dimethylsiloxane-*co*-diphenylsiloxane)). The liquid crystal is a eutectic mixture of cyanobiphenyl and cyanoterphenyl molecules and was provided by Merck (trade name 'E7'). Strictly speaking, since E7 is not a pure compound, the present mixtures may be viewed as quasibinary systems. Pure E7 shows a typical first order isotropic-to-nematic transition at about 62°C, as evidenced by microscopic observations in capillaries. The nematic phase of E7 is stable over a large temperature range around room temperature.

3.2
Sample Preparation

A typical mixture was heated to the isotropic phase (70°C) and stirred for a few minutes to ensure homogeneity. A thin cell, consisting of two glass slides separated by 20 μm-thick spacers, was filled by capillarity in the isotropic phase with the binary mixture. However, before the mixture was added, the glass slides had been previously coated with polyvinyl alcohol (PVA) and rubbed along a defined direction with a piece of velvet. This treatment causes the liquid crystal molecules to become homogeneously aligned parallel to the slides in the nematic phase (planar anchoring).

3.3
Methods

Temperature quenches of the binary mixture were performed from the pure nematic domain (N) to the diphasic domain (N+I) of the phase diagram (see Fig. 2) – quenches from the isotropic phase would have led to the formation of numerous and uncontrolled topological defects [28]. At 55°C, in the nematic phase, the liquid crystal was allowed to align for a few minutes. Once aligned, the system was quenched into the diphasic domain at 20°C by placing the sample onto a temperature-controlled stage of an optical microscope. At this temperature, phase separation occurs and as time evolves, the phenomenon can be easily pictured using a CCD camera attached to a microscope and connected to a frame grabber.

4
Demixing in a Nematic Liquid Crystal

In this section, we describe and explain some aspects of a thermally induced phase separation of a binary mixture containing a nematic liquid crystal forming the continuous phase. The first three subsections deal with a detailed description of the experimental observations. The phase diagram of the mixture, the scenario of the phase separation as well as the structures obtained after long times after the temperature quench are presented. The fourth subsection is devoted to a discussion of the experimental results on the basis of the concepts described in Sect. 2. In the fifth subsection, we report on the kinetic aspects of the phase separation.

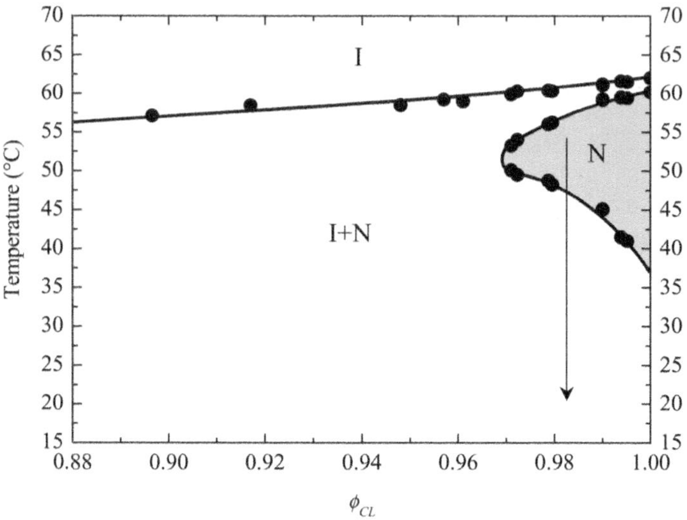

Fig. 2. Partial phase diagram of the binary mixture liquid crystal E7/silicone oil (Aldrich). ϕ_{LC} is the E7 weight fraction. The Fig. shows an isotropic phase (I) above 60°C, a coexistence domain between an isotropic and a nematic phase (I+N) below 60°C, and a nematic phase (N) in the extreme right part of the diagram. The systems are quenched from the nematic domain to 20°C as indicated by the *arrow*

4.1 Phase Diagram

Since the liquid crystal forms the continuous phase of the binary mixture, we are only interested in a small part of the total phase diagram. Weight fractions of the liquid crystal in the range 0.9 to 1 were used to determine the partial phase diagram of the mixture which is shown in Fig. 2. The system forms an isotropic (I) phase at high temperature, and a diphasic equilibrium between an isotropic and a nematic phase (N+I) at low temperature. A nematic domain (N) is found at intermediate temperatures and low silicone oil concentrations. As pointed out in the experimental section, the existence of this nematic domain has some importance prior to quenching the system to the diphasic region. The present mixture exhibits classical features usually observed in other mixtures of nematic liquid crystals and polymers or isotropic fluids [29, 30].

4.2 Scenario of the Phase Separation over a Short Time

Several mixtures with different silicone oil concentrations were prepared. Since the temperature quenches are performed from the nematic domain (N), the maximum weight fraction of silicone oil investigated was 0.035. When the sample is thermally quenched from the nematic domain (typically, 55°C) to the diphasic domain (20°C), the uniform single-phase mixture is thermodynami-

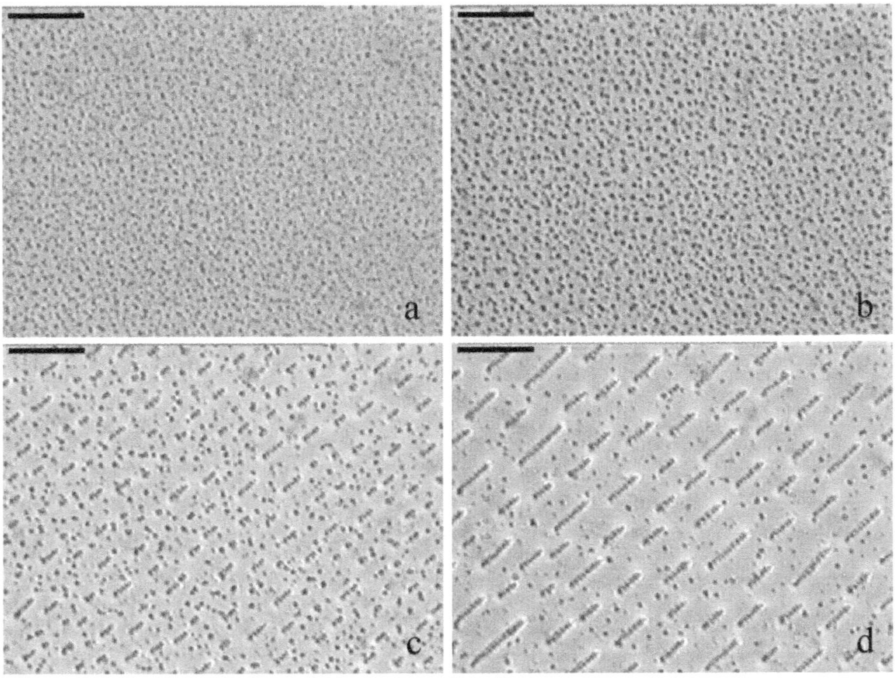

Fig. 3. The phase separation process in a nematic solvent. System composition: liquid crystal, 98%wt; silicone oil (Aldrich), 2%wt. **a** Some droplets form after the quench (picture: 20 s after the quench). **b** The droplets diffuse randomly and coalesce in the initial stages of the phase separation (picture: 42 s after the quench). **c** Coalescence stops once a critical size is reached. The droplets begin to form small linear aggregates oriented along the alignment direction of the liquid crystal (picture: 55 s after the quench). **d** The chains grow as time evolves (picture: 120 s after the quench). *Scale bar*: 60 μm

cally unstable and phase separation occurs. The first steps of the process are quite reminiscent of those taking place in mixtures composed of isotropic fluids [1, 19]. Small particles, mostly composed of silicone oil, form and grow through coalescence as they diffuse and collide each other. In classical mixtures, droplets varying noticeably in size are usually observed after a certain time. However, in our case, an unusual and striking phenomenon takes place when the coarsening silicone droplets reach a critical size, taken as R^*. This critical size is about a few microns in the considered experiments. The coalescence is then suddenly stopped and the droplets begin to attract one another like electrostatic dipoles to form small linear chains aligned along the rubbing direction. This situation is depicted in Fig. 3 which shows the system at different times after the temperature quench. We note that the use of a different silicone oil leads to a different critical size. Indeed, the critical radius R^* is about 2 μm for the Aldrich oil while critical radii of up to 4 μm are observed with the Fluka oil.

We would like to note here that the silicone oils used have a lower density than the liquid crystal. As a result, the oil particles tend to cream and lie closer to the

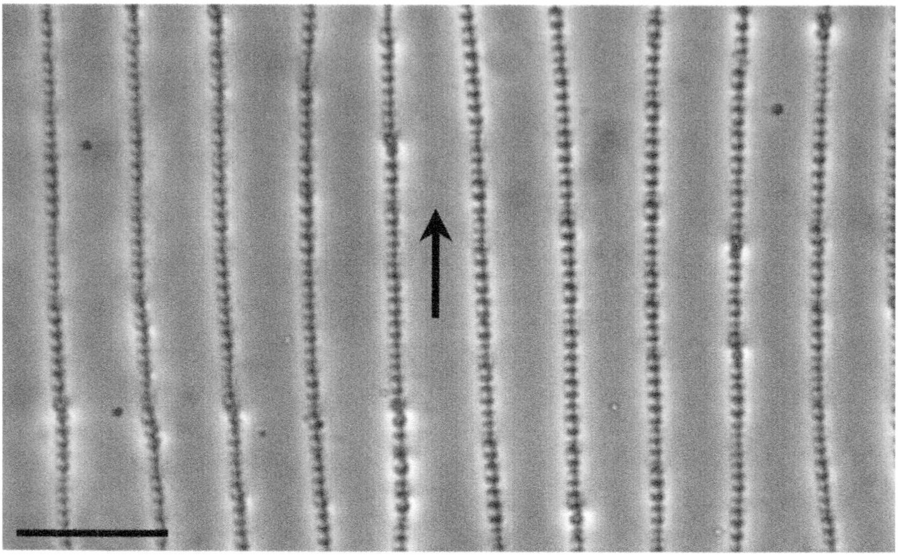

Fig. 4. Very long chains are obtained over long times. They form highly ordered arrays made of monodisperse droplets. System composition: liquid crystal, 98%wt; silicone oil (Aldrich), 2%wt. *Black arrow*: direction of polymer rubbing. *Scale bar*: 50 μm

upper slide than to the lower one of the cell. Consequently, the considered phase separation is a quasi two-dimensional process.

4.3 Structures Obtained over a Long Time

As time elapses, the chains grow through 'tip to head' aggregation and self-organize with respect to each other. Over long time periods, typically 45 minutes after the quench, they form highly ordered arrays of macroscopic chains (several hundred microns long) made of monodisperse droplets which do not coalesce. An example of the resultant system is shown in Fig. 4. The samples remain in this state for months and even years. These observations are in marked contrast to the long-time behavior of classical mixtures which usually fully phase separate [1]. The whole sample is filled with chains and the interchain distance d is uniform and decreases when the silicone oil concentration increases as shown on the plot in Fig. 5. This graph shows the variations of R^2/d, where R is the mean radius of the particles, as a function of ϕ, the weight fraction of silicone oil. Assuming that the weight fraction of the dispersed phase roughly corresponds to that of the silicone oil and considering that the chains span the whole surface area, R^2/d is expected to scale as ϕ. This is indeed what we observe in Fig. 5. This behavior proves the existence of a long-range repulsion between the chains. This is further confirmed by the formation of edge-like dislocations of chains in some parts of the sample (Fig. 6). These were the subject of a specific work [31]

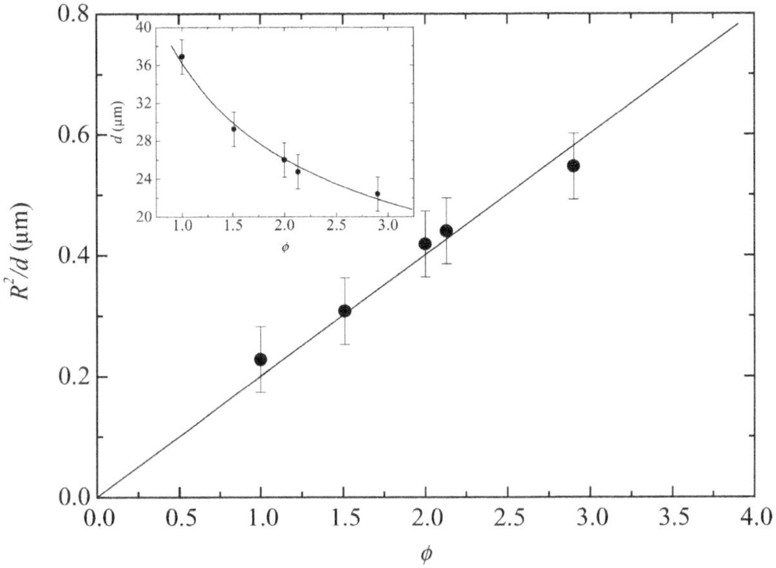

Fig. 5. R^2/d against weight fraction of silicone oil, ϕ; R is the particle radius and d the separation between the chains. From mass conservation (see text), R^2/d is expected to scale as ϕ. The present data seem to confirm this behavior which is an indication of a long-range repulsion between the chains. The *solid line* is a linear fit. *Inset*: values of d as a function of ϕ

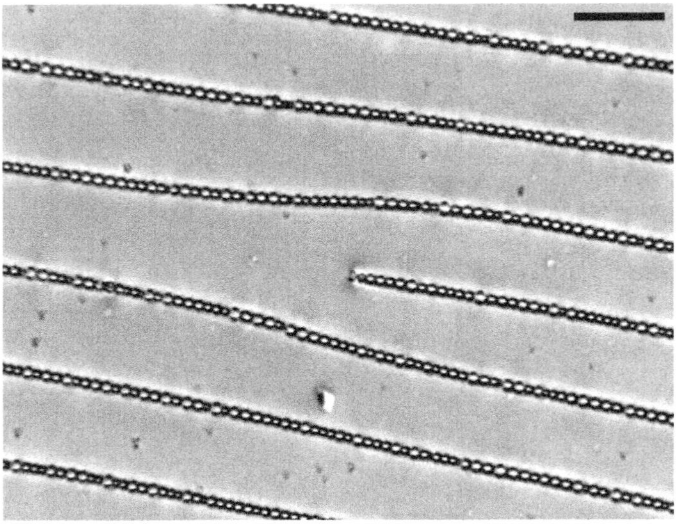

Fig. 6. Isolated edge dislocation. System composition: liquid crystal, 98.4%wt; silicone oil (Aldrich) 1.6%wt. *Scale bar*: 30 μm

as an example of dislocations where rotational invariance is broken, in contrast to more classical dislocations of smectics or cholesterics [19, 23].

4.4 Discussion

In this subsection, our aim is to explain the main features of the phase separation described above. We first address the issues of stability, monodispersity of the particles as well as the formation of chains. We then focus on the repulsion of the chains mentioned above. A short comparison with the long-time behavior of electrorheological fluids is presented.

4.4.1 *Coalescence Inhibition and Monodispersity of the Droplets*

In Sect. 2, we saw that a spherical particle suspended in a liquid crystal tends to distort the long-range ordering of the director. Assuming a single elastic constant K for the nematic phase, the elastic cost of the distortions around a single particle should be of the order of KR, where R is the particle radius (see Sect. 2). These distortions, which mediate elastic interactions between the particles, depend critically on the boundary conditions at the surface of the particles. These boundary conditions are spontaneously set by molecular interactions between the liquid crystal molecules and the silicone oil interface. In the present experiments, the preferential orientation of the liquid crystal molecules is normal to the oil interface. The characteristic energy specifying a deviation from this preferential alignment is given by WR^2 (see Sect. 2). We see that the surface energy scales as R^2, whereas the bulk elastic energy scales as R. Thus, surface energy dominates over elastic energy for large drops whereas it is the opposite for small ones. This means that for drops typically larger than $R^*=K/W$, the liquid crystal molecules will preserve a normal orientation at the surface of the particle regardless the cost of elastic energy. However, for drops typically smaller than R^*, we can expect the surface director to deviate from its preferred alignment and adopt different orientations to minimize the elastic cost of the distortions. From these considerations, we thus expect the distortions of the director to be size dependent. This is exactly what is happening in the considered experiments. At the early stages of the phase separation, the small droplets all exhibit the Saturn or surface ring configuration as can be seen in Fig. 7. We would like to point out here that we will use in the followings the term "Saturn ring" to describe the quadrupolar distortions around drops. Although a surface ring has been predicted, it is rather difficult to discriminate between these two configurations because of the limited experimental resolution. It seems to us that the ring does not sit directly at the surface, at least for the biggest quadrupoles. These floating quadrupoles can freely contact and coalesce since there is no elastic repulsion and no surfactant in the system. The coalescing quadrupoles therefore grow in size but the coalescence suddenly stops when they reach a well-defined characteristic size. We then observe that the Saturn ring shrinks continuously to the hedgehog defect and transforms into the dipole configuration. The coalescence

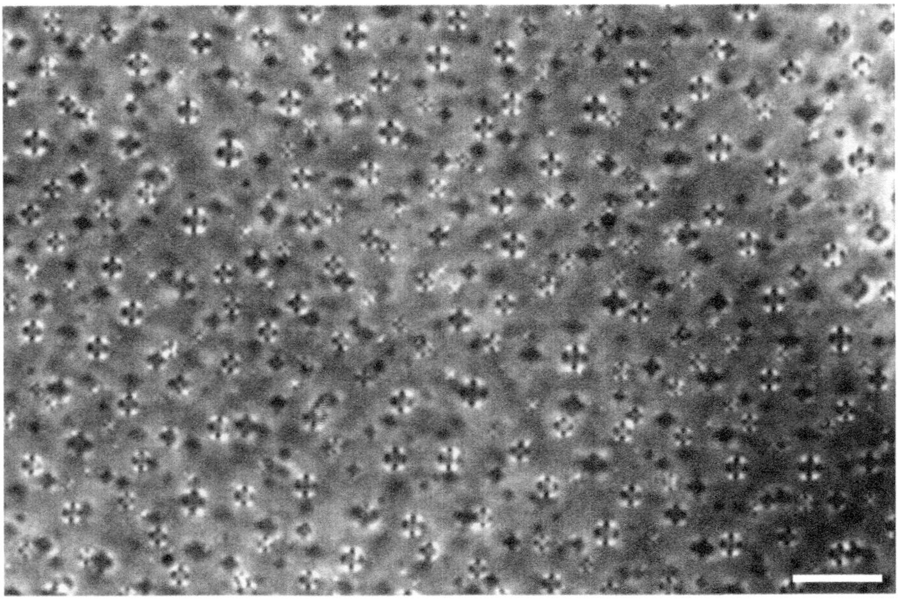

Fig. 7. Early stage of the phase separation. The droplets exhibit four bright regions between crossed polarizers which is characteristic of quadrupolar distortions. All the droplets have the Saturn or surface ring configuration. System composition: liquid crystal, 96.5%wt; silicone oil (Fluka), 3.5%wt. *Scale bar*: 9.9 μm

is inhibited as soon as the dipoles form because the companion hedgehog provides a short-range elastic repulsion as mentioned in Sect. 2. The resultant elastic dipole-dipole interactions between drops lead to the formation of chains in which the droplet-defect pairs point in the same direction (Fig. 8). The chains follow the nematic director alignment imposed by the surface treatment. This is why they are all oriented along a well-defined direction. We stress that the transition between the Saturn ring and dipole configurations is sharp enough to lead to remarkably uniform dispersions [2]. Assuming that W is of the order of $5 \cdot 10^{-6}$ J/m^2, a typical surface anchoring energy for cyanobiphenyl molecules and silicone material [32], and that K is of the order of 10 pN [19, 23], we expect the droplets to have a diameter D close to 4 μm, a value very close to that experimentally observed. This rough estimate provides additional support for the described mechanism.

From the above considerations, we expect the critical radius $R^*(=K/W)$ to change if either K or W varies. Keeping the same liquid crystal, changing the silicone oil is likely to affect the value of W. This is indeed what is observed with the Fluka oil. Quadrupoles with diameters approaching 8 μm can be observed; meaning that W is weaker in this case than with the Aldrich oil. These observations prove that decreasing W tends to favor the Saturn ring over the dipole configuration, in agreement with previous experimental [33] and theoretical works.

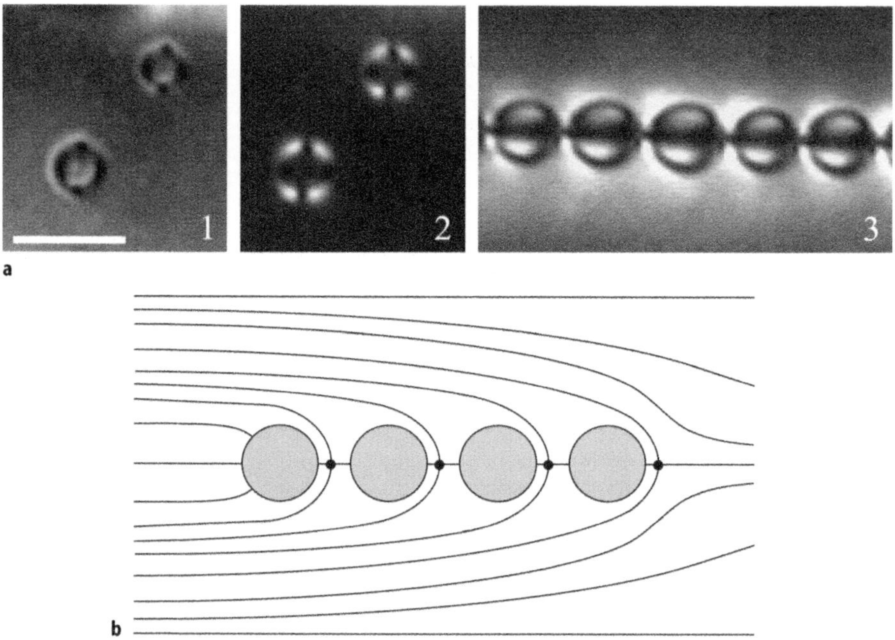

Fig. 8. a Elastic dipoles and quadrupoles are easily recognizable between crossed polarizers since the director distortions regions differ in the two configurations. **1**: two quadrupoles between uncrossed polarizers. **2**: the same two quadrupoles between crossed polarizers. **3**: dipoles in a chain. *Scale bar*: 6.3 μm. **b** Schematic representation of a chain of dipoles

4.4.2
Lateral Repulsion of the Chains

The linear chaining of the particles is reminiscent of other dipolar fluids such as electrorheological (ER) or magnetorheological (MR) fluids [17, 18]. In such fluids, the particles are polarized by an external electric or magnetic field and behave as dipoles subject to an attraction along the direction of the external field. The particles form long chains parallel to the applied field, exactly like in our experiments. However, a significant difference is observed over long time periods. In ER or MR fluids, long chains aggregate laterally and assemble into larger columns [17, 18, 34]. The attractive interaction between the chains is presumed to arise from thermal fluctuations which develop local concentrations of dipole moments [34]. In the present experiments, the chains never aggregate laterally but form perfectly ordered arrays than span the whole surface area of the sample. The chains remain quite straight and their thermal fluctuations are too small to be probed using optical microscopy. As mentioned in Sect. 2, the structures formed due to elastic attractive interactions are stable against thermal fluctuations. This is maybe the origin of the different behavior observed for ER fluids and nematic colloids. As aforesaid, there is rather a long-range repulsion between the chains which sometimes form edge-like dislocation patterns (see

Fig. 6). The details of this lateral repulsion are not known. A simple elastic model based on a phenomenological approach [31] is unable to gain precise quantitative information on this interaction.

4.5 Kinetic Aspects

This section is devoted to the kinetics of chain formation. Experimental studies and computer simulations on the irreversible growth of field-induced chains of dipolar particles, such as ER or MR fluids, suggest an asymptotic power-law behavior for the mean cluster size $S(t)$:$S(t) \propto t^z$ where z is the dynamic exponent [35–39]. In general, z depends on the dimension d of the space as well as on the nature of the aggregation process. Chain growth in a nematic environment may be somewhat different from the well-known kinetic growth of ER or MR fluids because of the anisotropy of the host fluid. However, we can also expect some direct analogies because the coarsening is dominated by the long range attractions which have been theoretically and experimentally shown to exhibit a dipolar character. We first present the experimental data for three oil concentrations and then compare the results with the literature on ER and MR suspensions.

4.5.1 *Experimental Results*

Three mixtures with weight fractions of silicone oil of 0.01, 0.02 and 0.03 were prepared. The temperature quenches were performed in the same way as described earlier. The kinetic evolution of the phase separation is recorded on videotapes from which series of digitized pictures taken at different times can be extracted. The time origin t=0 is chosen when the initial coalescence stops and uniformly-sized drops start to aggregate. In each picture, the mean chain length, $S(t)$, is calculated. It is defined as

$$S(t) = \frac{\sum_k k n_k(t)}{\sum_k n_k(t)} = \frac{N}{N_{CH}(t)} \tag{6}$$

where $n_k(t)$ is the number of chains containing k particles, $N_{CH}(t)$ the total number of chains and N the total number of particles once the initial coalescence stops. N is not quite constant at the very beginning of chain formation. Small scattered quadrupolar droplets still exist and coalesce. However, this weakly affects the value of $S(t)$.

Figure 9 shows a log-log plot of $S(t)$ for the three oil concentrations studied. The black lines are linear fits to the data and suggest that the chain growth has a power-law behavior just like in ER fluids. The dynamic exponents range from 0.664 to 0.695, i.e., they do not seem to depend a lot on the oil amount, at least in the concentration range investigated. Furthermore, since the linear fits never cover more than two decades in time, we can reasonably consider an average value $\bar{z}$=0.68. Thus, we find that $S(t) \propto t^{0.68}$ for a two-dimensional aggre-

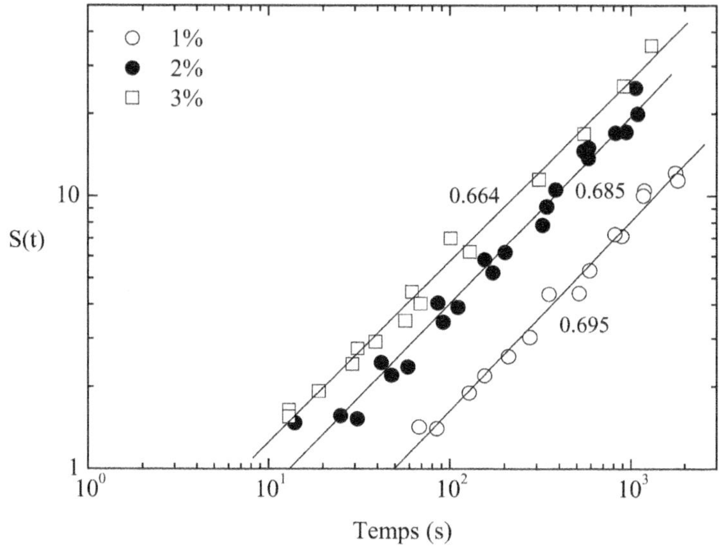

Fig. 9. Log-log plot of the mean chain size, $S(t)$, as a function of time for three oil concentrations (%wt). The *black lines* are linear fits. It seems like the data follow a power law with time. The values of the exponents do not vary much with the oil concentration and approach a common value of 0.68. Then, $S(t) \propto t^{0.68}$, a growth rate slightly higher than that found in most ER and MR fluids (see text)

gation, a growth rate a bit greater than that found in most ER and MR fluids as discussed below.

4.5.2
Comparison with ER and MR Fluids

A great number of experimental and theoretical studies as well as computer simulations have been devoted to the understanding of structural as well as dynamic properties of ER and MR fluids [17, 18]. Chain growth kinetics depends, among other things, on the relative importance of dipolar interactions versus k_BT. For a diffusion-limited aggregation, the temperature in the system is high enough so that dipolar interactions are weak relative to k_BT, for particles separated by a distance that corresponds to the average distance set by the concentration of the particles. Thermal energy dominates and particles and clusters diffuse randomly via Brownian motion. However, dipolar interactions are predominant at short distances; and when two particles come close enough to each other so that the dipolar interactions are greater than k_BT, a transition between random and ballistic motion occurs and the particles stick irreversibly. Most of the experimental and theoretical studies presented so far deal with diffusion-limited aggregation. For an isotropic diffusion of the chains in two or three dimensions, the well-known value of 0.5 is found for z [40, 41]. If the anisotropy of the drag coefficients for a moving chain is taken into account, the simulated reported values of z are close to 0.6 [39], in very good agreement with the experiments of Fraden et al. [37]. In their numerical simulation, See and Doi [38] found z=0.4 for an isotropic diffusion of the chains and when dipolar interactions dominate at all distances over thermal diffusion. Klingenberg et al. [36] reported values in the range 0.31-0.56 depending on the particle concentration.

In the present experiments, the chaining of the particles is entirely dominated by elastic interactions. The particles and chains do not diffuse randomly but rather have straight trajectories even when they are several particle diameters away from each other. Using previous experimental [26] and theoretical [8, 10] results, we can roughly estimate the strength of the elastic interactions versus k_BT. The long-range dipolar elastic pair potential, $U(r,\theta)$, between two elastic dipoles located at $\mathbf{r}_i$ and $\mathbf{r}_j$ is given by $U(r,\theta)=CKR^4(1-3\cos^2\theta)/^3r^3$ where r is the center-to-center distance between the droplets, θ is the angle between $\mathbf{r}=\mathbf{r}_i-\mathbf{r}_j$ and the z-axis, K is the elastic constant of the nematic phase in the isotropic approximation, and C is a unitless constant ranging from 70 to 300 [26]. Taking $K=10^{-11}$ N and R=1 μm, the distance at which $U(r,\theta)/k_BT|=1$ in the θ=0 direction is comprised between 49 μm (C=70) and 79 μm (C=300). This estimate agrees with the ballistic nature of the observed aggregation since the average separation between droplets is much smaller in the concentration range of these experimental studies.

As previously mentioned, the chains seem to grow faster in a nematic host than in most ER or MR fluids in two dimensions. This behavior might arise from the anisotropy of the continuous phase itself which facilitates the motion of anisotropic objects in a particular direction. Since the chains are oriented along the nematic director, their motion may be easier in this direction rather than along the perpendicular one. Recent theoretical predictions about a single particle moving in an aligned nematic [42–44] indeed predict that the droplet moves more easily along the director alignment rather than along the perpendicular direction. However, the motion of chains of particles has not yet been theoretically predicted but we can intuitively expect a similar tendency for them.

5 Application of an External Electric Field

In this section, we discuss the behavior of liquid crystal suspensions under the action of an external electric field. The behavior of colloidal suspensions in electric fields is of considerable technological interest with the so-called Electro-Rheological (ER) fluids [17, 18]. The main features of this behavior are now rather well understood. When an external field is applied, particles suspended in an isotropic fluid become polarized. Resultant dipole–dipole interactions between the particles lead to their chaining along the direction of the applied field. When suspended in a liquid crystal host, colloidal particles are also expected to be polarized upon the application of an electric field. However, new phenomena may take place because of the specific response of the liquid crystal. In this case, the external field is likely to alter the distortions of the liquid crystal alignment around the particles. These distortions induce elastic interactions between the particles and therefore play a crucial role in their stability and the formation of particular structures. In the absence of any external field, we know, from Sects. 2 and 4, that elastic distortions in a nematic liquid crystal can have dipolar or quadrupolar symmetry depending on surface boundary conditions and droplet size. Different kinds of distortions and symmetries lead to completely different behaviors and structures. It is thus of critical importance to determine how the electric field affects the distortions around the particles. From a theoretical point of view, Stark has predicted that an elastic dipole could be transformed into the quadrupolar Saturn ring configuration in the presence of an electric field [11]. Experimentally, Gu and Abbot, using cylindrically confined samples, have recently observed stable Saturn rings in the absence of a field and the expansion of the ring when the field is applied [45]. Nevertheless, the transition predicted by Stark was not observed in these experiments. Here, we focus on more recent experimental results which agree with Stark's theoretical predictions [16]. These experiments were performed in emulsions obtained from bulk demixing in liquid crystals as described in the previous sections. The systems used have a planar geometry, which may explain the differences from Gu and Abbot's experimental observations.

5.1 Elastic Dipole to Elastic Quadrupole Transition

The system we are interested in still consists of the binary mixture E7/silicone oil (Alrich or Fluka). As before, the experimental systems are all prepared by a thermally induced phase separation of this mixture. The experiments are performed in thin glass cells fitted with 50-μm-thick spacers which served as electrodes. Their spacing is 1 mm and the field is applied in the plane of the cell. As the phase separation proceeds, it is possible to focus on an isolated particle before it interacts with its neighbors. We are primarily interested in the behavior of the dipoles in the presence of a field. If a high electric field (3–4 V/μm) is applied along the axis of an isolated elastic dipole, we observe a transition from the dipole to the quadrupole. Figure 10a shows the dipole point defect opening up into an equatorial ring which consists of a −1/2 disclination [19]. This opening is very quick and lasts less than a second (see Fig. 10b). It corresponds to a discontinuous transition between the dipole and the quadrupole and occurs only above a certain threshold field, E_S, which depends on the droplet size as well as on the anchoring strength [16]. These observations are in qualitative agreement with the simulations made by Stark [11]. Interestingly, a similar transition has been reported when an electric field is applied to nematic droplets suspended in isotropic phases [46]. A hedgehog defect located at the center of the drop loses its stability in the presence of the field and transforms into a ring at the surface of the drop.

While the field is on, the Saturn ring configuration is maintained and appears to be the most stable configuration in the presence of the field. However, when the field is turned off, the ring continuously shrinks back to the hyperbolic hedgehog defect within a time scale of a few tens of seconds on average [47]. This

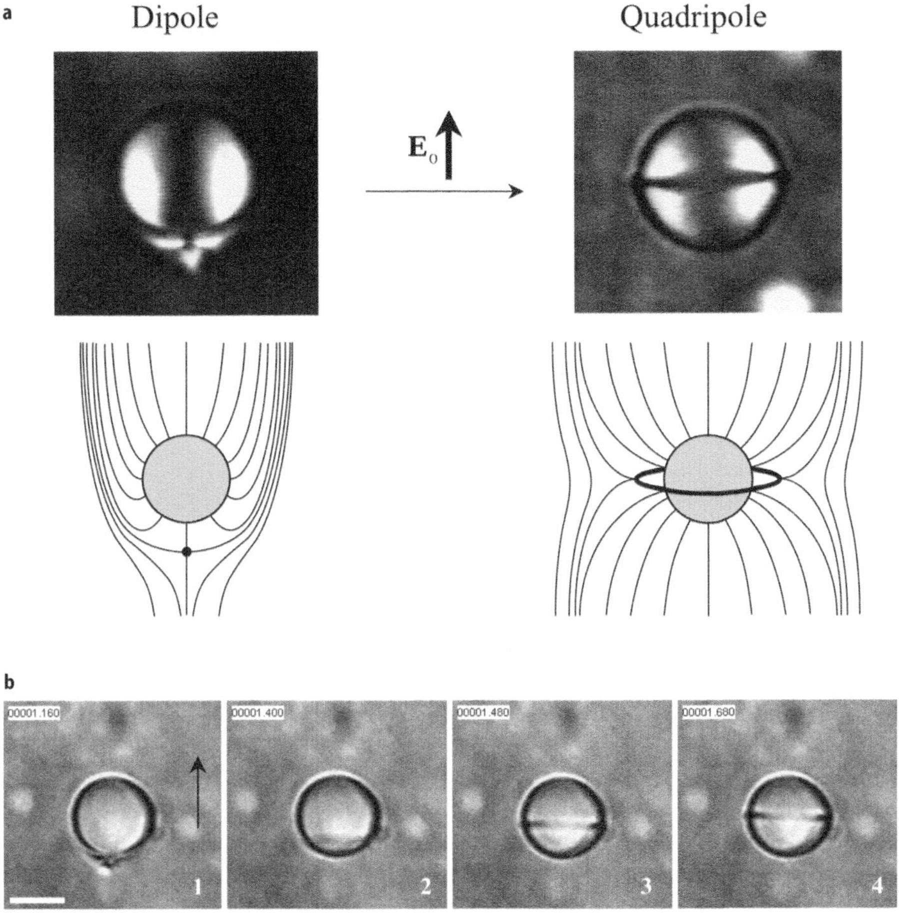

Fig. 10. **a** Dipole to quadrupole transition in the presence of an electric field $\mathbf{E}_0$ applied along the dipole axis. The pictures were taken between crossed polarizers and the schematics below each picture show the associated director distortions. Drop diameter: 8 μm. **b** Sequence of images illustrating the opening of the hedgehog defect into a disclination ring: **1** dipole; **2** right after the opening; **3** off-centered ring; **4** Saturn ring. *Black arrow*: applied field direction. The *black digits* on each picture indicate the time in seconds. *Scale bar*: 7.6 μm

relaxation phenomenon, which is illustrated on Fig. 11, is hence much slower than the defect opening. The ring closing velocity is not constant but strongly increases whilst approaching the hedgehog defect [47]. In the absence of field, the dipole configuration is always recovered. This observation suggests that the dipole is actually stable against the quadrupole in these experimental conditions. Moreover, this behavior shows that the predicted metastability barrier between the two configurations is weak or does not even exist in the present experiments [10, 11]. From these experimental data, it is possible to estimate the energy dissipation numerically during the ring relaxation. Within appropriate approxima-

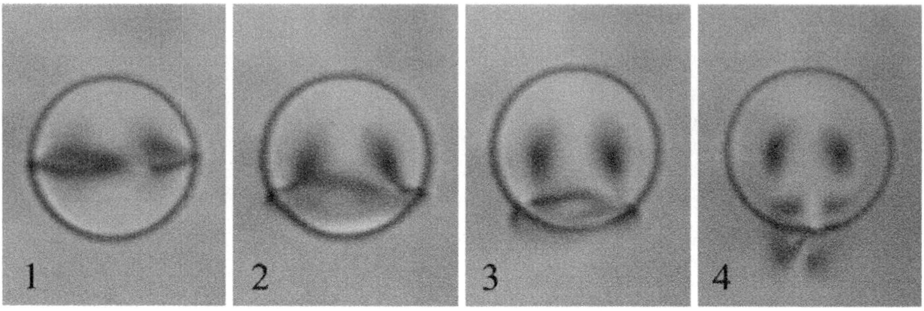

Fig. 11. Series of pictures showing the ring defect relaxation once the electric field has been turned off. **1** Saturn ring; **2** and **3** intermediate configurations; **4** dipole. Drop diameter: 35 μm

tions, this energy is expected to be directly related to the previously calculated Frank elastic free energy when going from the quadrupole to the dipole. The calculation yields to a reduced dissipated energy $\bar{E} = E/\pi KR$ ranging from about 10 to 14 [47] while the reduced Frank energy variation, $\Delta\bar{F} = \Delta F/\pi KR$ (K is the nematic elastic constant and R the particle radius), is expected to be close to 2 [10, 11, 13].

5.2 Elastic Quadrupolar Repulsion

At the very beginning of the phase separation, we saw that the droplets exhibit the Saturn ring configuration. The system is strongly unstable since these floating quadrupoles attract one another due to attractive elastic interactions and freely coalesce. Application of a weak electric field $\mathbf{E}_0$ (E_0<1 V/μm) at this moment polarizes the droplets and induces electrostatic dipole–dipole interactions between the quadrupoles which assemble into chains; exactly like in ER fluids. In those chains, the vector joining the centers of two drops, **r**, is parallel to the director alignment far from the drop. In such a configuration, theory predicts that the quadrupoles experience an elastic repulsion [4, 7]. This is indeed verified experimentally. There is a competition between the attractive electrostatic force, which forces the drops to come in close contact, and the repulsive elastic force, which tends to drive them apart. This elastic repulsion is clearly evidenced on the sequence of images of Fig. 12. On picture **a**, the quadrupoles are close to one another but they do not touch each other. When the electric field intensity is decreased, these quadrupoles begin to move away from each other because of the elastic repulsion. For these field values (reported in Fig. 12), the elastic repulsion balances the electrostatic attraction. For each field value E_0, an equilibrium distance between drops is fixed by the condition $F_{elastic} = -F_{electro}$. It is then possible to directly measure this elastic quadrupolar repulsion by measuring the center-to-center distance r between two drops of a chain for each E_0. A typical force profile is presented in Fig. 13. This graph illustrates how elastic quadrupoles actually repel each other when they are oriented in this configuration. On the y-axis

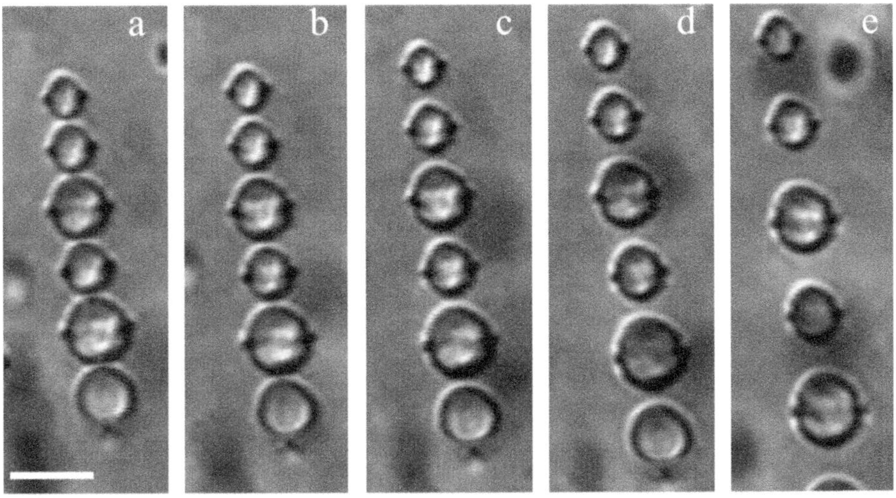

Fig. 12. Evidence of the elastic quadrupolar repulsion. When the field intensity decreases, the droplets begin to be driven apart. The *digits* below each picture indicate the values of E_0 in V/μm. *Scale bar*: 6.3 μm

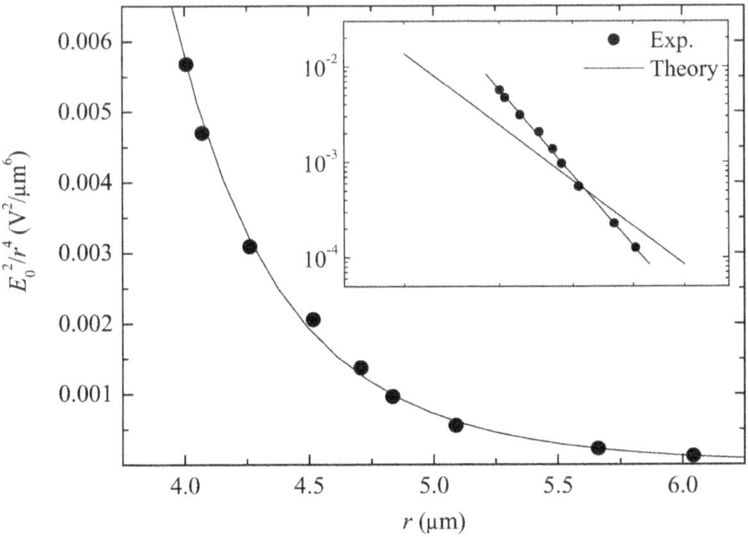

Fig. 13. Quadrupolar elastic force $F_{elastic} \propto E_0^2/r^4$ as a function of the center-to-center distance r in the presence of an electric field. The force profile is repulsive and steeper than a long-range prediction in the absence of field (*inset*). A power law fit leads to a scaling close to $1/r^9$ (solid line)

of this graph, the quantity E_0^2/r^4 corresponds to the electrostatic force expression between two aligned electrostatic dipoles [36]

$$E_{electro} = 2F_0 \left(\frac{D}{r}\right)^4 \tag{7}$$

with $F_0=3\pi\varepsilon_0\varepsilon_s D^2\beta^2 E_0^2/16$ and $\beta=(\varepsilon_p-\varepsilon_s)/(\varepsilon_p+2\varepsilon_s)$. D is the diameter of the drops, ε_0 the permittivity of free space and ε_p the permittivity of the silicone oil particles. As a first approximation, we set $\varepsilon_S=\bar{\varepsilon}_{LC}$ where $\bar{\varepsilon}_{LC}$ is the mean dielectric constant of the liquid crystal [23]. Therefore, we deduce $R^*(=K/W)$. Taking $\varepsilon_p=2.24$ [Aldrich source] and $\bar{\varepsilon}_{LC}\approx 8.7$ [48], a crude estimate of the repulsive elastic interaction leads to force values around a few pN.

We stress that, in the absence of field, the system is highly unstable since the quadrupoles freely coalesce. Thus, the reported observations lead to the surprising conclusion that initially unstable particles can be stabilized by application of an electric field, provided that the particles are driven in a particular direction along which they repel each other. In classical isotropic fluids, the electric field would tend to destabilize the system following an attraction and a subsequent coalescence of the drops.

In the absence of a field, previous theoretical predictions based on a long-range calculation predicted that the elastic quadrupolar repulsion force should follow the power law $F_{repulsion}\propto 1/r^6$ [4, 7]. Under the present experimental conditions, i.e., in the presence of an electric field, we observe a steeper repulsion as shown by the log–log plot in the inset of Fig. 13. Two reasons might explain the discrepancy between the experimental measurements and the theory. First, since the electric field is likely to distort the ordering of the liquid crystal molecules in the vicinity of the drops, the measured quadrupolar repulsion may intrinsically depend on E_0. Second, short-range effects, not considered in the theoretical approach, may come into play in the experiments. Indeed, the maximum measured separation between two drops is of the order D.

Let us now consider again picture **a** in Fig. 12. If E_0 is increased to values ranging from 2–3 V/μm, we observe that the quadrupoles come into contact and can coalesce despite the elastic repulsion evidenced above. Indeed, for these field values, the attractive electrostatic force completely overwhelms the elastic repulsion force which eventually leads to a one-dimensional coalescence of the drops as shown in Fig. 14. Thus, the system is again unstable in the presence of high electric fields. Depending on the field intensity, we can then control the droplet coalescence and thereby their size.

6 Conclusion

The results reported in this paper demonstrate that colloidal dispersions in thermotropic liquid crystals are challenging systems for discovering and studying new physical effects and structures. The anisotropy of the continuous phase leads to the observation of phenomena that are markedly different from that known in isotropic solvents. We have shown that a thermally induced phase sep-

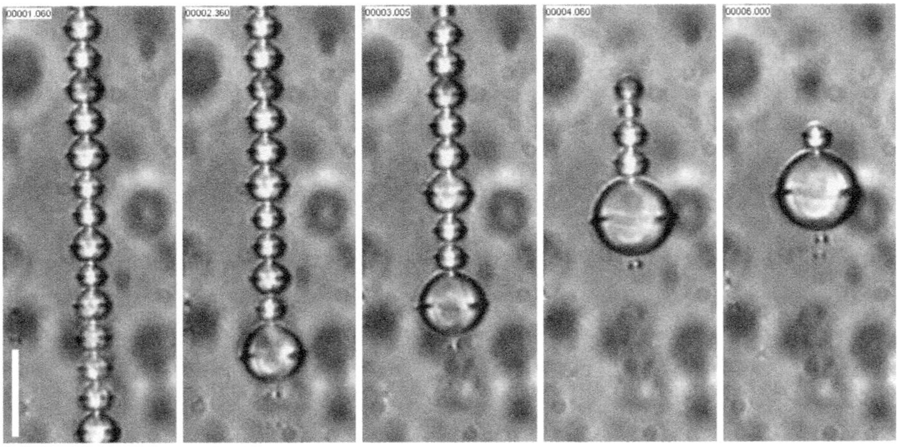

Fig. 14. For E_0>1 V/μm, the electric field dominates the elastic repulsion and the quadrupoles are forced to coalesce within the chains. *Scale bar*: 14.1 μm

aration of a binary mixtures containing a liquid crystal solvent and an isotropic oil, do not lead to full phase separation. In a nematic liquid crystal, the mixture self-stabilizes and self-organizes into highly ordered arrays of monodisperse colloidal droplet chains. In such systems, the size and spatial organization of the particles are controlled by the orientational elasticity and the topological defects of the continuous phase. We have also discussed the basics of the influence of an external electric field on these systems. The observed behaviors originate from the specific response of the liquid crystal to the applied field which alters the orientation of the molecules. Among the most striking results, we first described an electric field-induced transition between an elastic dipole and an elastic quadrupole; followed by a relaxation phenomenon once the field is switched off. Surprisingly, we have seen that it is possible to stabilize an initially unstable dispersion by applying an electric field, a behavior in marked contrast to that of isotropic dispersions.

Monodispersity, spatial ordering and absence of coalescence from phase separations in liquid crystals provide new and potentially helpful tools for the design of ordered composites and functional materials. The unusual behaviors of inclusions in the presence of an electrical field could also be useful for the development of a new class of field-responsive fluids. To practically estimate the potential of these systems for future applications, it is now time to start exploring the physical properties of these materials. Little is still known about their optical or rheological properties. They will presumably differ from that of classical emulsions; opening thereby the possibility for the development of novel emulsion applications.

References

1. Gunton JM, San Miguel M, Sahni PS (1983) In: Lebowitz JL (ed) Phase Transition and Critical Phenomena, vol 8. Academic Press, London, p 267
2. Loudet JC, Barois P, Poulin P (2000) Nature 407:611
3. Terentjev EM (1995) Phys Rev E 51:1330
4. Ramaswamy S, Nityananda R, Raghunathan VA, Prost J (1996) Mol Cryst Liq Cryst 288:175
5. Kuksenok OV, Ruhwandl RW, Shiyanovskii SV, Terentjev EM (1996) Phys Rev E 54:5198
6. Ruhwandl RW, Terentjev EM (1997) Phys Rev E 55:2958
7. Ruhwandl RW, Terentjev EM (1997) Phys Rev E 56:5561
8. Poulin P, Stark H, Lubensky TC, Weitz DA (1997) Science 275:1770
9. Poulin P, Weitz DA (1998) Phys Rev E 57:626
10. Lubensky TC, Pettey D, Currier N, Stark H (1998) Phys Rev E 57:610
11. Stark H (1999) Eur Phys J B 10:311
12. Andrienko D, Germano G, Allen MP (2001) Phys Rev E 63:041701
13. Stark H (2001) Physics Reports 351:387
14. Lev BI, Chernyshuk SB, Tomchuk PM, Yokoyama H (2002) Phys Rev E 65:021709
15. Fukuda J, Yoneya M, Yokoyama H (2002) Phys Rev E 65:041709
16. Loudet JC, Poulin P (2001) Phys Rev Lett 87:165503-1
17. Bullough WA (1996) In: Proceedings of the 5th international conference on ER fluids, MR suspensions and associated technology, World Scientific, Singapore
18. Parthasarathy M, Klingenberg DJ (1996) Mater Sci Eng R 17:57
19. Chaikin PM, Lubensky TC (1995) In: Principles of condensed matter Physics. Cambridge university Press, Cambridge
20. Mermin N (1977) In: Trickey SB, Adams ED, Dufty JW (eds) Quantum Fluids and Solids. Plenum Press, New York
21. Kurik M, Lavrentovich OD (1982) JETP Lett 35:444
22. Volovik GE, Lavrentovich OD (1983) Sov Phys JETP 58:1159
23. P. G. De Gennes PG, Prost J (1994) In: The Physics of Liquid Crystals. Oxford University Press, London
24. Rapini A, Papoular M (1969) J Phys (Paris), colloq. 30:C4-54
25. Blinov LM, Kabayenkov AY, Sonin AA (1989) Liq Cryst 5:645
26. Poulin P, Cabuil V, Weitz DA (1997) Phys Rev Lett 79:4862
27. Poulin P, Francès N, Mondain-Monval O (1999) Phys Rev E 59:4384
28. Yurke B, Pargellis AN, Chuang I, Turok N (1992) Physica B 178:56
29. Shen C, Kyu T (1995) J Chem Phys 102:556
30. Benmouna F, Bedjaoui L, Maschke U, Coqueret X, Benmouna M (1998) Macromol Theory Simul 7:599
31. Loudet JC, Poulin P, Barois P (2001) Europhys Lett 54:175
32. Lavrentovich OD, Nazarenko VG, Sergan VV, Durand G (1992), Phys Rev A 45:R6969
33. Mondain-Monval O, Dedieu JC, Gulik-Krzywicki T, Poulin P (1999) Eur Phys J B 12:167
34. Halsey TC, Toor W (1990) J Stat Phys 61:1257
35. Miyazima S, Meakin P, Family F (1987) Phys Rev A 36:1421
36. Klingenberg DJ, Zukoski CF, Hill JC (1993) J Appl Phys 73:4644
37. Fraden S, Hurd AJ, Meyer RB (1989), Phys Rev Lett 63:2373
38. See H, Doi M (1991 J Phys Soc Jpn 60:2778)
39. Carmen Miguel M, Pastor-Satorras R (1999) Phys Rev E 59:826
40. Jullien R, Botet R (1987) In: Aggregation and Fractal Aggregates. World Scientific, Singapore
41. Vicsek T (1992) In: Fractal Growth Phenomena, 2nd edn. World Scientific, Singapore
42. Ruhwandl RW, Terentjev EM (1996) Phys Rev E 54:5204
43. Stark H, Ventzki D (2001) Phys Rev E 64:031711
44. Stark H, Ventzki D (2002) Europhys Lett 57:60
45. Gu Y, Abbott NL (2000) Phys Rev Lett 85:4719
46. Bodnar VG, Lavrentovich OD, Pergamenshchick VM (1992) Sov Phys JETP 74:60
47. Loudet JC, Mondain-Monval O, Poulin P (2002) Eur Phys J E 7:205
48. Lippens D, Parneix JP, Chapoton AJ (1977) J Phys (Paris) 38:1465

Top Curr Chem (2003) 226: 197–223
DOI 10.1007/b10829

Templating Vesicles, Microemulsions, and Lyotropic Mesophases by Organic Polymerization Processes

Hans-Peter Hentze, Carlos C. Co, Craig A. McKelvey, Eric W. Kaler

Center for Molecular Engineering and Thermodynamics, Department of Chemical Engineering, University of Delaware, Newark, DE 19716, USA. *E-mail: 57116@udel.edu*

The self-organized templates that form in surfactant solutions can direct the synthesis of organic polymers. Several examples of such syntheses are reviewed here, beginning with discussion of the use of surfactant vesicles to guide the formation of spherical polymer shells. This process can produce a variety of morphologies including, under certain conditions, the desired shells. Microemulsions, defined as equilibrium microstructured fluids containing oil, water, and surfactant, can be used to produce latex dispersions of high molecular weight polymers having a relatively narrow size distribution. The experimental and theoretical aspects of this process are now fairly well understood and are summarized. Finally, recent efforts to use lyotropic phases to template polymers are reviewed. For each of these morphologies, the final product materials reflect a delicate balance between phase behavior and the reaction and mass transfer parameters controlling structure formation during template synthesis.

Keywords. Self-organization, Nanoparticles, Hollow spheres, Mesoporous materials, Liquid crystals, Morphosynthesis, Kinetic models, Surfactant templating

1 Introduction

Self-organization is a general and important natural principle guiding the structuring of materials, and describes spontaneous arrangement into states of higher order. Depending on the range of interactions and the size of the building blocks the length scales of self-organization range from angstroms to light years. On the nanometer scale, examples of self-organization abound in surfactant solutions, where the surfactant molecules arrange spontaneously into ordered aggregates, such as micelles, vesicles, or lyotropic mesophases. Such self-organized surfactant solutions are an excellent media for templated synthesis of nanostructured materials.

The most general definition of a template is as a structure-directing agent. In surfactant solutions the final templated polymers can be either discrete nanoparticles or mesostructured bulk materials as a consequence of polymerization, respectively, in the non-continuous or continuous domains of the template. Thermodynamically stable media, such as microemulsions, equilibrium vesicles, or lyotropic mesophases are especially useful as templates because of their structural definition and reproducible morphologies. The mesostructure of a thermodynamically stable template is defined by composition and temperature, but this same feature makes the structure unstable to changes in temperature, pH, or concentration. The aim of template synthesis is to transfer the self-organized template structure into a mechanically and chemically stable, durable, and processable material.

Two classes of templating can be distinguished. The first class is direct templating, which means that the morphology and length scale of the templated polymer is retained during the course of reaction. If the self-organizing species is itself polymerized, the result is a cured template. This direct templating is called synergistic synthesis. If instead the reaction is caused by polymerization at the surface of the assembly, a one-to-one replica is obtained by a process called transcriptive synthesis [1]. The second broad class, indirect templating, involves structuring of matter without retention of the parental template structure. Here the template directs the growth of the polymer within a confined geometry in a process called reconstructive synthesis. Materials templated indirectly are usually ordered on a larger length scale than the template. The following discussion provides examples of synergistic, transcriptive, and reconstructive synthesis by polymerization within thermodynamically stable, self-organized media. Such materials have many potential applications.

2 Templating from Vesicular Solutions

Vesicles form when a surfactant bilayer encapsulates an aqueous core (Fig. 1). The microstructure resembles that of a biological cell in which the plasma membrane has been replaced by a surfactant bilayer. These structures can form either spontaneously or as a result of shear or other processing of lamellar liquid crys-

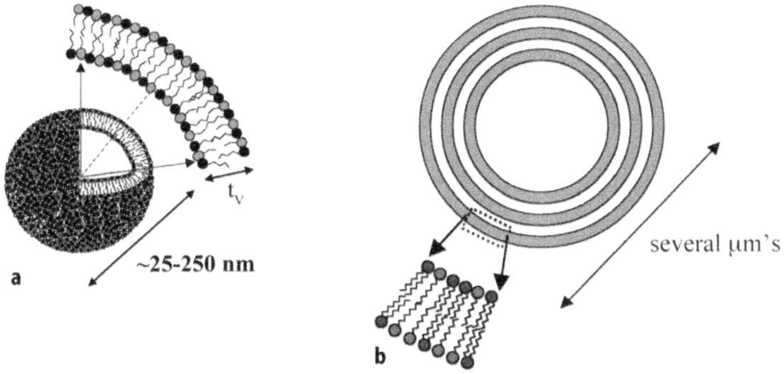

Fig. 1a, b. Schematic of vesicle architectures: **a** unilamellar vesicle, typical radii range from 25 to 250 nm, bilayer thicknesses are about 2–4 nm; **b** multilamellar vesicle, composed of several concentric surfactant bilayers surrounding an aqueous core

talline phases. Typical dimensions for the bilayer thickness are 2–4 nm and the vesicle diameter can range from 20 nm to over 250 nm.

The nomenclature used to describe vesicular structures is sometimes variable, but the following definitions are used here. The term "vesicle" is restricted to the unilamellar vesicle (ULV) architecture depicted in Fig. 1a. Liposomes are vesicles formed from biologic surfactants such as lecithin and cholesterol, and were first studied by Bangham et al. over 35 years ago [2]. Multilamellar vesicles (MLVs or "onion" phases) consist of multiple surfactant bilayers forming concentric shells around an aqueous core (Fig. 1b). MLVs are generally much larger than ULVs, and can be as large as several microns.

2.1
Polymerization and Vesicle Phases

Polymerization offers an approach to making vesicle formulations suitable for applications. The major benefits of polymerization include increasing the chemical-mechanical strength of the vesicle architecture, and the potential for performing subsequently a variety of reactions to create a highly functionalized surface. The most common approach to polymerization in vesicles is to use polymerizable surfactants (Fig. 2a). The use of polymerizable surfactants is best described as the polymerization of vesicles or fixation of vesicles, and so is a synergistic template synthesis. Typically, unsaturated biological surfactants have been specifically synthesized for these types of polymerizations, and there are a number of excellent reviews of this subject [3–6].

An alternative to the polymerization *of* vesicles with polymerizable surfactants is the polymerization *in* vesicles, or vesicle templating, in which a hydrophobic monomer swells the surfactant bilayer and is subsequently polymerized (Fig. 2b). In this case the vesicle is used like a mold that directs the polymerization process, and the surfactant molecules themselves are not incorporated into the final polymer. This is a transcriptive synthesis. The polymeric

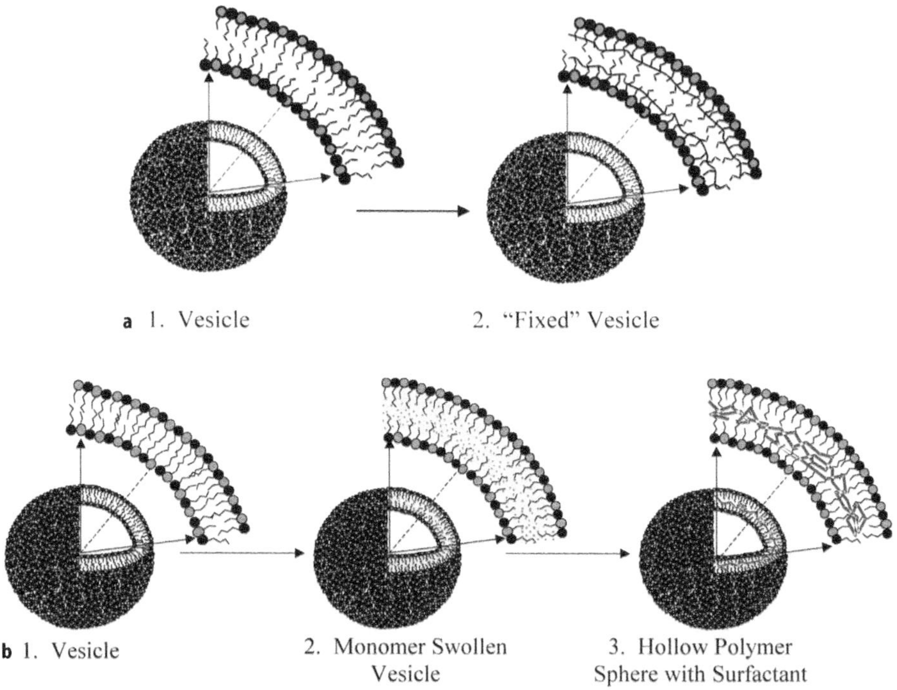

Fig. 2a, b. Templating vesicles: **a** vesicle-forming polymerizable surfactants are fixed into place by polymerization (synergistic approach); **b** a hydrophobic monomer swells the interior of a vesicle bilayer and is subsequently polymerized to form a hollow polymer shell (transcriptive approach)

properties could ideally be controlled through careful monomer selection, so that hydrolysable, charged, or cross-linked polymers could be made. Following polymerization, the polymer shells can be isolated from the templating surfactants and modified further to suit the needs of a particular application.

2.2
Approaches Toward Vesicle Templating

A myriad of polymerization conditions have been employed to polymerize monomers held in vesicle bilayers. Research groups have claimed both success and failure of morphosynthetic templating in these studies – even for nearly identical reaction conditions!

At least five groups have attempted polymerization *in* vesicles [7–11]. Pertinent experimental conditions, results, and appropriate references are summarized in Table 1. The polymerized morphologies observed are hollow polymer shells, "parachutes", "matrioshka" structures, and "necklaces" as shown in Fig. 3. ("Matrioshka" refers to the concentric sphere morphology that resembles the nesting Russian dolls bearing the same name.) Structures of the surfactants and monomers used in the experiments are shown in Fig. 4.

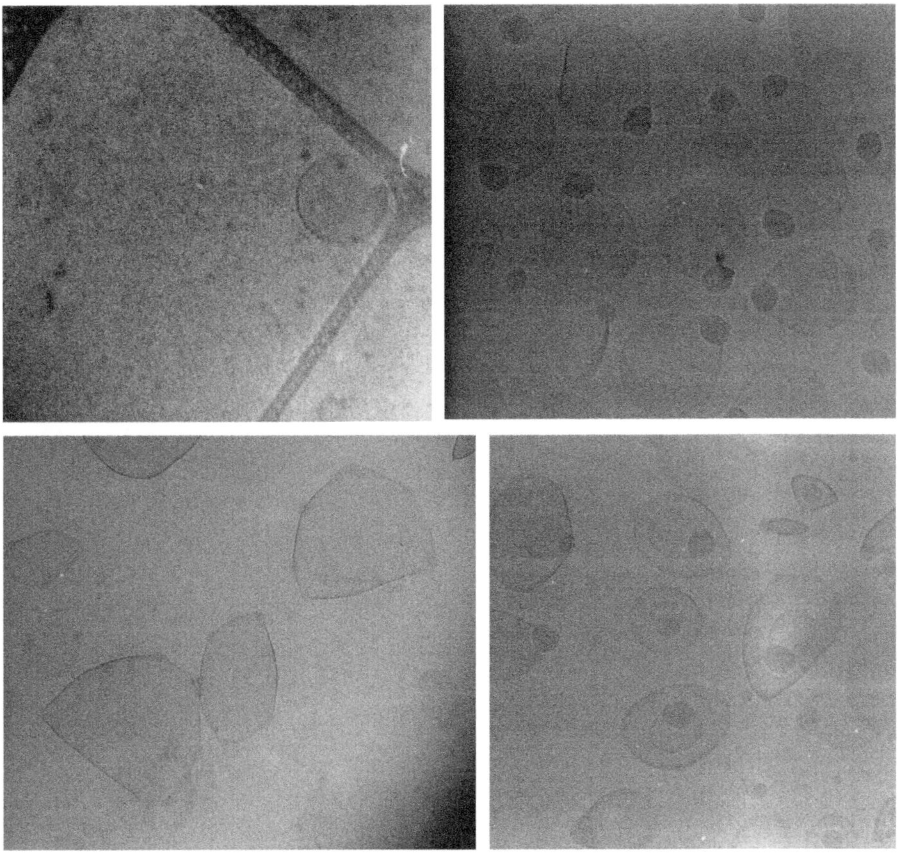

Fig. 3. Polymer morphologies obtained by vesicle templating: Hollow polymer spheres (*top left*) [24], parachutes (*top right*) [18], necklaces (*bottom left*) [17], and matrioshka structures (*bottom right*) [17]. Adapted with permission from [17, 18, 24]

Hollow polymer shells result when the monomer initially swells the surfactant bilayer and reacts rapidly to form a polymer shell. For a polymer shell to form, no significant polymer or monomer rearrangement can occur during the polymerization. All five groups have claimed success in forming hollow polymer spheres [7–11]. The most complete studies are from German et al., although more recently this group has concluded that hollow polymer shells did not result from any of their polymerizations, in contrast to their first report [12–22]. These authors propose that phase separation of polymer from the vesicle template is ubiquitous.

Parachutes, matrioshka structures, and necklaces result when the polymer phase separates from the surfactant bilayer during or after the polymerization process. Two mechanisms have been proposed to describe this phenomenon: polymer rearrangement and preferential polymerization [12]. Polymer rearrangement occurs when growing or terminated polymer molecules initially dis-

Table 1. Experimental conditions for polymeric vesicle templating (e.g., the polymerization of monomers in vesicles)

Monomer	Surfactant(s)	Surfactant: monomer ratio (molar)	Vesicle preparation	Polymerization method	Observed morphology	Characterization methods	References
Styrene, DVB, and styrene/DVB	DODAB (DMPC also mentioned)	1:1	Sonication	AIBN, 60 °C, 300 nm photo-polymerization	Hollow polymer spheres	QELS, electron mic., lateral and transmembrane diffusion measurements	Murtagh and Thomas (1986) [7]
ISODAC	SEHP	3.5:1–5:1	Sonication	KPS or AIBN, 60 °C	Hollow polymer spheres	QELS, FF TEM, probe trapping, stability tests	Poulain et al. (1996) [9]
Styrene, DVB, and styrene/DVB	CTAT/SDBS and others	3:1	Spontaneous	V-50, 60–65 °C	Hollow polymer spheres	QELS, Cryo TEM, SANS, AFM, GPC, stability tests	Morgan et al. (1997) [11] McKelvey et al. (2000) [24]
BMA/EGDMA	DODAC	1:1	Sonication or Extrusion	AIBN, 55 °C, then 254 nm UV	Hollow polymer spheres	Confocal mic. and SEM	Meier et al. (1998) [10]
Styrene	DODAB	1:2	Sonication	CHP/Fe(II), 50 °C	Hollow polymer spheres	TEM, SEM, paramagnetic probe diffusion	Kurja et al. (1993) [8]
Styrene	DODAB	1:2	Extrusion	DMPA, UV lamp or pulsed laser	Parachutes	Cryo TEM, QELS, μDSC, stability tests, lateral and transmembrane diffusion measurements	Jung et al. (1997, 2000) [17, 19]

Table 1 (continued)

Monomer	Surfactant(s)	Surfactant: monomer ratio (molar)	Vesicle preparation	Polymerization method	Observed morphology	Characterization methods	References
Styrene	DODAB	1:2	Extrusion	V-50, UV lamp or pulsed laser	Matrioshkas	same as above	Jung et al. (2000) [19]
DVB	DODAB	1:2	Extrusion	V-50 or amphiphilic initiator, 50–70 °C	Broccoli-type matrioshkas or parachutes	Cryo TEM	Jung et al. (2000) [14]
Styrene/ EGDMA or BMA/EGDMA	DODAB	2:1–1:1	Extrusion	DMPA, UV lamp or pulsed laser	Necklaces	Cryo TEM	Jung et al. (2000) [16]
Styrene, DVB, and surfactants	A1, A2, A1/A2	1:2–2:1	Extrusion	DMPA, UV lamp or pulsed laser	Hollow polymer spheres and phase separated structures	Cryo TEM, QELS, GPC, stability tests	Jung et al. (2000) [15]

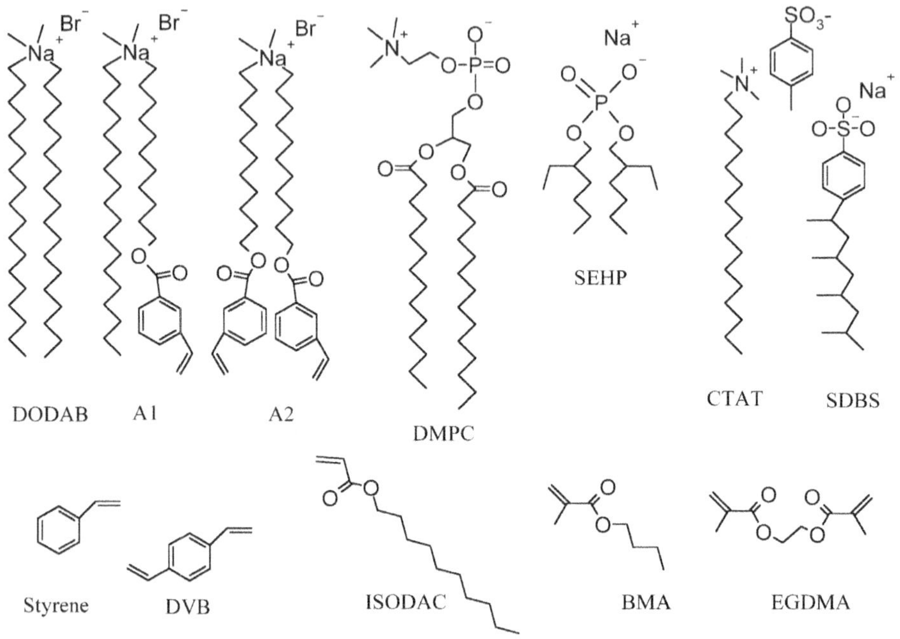

Fig. 4. Surfactants (*top*) and monomers (*bottom*) used for vesicle templating. See Table 1 for experimental conditions

tributed throughout the surfactant bilayer migrate to each other and coalesce due to unfavorable polymer-surfactant bilayer interactions. Preferential polymerization occurs when a locus of polymerization develops (presumably by radical entry or initiation) and the growing polymer is swollen by monomer. In this case, monomer diffuses to the polymerization site due to favorable polymer-monomer interactions and less favorable monomer-surfactant bilayer interactions.

Either or both mechanisms may be responsible for phase separation. Table 2 summarizes the anticipated effect of various polymerization conditions on polymer phase separation. Cross-linking monomers coupled with polymerization rates fast relative to lateral polymer diffusion are obvious ways to combat phase separation caused by polymer aggregation. Anchoring the growing polymer chains by employing polymerizable surfactants is also an alternative that should slow lateral polymer diffusion and thereby suppress phase separation. Similar approaches govern some of the features of microemulsion polymerization discussed below.

Suppressing phase separation in the case of preferential polymerization is difficult. Choosing surfactants with favorable interactions for the chosen monomers may decrease the driving force for monomers to leave the vesicle bilayer and swell the growing polymer radicals. Again, increasing the polymerization rates and cross-linking should also discourage phase separation. Other factors likely to influence phase separation include the polymer and vesicle bilayer glass or gel transition temperatures since the relative fluidity of both the bilayer

Table 2. Anticipated affect of experimental conditions on polymer phase separation and referenced experimental results

Condition	Anticipated effect on polymer phase separation	Actual observation
Cross-linking monomers	Suppress	Hollow shells [8, 10, 11, 25], brocolli parachutes [14], and necklaces [16]
Fast polymerization rate	Suppress	Necklaces/raspberries [15, 17]
Polymerizable surfactant anchoring	Suppress	Hollow shells and necklaces [16]
Improve surfactant-monomer interaction using alkyl monomers/surfactants	Suppress	Hollow shells [9, 10] and necklaces [12, 15, 17]
Improve surfactant-monomer interaction using aromatic monomers/surfactants	Suppress	Hollow shells [11, 16, 24]
Raise temperature/decrease polymer T_g	Encourage	Parachutes/complete phase separation [20]
Raise temperature/decrease surfactant bilayer T_m	Encourage	Complete phase separation [17]
Decrease temperature	Suppress	Necklaces [15]
Initiation mechanism	Unclear	Wide variety of structures including matrioshka [15, 20]

and polymer domains will affect the lateral diffusion of both monomer and polymer molecules. Manipulation of the reaction temperature (e.g., for thermal initiation) around this transition temperature could drastically affect the transport properties of the growing polymer and surfactant bilayer.

Distinguishing phase-separated morphologies from polymer shells experimentally is not trivial, and several complimentary techniques are generally required to ascertain fully the polymer morphology that results from polymerization. Simply measuring the hydrodynamic diameter by quasielastic light scattering (QLS), for example, is not sufficient to distinguish between hollow polymer shells or parachutes since the apparent hydrodynamic diameters of both structures can be identical. Even measurement of trans-bilayer and lateral diffusion rates with fluorescent probes can lead to erroneous conclusions about the particle morphology as described by German et al. [20].

The utility of cryo TEM is clear in obtaining a qualitative image of the apparent vesicle morphology in aqueous dispersions (Fig. 3). This technique avoids artifacts of drying (e.g., aggregation) that regular TEM methods require and allows characterization of aqueous dispersions in their native state. Scattering techniques compliment TEM techniques by yielding statistically meaningful information about the population as a whole.

Murtagh and Thomas made the first attempts to polymerize a hydrophobic monomer in surfactant vesicle bilayers [7]. These authors primarily relied on

QLS and electron microscopy to monitor particle diameters and fluorescence spectroscopy to measure both trans-membrane and lateral diffusion in the vesicle bilayers before and after polymerization. The authors concluded that hollow polymer shells were formed based on an apparent decrease in the lateral diffusion rate of a probe molecule. No changes in the vesicle diameter or gel transition temperature were observed following polymerization, but it is not clear from the data that phase separation was avoided. German et al. described similar fluorescence results in their phase-separated parachute samples and suggest that the hydrophobic pyrene probe used for lateral membrane diffusion studies is compartmentalized in the polymer latex beads of parachute aggregates [20]. Thus, measurement of the lateral diffusion rate with pyrene is not sufficient evidence that hollow polymer spheres have been synthesized. Murtagh and Thomas did not show electron micrographs or use scattering techniques to indicate their polymer structures are hollow, and the QLS measurements simply show that the outermost dimension of the vesicles is not altered by polymerization.

Nakache et al. attempted the polymerization of isodecyl acrylate (ISODAC) in anionic sodium di-2-ethylhexyl phosphate (SEHP) vesicles [9]. QLS and freeze fracture TEM were used to characterize the morphologies and no polymer phase separation was reported. Entrapment of hydrophobic moieties was possible in the polymerized vesicles, but hydrophilic moieties could not be entrapped. Polymerized vesicles were apparently more temperature stable than unpolymerized structures and the addition of electrolytes induced polymer precipitation.

QLS and freeze fracture TEM may not be capable of differentiating between parachutes and hollow polymer spheres. QLS only measures the outermost aggregate dimension and neither QLS or freeze fracture TEM can distinguish surfactant aggregates from polymeric ones. Freeze fracture micrographs of either the vesicle template or polymeric product only depict the structure of the surface of an aggregate (i.e., typically they are images of the interface between the aggregate and the external water domain). Freeze fracture images of parachutes containing relatively flat latex beads would appear quite similar to those of vesicle templates without polymer. Additionally, it is possible that the latex bead of a parachute may lie outside of the plane along which the fracture is generated and thus be obscured from view in a freeze fracture TEM image.

Probe diffusion rates observed from parachutes or hollow polymer spheres may be indistinguishable since the probe could be bound in either a polymer bead or polymer shell with similar release characteristics. For example, only hydrophobic probes could be trapped in the polymerized vesicles synthesized by Nakache et al. Here, "trapping" refers to a decrease in the rate of probe release after vesicle polymerization. The trans-membrane diffusion rates of hydrophilic probes should decrease following polymerization if a polymer shell is successfully formed in the surfactant bilayer. Nakache et al. only observed a decrease in the trans-membrane diffusion rate of hydrophobic probes. This is important since the hydrophobic probe may be released from both hollow polymer spheres and polymer latices with similar release kinetics. Again, caution should be taken, as already shown by German et al. [20] in the case of the fluorescence studies, since diffusion rate data interpretation and particle morphology are inherently coupled.

The increased temperature stability of the polymerized vesicles is, however, stronger evidence suggesting that complete phase separation has not occurred in the particles synthesized by Nakache et al. In a later paper [23], the authors noted that only a fraction of the vesicles appear to polymerize. If the monomer is evenly distributed among all the vesicles initially, one could expect monomer from non-reacting vesicles to diffuse to vesicles containing a growing radical, much like a typical emulsion polymerization experiment in which growing radicals in micelles are fed by monomer droplets.

Kaler et al. have attempted polymerization of styrene and divinyl benzene in catanionic vesicles that form spontaneously [11]. The polymer particles formed have been characterized with QLS, AFM, cryo TEM, and SANS. Each step of the polymerization process was characterized from the initial vesicle template to the monomer-swollen vesicles, and finally to the polymer product. The polymer spheres were isolated from the templating surfactants by methanol dialysis and completely dried. Redispersion of the hollow polymer spheres was facilitated either by sulfonating the aromatic rings of the cross-linked polystyrene to provide electrostatic stabilization, or by the addition of non-ionic ethoxylated alcohol surfactants to provide steric stabilization to the otherwise hydrophobic particles [24]. Even the re-dispersed particles appear hollow by cryo TEM and SANS measurements [25]. The size of the polymer spheres measured with QLS was independent of concentration, while the sizes of unpolymerized vesicles changed with concentration. Electrolyte addition destroyed surfactant vesicles and caused precipitation of the polymer particles, as Nakache et al. also observed [9]. Addition of the surfactant, Triton X-100, destroyed unpolymerized vesicles completely and yielded a clear solution. The polymerized solutions remained stable and exhibit the Tyndall effect with the addition of Triton X-100 (e.g., the dispersions scattered blue light due to Rayleigh scattering) [26]. The highest polymer molecular weights measured (2600 monomer units/chain average) were reported in these studies with polystyrene [11]. These are the only reports of vesicle templating using a spontaneously forming vesicle system composed of cationic and anionic surfactants.

Hotz and Meier polymerized butyl methacrylate (BMA) with ethylene glycol dimethacrylate (EGDMA) in dimethyldioctadecylammonium chloride (DODAC) vesicles [10, 27]. This group also concluded that hollow polymer spheres formed and characterized the structures with QLS, SEM, AFM, and confocal laser scanning microscopy. Isolation of the polymer spheres was performed by repeated precipitation of the polymer in methanol and re-dispersion in tetrahydrofuran (THF). An enormous increase in the vesicle bilayer thickness was observed from 3–5 nm in the surfactant vesicles to 200–300 nm for the polymer shells. This type of increase in shell thickness is explained by vesicle shrinkage, but could also be explained by only a fraction of vesicles polymerizing. Particles were apparently broken by shear between two glass plates and imaged by SEM to demonstrate their hollow nature. These particles are quite large (at least 100 μm in diameter) and only pictures of single particles have been published.

German et al. have reported the most complete study of polymerization in vesicles. They have systematically investigated polymerization *in* and *of* vesicles. This group has carefully characterized the morphologies resulting from the polymerization of a variety of monomers in several surfactant systems. With the exception of their first publication in this area [8], their characterization results clearly demonstrate that phase separation occurs for nearly every set of polymerization conditions attempted. The authors conclude from these observations that phase separation in one form or another is a general phenomenon of polymerization in vesicles.

One of the exceptionally useful aspects of this group's work is not only the wide variety of monomers and polymerization conditions attempted but also the wide variety of characterization methods employed. Like Kaler et al., this group has characterized the initial surfactant vesicle template, the monomer-swollen vesicles, and the final polymeric product. Particular attention was paid to the effect of monomer and polymer on the surfactant bilayer properties. Cryo TEM experiments revealed subtle changes in the vesicle morphology, which were investigated further with micro DSC experiments [21].

A number of other aspects of polymerization in vesicles studied systematically by this group include initiation chemistry, cross-linking, monomer-surfactant compatibility, polymerization kinetics, and incorporation of polymerizable surfactants. Although phase separation was observed under all conditions, the degree of the phase separation appears to depend heavily on each of these polymerization conditions in a surprisingly intuitive way.

Incorporation of monomers with similar characteristics to the hydrophobic tails of the surfactants involved (typically alkane chains of DODAB and DMPC) tends to suppress phase separation somewhat, and results in either multi-polymer bead aggregates (e.g., necklaces) or parachutes containing an elliptical rather than a spherical latex bead. Copolymerization of butyl methacrylate with ethylene glycol dimethacrylate in DODAB vesicles resulted in polymer necklaces where the polymer beads appear randomly dispersed in the vesicle bilayer [15] in contrast to the polymer shells observed by Hotz and Meier [10] for the same reaction in DODAC vesicles. Similarly, polymerization of octadecylacrylate, another straight-chain monomer, in DODAB vesicles produced parachutes with extremely ellipsoidal polymer beads in contrast to the rather spherical beads observed commonly for the polymerization of aromatic monomers such as styrene in DODAB [12]. Presumably these differences are caused by an increased compatibility between the surfactant bilayer and the monomer chosen.

Polymerization of divinyl benzene in DODAB vesicles also resulted in phase separation where "broccoli parachutes" were observed [15]. The latex beads in these cases appear as clusters of fuzzy latex particles. Phase separation appears to be slightly suppressed by the cross-linking, but the incompatibility of the aromatic monomer with the surfactant bilayer apparently dominates the final morphology as phase separation is obvious. Interestingly, when the alkyl chain surfactant is replaced with an aromatic polymerizable one, phase separation is completely suppressed [16]. With the A1 and A2 surfactants illustrated in Fig. 4, German et al. were able to successfully copolymerize styrene and divinyl benzene. The authors concluded that using a polymerizable amphiphile is a prerequisite for successful vesicle templating (i.e., avoiding phase separation). The introduction of these surfactants not only potentially anchors the growing polymer chains, but also increases the compatibility of the surfactant bilayer and the monomers used as they now both contain aromatic rings. It is unfortunate that experiments with a non-polymerizable aromatic surfactant vesicle system and divinyl benzene were not attempted. This would most directly compare to the successful polymerization of styrene and divinyl benzene in the catanionic vesicles investigated by Kaler et al. where both of the surfactants used contain aromatic rings [24].

The local concentration of monomer at the polymerization locus has also been estimated by German et al. using pulsed excimer laser polymerizations (PLP) [22]. The estimated local concentration of monomer at the polymerization site was relatively independent of the monomer loading, varied linearly with the vesicle size, and was inversely proportional to the temperature. Larger vesicles tended to have the highest local monomer concentrations at the polymerization site, and the authors suggested this is related to the extra room available for the polymer bead to swell with monomer. The inverse temperature dependence is unexpected. The solubility of the monomer in the aqueous phase and the bilayer presumably increases with temperature, resulting in a decrease in monomer at the reaction locus. If the monomer is more evenly distributed throughout the vesicle instead of concentrating at the reaction site, phase separation should be suppressed. Cryo TEM indicates, however, that phase separation was most obvious at higher reaction temperatures, and in this case the polymer latex beads were often completely ejected from the vesicle template.

The number of estimated polymer chains for a 55-nm latex bead in these parachute morphologies was 19,000 with M_n=2850 [20]. This number is much larger than the estimates of others of the number of polymer molecules in their own polymeric products [11].

Finally, the "matrioshka" structures observed by German et al. are perhaps the most surprising and intriguing. Their formation appears connected with the use of ionic initiators and monomers like V-50 and (3-methacryloylamino)-propyl trimethylammonium chloride respectively. A uni-vesicular transition process resulting in these structures has been proposed [13]. It is surprising that such a subtle change in the reaction conditions can cause such a dramatic morphological change.

2.3 Templating Vesicles – Concluding Remarks

Of all the characterization methods employed to study the morphologies of polymerized vesicles, cryo TEM is perhaps the most powerful. SEM, freeze fracture TEM, QLS, and probe diffusion studies alone cannot adequately distinguish between the polymer morphologies that have been proven to occur such as between hollow polymer shells and phase separated polymer-vesicle systems.

Characterization of polymer isolated from the templating surfactant by precipitation, vesicle lysis, or solvent washing/dialysis are other methods to characterize unambiguously the resulting polymer morphologies. The surfactant-induced vesicle lysis experiments (e.g., with Triton X-100) are a rapid and simple way to separate potentially latex particles from unpolymerized vesicles by micellizing any fragile surfactant structures. QLS can then be used to measure the approximate hydrodynamic diameter of the remaining rigid polymerized spheres (latex or hollow shells). The addition of electrolytes also destroys vesicles, but this technique generally causes the polymer to precipitate, making subsequent characterization more difficult.

Success in vesicle templating depends on control of reaction kinetics, monomer partitioning, membrane and polymer fluidity (T_g vs polymerization temperature), the use of cross-linking monomers, and anchoring polymerizable monomers. In practice, phase separation of the newly forming polymer and the surfactant bilayer has been observed under a variety of conditions. The extent of phase separation appears to follow intuitive guidelines, with the compatibility of the monomer and polymer with the surfactant of central importance.

All the observed morphologies (including the phase separated parachutes and necklaces) are potentially useful in a variety of applications. German et al. have proposed that necklaces composed of a variety of different polymeric beads could be useful as controlled release materials [20]. Most importantly, the implications of polymerization in vesicles can reveal more about the fundamental properties of vesicles as well as provide information about polymerization reactions in confined media.

3 Microemulsion Polymerization

Microemulsions are thermodynamically stable mixtures of water, oil, and surfactant that exhibit either a discrete droplet or bicontinuous structure on the length scale of nanometers. The simple nature of microemulsions or swollen micelles makes it easier to study the details of monomer/polymer/surfactant interactions and the importance of nucleation and monomer transport processes than is the case for reactions in lyotropic phase structures. Polymerization is initiated using the same free-radical sources used in vesicle and liquid crystalline template polymerizations, but usually only a very small fraction (~1/1000) of the micelles are initiated and grow into polymer particles [28]. The resulting polymer particles are typically much larger than the original microemulsion droplets. Thus such syntheses are reconstructive morphosyntheses. Monomer partitioning between polymer particles and the uninitiated micelles via diffusion through the aqueous phase plays an important role in controlling polymer particle formation and growth.

3.1 Factors Determining Polymer Growth: Monomer Partitioning, Swelling Behavior, and Side Reactions

Mechanically, surfactant monolayers are insufficiently rigid to restrict polymerization and preserve the original micelle size. Typical surfactant lifetimes on a micellar surface are on the order of microseconds, which are short relative to the milliseconds required for adding a monomer unit to the radical end of a polymer [29]. Instead, the microemulsion "template" exerts significant thermodynamic control over the outcome of the polymerization via monomer/surfactant interactions that limit the concentration of monomer in the polymer particles. SANS measurements of monomer partitioning between polymer particles and monomer-swollen micelles show that the monomer concentration profile in the polymer particles over the course of polymerization depends strongly on the properties of the starting microemulsions (Fig. 5) [30]. As the initial microemulsion approaches the phase-boundary, the concentration of monomer in the polymer particles increases and the concentration profile becomes more non-linear.

The solid curves in Fig. 5 are the results of a thermodynamic analysis based on a Flory-Huggins model for the interaction between the monomer and polymer and a phenomenological curvature energy model to describe the chemical potential of the monomer in the micelle [30].

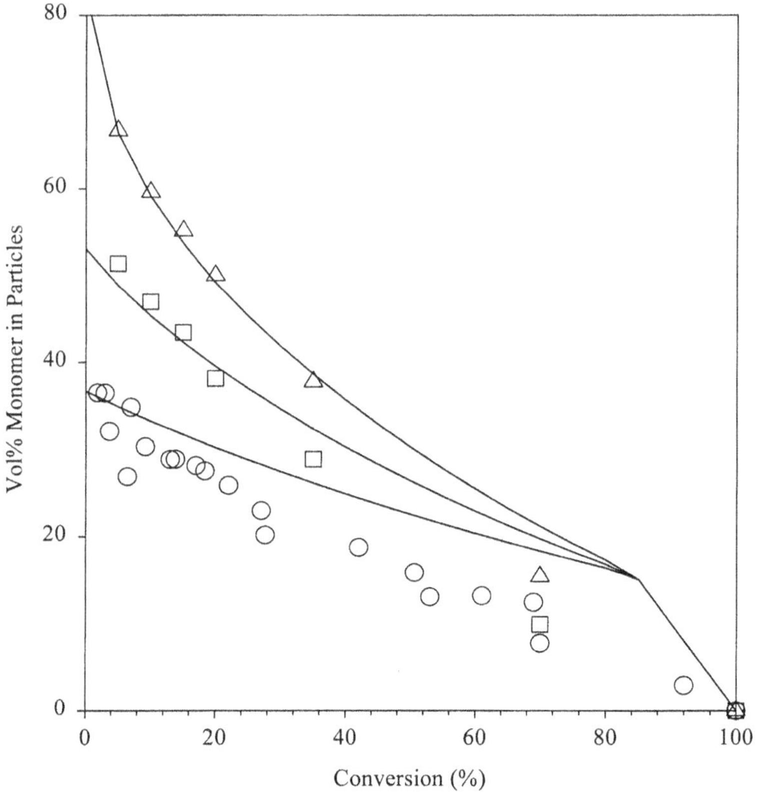

Fig. 5. SANS measurements of monomer volume fraction in polymer particles over the course of styrene microemulsion polymerizations at 60°C [30]. Microemulsions consist of 12 wt% dodecyltrimethylammonium bromide surfactant in D_2O and (*circles*) 3 wt%, (*squares*) 5 wt%, and (*triangles*) 7.5 wt% monomer on a surfactant-free basis

For this particular example of styrene/DTAB microemulsions, the strong affinity of the micelles for monomer causes the polymer particles to be depleted of monomer at low conversions. Thus, diffusion limitations due to glass transition may set in at unexpectedly low conversions. As an example, styrene/DTAB microemulsions containing 3 wt% monomer will become glassy at less than 40% conversion at 60°C. This observation suggests a method for preparing exceptionally small and monodisperse particles by feeding styrene monomer slowly and homogeneously. By maintaining low styrene loadings, existing polymer particles are starved of monomer and become glassy. Nascent particles containing low molecular weight polymer grow rapidly until they too become glassy. Adding a cross-linker would further minimize the swelling of the polymer particles. The presence of monomer-swollen micelles makes it possible to maintain a finite, albeit still very low, overall loading of monomer.

The idea of gradual addition of monomer to polymerizing microemulsions is not new and has been attempted for methylmethacrylate [31] and styrene [32–34] to make more efficient use of the surfactant. For example, Gan et al. have performed styrene microemulsion polymerizations where monomer was added gradually either dropwise or using a hollow-fiber membrane [32, 35]. They observed relatively large (~40 nm) and uniform particle sizes with no indications of continuous particle nucleation. However, this result is more likely due to a combination of fast monomer addition relative to the polymerization rate and depletion of the redox initiator system employed rather than to the growth of glassy particles.

3.2 Polymerization of Bicontinuous Microemulsions

The polymerization of bicontinuous microemulsions has been attempted frequently but has not succeeded in transcriptive synthesis without the use of polymerizable surfactants. Although the microemulsion length scales are ideal for preparing nanostructured membranes with diverse separation [36–41] and electrochemical [35, 42] applications, the usual result is that polymerization yields macroporous gels. Rapid reorganization of the microemulsion template apparently makes it unable to arrest the reorganization of the polymer to its thermodynamically favored configuration. The only reports describing successful capture of bicontinuous microemulsion structures have employed polymerizable surfactants. Elements controlling polymer morphology include nucleation and growth of the polymer and the phase behavior of the parent bicontinuous microemulsions, which depends strongly on the choice and quantity of monomer [43]. Clearly the curvature of the surfactant film, which governs the phase behavior, reflects the penetration of hydrophobic monomer into the surfactant hydrocarbon tails. Consumption of monomer by polymerization will then unavoidably and substantially change the curvature of surfactant monolayers forming the template, thereby frustrating attempts to replicate the bicontinuous structure. The use of co-surfactants, co-oils, and hydrophobic monomers of high molecular weight that minimize the impact of polymerization on the curvature of the surfactant film thus offer the beginnings of routes to successful polymerization.

Even for polymerizable surfactants, curvature control is still very important, and Eastoe et al. describe mixtures of polymerizable surfactants that retain the same preferred curvature before and after cross-linking [44, 45]. Current approaches to the one-to-one replication of bicontinuous or other surfactant mesophases without the use of polymerizable surfactants involve the use of surfactants of high molecular weight [46] or highly viscous cubic phases featuring slower rearrangement dynamics.

3.3
General Aspects of Templating Within Microemulsions

Accurate replication of the original template structure during polymerization of a microemulsion is complicated by initiation that is spatially non-uniform relative to the length scales of the microstructure, and further confounded by the inability of the "soft" surfactant template to prevent the reorganization or swelling of the polymer. Nonetheless, if surfactant/monomer interactions are sufficiently strong, the polymer can be depleted of monomer and become glassy at early stages of polymerization. Addition of cross-linkers can be used to reduce further the swelling and mobility of the polymer. Under these conditions, the reactive radical ends migrate via reaction-diffusion to unpolymerized regions rich in monomer. Meanwhile, the cross-linked and glassy polymerized regions do not reorganize. This picture suggests that when a monomer yielding a mechanically rigid and non-swelling polymer is used for replication, the demands on the template can be reduced to guiding the direction of polymerization at the outermost edges. These ideas are also relevant to the successful templating of comparatively "soft" equilibrium catanionic vesicles discussed above [11, 24, 25] with styrene and divinylbenzene, wherein the high degree of cross-linking would likely lead to the phase separation of a mechanically rigid structure, and may be partly due to this mechanism of polymerization.

4
Polymerization in Lyotropic Mesophases

Liquid crystals combine properties of both liquids (fluidity) and crystals (long range order in one, two, or three dimensions). Examples of liquid crystalline templates formed by amphiphiles are lyotropic mesophases, block copolymer mesophases, and polyelectrolyte-surfactant complexes. Their morphological complexity enables the template synthesis of particles as well as of bulk materials with isotropic or anisotropic morphologies, depending on whether the polymerization is performed in a continuous or a discontinuous phase. As the templating of thermotropic liquid crystals is already described in other reviews [47] the focus here is the template synthesis of organic materials in lyotropic mesophases.

Potential applications of liquid crystalline templated polymer gels range from separation media (membranes, chromatography columns, or electrophoresis gels) to low dielectric constant insulators for microelectronic devices, to nanostructured optoelectronic devices, to catalysts supports, drug carriers, or materials for controlled release.

4.1
Transcriptive and Reconstructive Synthesis Within Lyotropic Mesophases

Transcriptive synthesis aims at a 'casting' of a self-assembled, non-polymerizable template under retention of the parental order (Table 3). In many cases during the course of the reaction the template structure itself is retained, but newly formed polymer phase separates by a nucleation and growth mechanism. Hence, the obtained morphology is usually not a one-to-one copy of the template. Nevertheless highly ordered polymer morphologies can be obtained by the structure-directing effect of the template, although they usually display characteristic lengths much larger than those of the original template structure (i.e., this is a reconstructive morphosynthesis [1]).

The photoinitiated polymerization of divinylbenzene (DVB) within four separate cubic phases of the system DVB/didodecyldimethylammonium bromide (DDAB) is reported to yield retention of the lyotropic order during the course of the reaction [48], although the structure of the pure polymer matrix after removal of the template was not investigated. Similarly, polymerization of acrylamide within lamellar, hexagonal, and cubic phases of the surfactant Aerosol OT led to preservation of the parental mesostructure [49]. SAXS measurements showed similar diffractograms before and after polymerization, but again there was no report of characterization of the polymer matrix after surfactant removal. Hence, at least in these cases, the formation of a polymer phase within a lyotropic mesophase does not prevent the formation of lyotropic surfactant phases.

Recently the polymerization of styrene within lamellar and cubic phases of the surfactant DODAB (dioctadecyldimethylammonium bromide) was studied [50]. After polymerization, the polystyrene/water/DODAB system showed the same phase behavior as the binary water/DODAB system, a result suggesting a phase separation during polymerization into a polymer-rich (with M_w~400,000) and a lyotropic phase.

There are other reports of similar phase separations into polymer-rich and surfactant-rich phases [51]. A detailed investigation of structure formation during the cross-linking polymerization of acrylamide within hexagonal phases of cetyltrimethylammonium surfactants indicated a nucleation and growth mechanism [52]. There are no significant changes in the X-ray scattering pattern before and after polymerization, but after removing the template by Soxhlet extraction no nanostructure was detected by SAXS. Instead SEM showed a homogeneous, layer-like morphology on a micrometer scale (Fig. 6). A mechanism consistent with these products is that during the reaction polymer particles are formed which separate from the lyotropic phase and assemble towards sheet-like structures. This assembly is directed by the anisotropy of viscosity and transport properties within the single domains of the lyotropic template. The particular shape and extension of the single layers depends on their degree of cross-linking. Chemical and mechanical properties can be adjusted by copolymerization with functional co-monomers like acrylic acid.

Other lyotropic mesophases of amphiphilic block copolymers, such as polystyrene-*b*-polyethylene oxide have been used as templates, with the expectation

Table 3. Examples for transcriptive and reconstructive approaches towards organic polymer synthesis within lyotropic mesophases

Surfactant	Monomers	Morphology before polymerization	Morphology after polymerization	Characterization of the isolated polymer	Remarks	Reference
DDAB	Styrene/DVB	Cubic	No phase transition	Not characterized	–	[48]
AOT	Acrylamide/*N,N'*-methylenebisacrylamide	Lamellar, hexagonal, cubic	No phase transition	Not characterized	–	[49]
DODAB	Styrene	Lamellar, cubic	Phase transition cubic→lamellar	Linear polymer (M_w~400000) M_w/M_n~4)	Phase separation between template and polymer	[50]
PS-*b*-PEO	Styrene, *m*-diisopropenylbenzene	Hexagonal, lamellar	No phase transitions	Macroporous polymer gels, elongated structure elements	Phase separation between template and polymer	[53]
CTMA	Acrylamide/*N,N'*-methylenebisacrylamide, carboxyethyl acrylate, acrylic acid, etc.	Hexagonal	No phase transition	Macroporous layer-like morphology	Phase separation between template and polymer	[52]
DTAB	Hexanediol diacrylate, *n*-decylacrylate, PEG400-dimethaccrylate, 2-hydroxyethylmethacrylate	Cubic, hexagonal, lamellar	Phases retained to a great extent	Not characterized	Investigation of polymerization kinetics	[54]

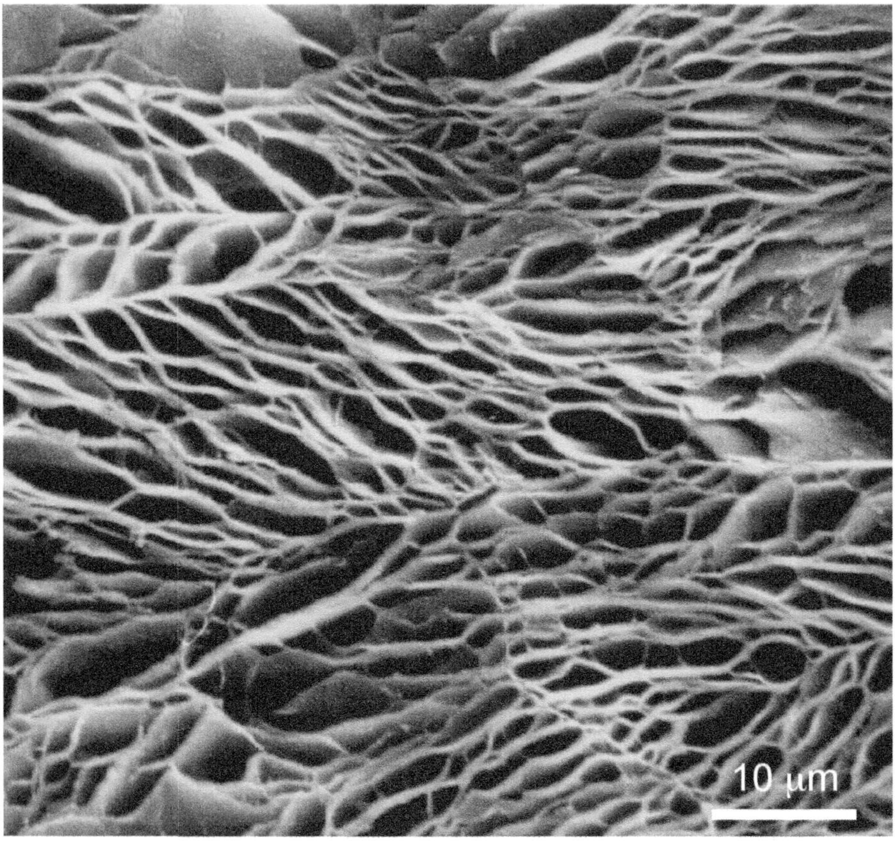

Fig. 6. Scanning electron micrograph of a poly(acryl amide) gel templated in a lyotropic mesophase of CTMA-tartrate. The formation of a layer-like morphology is directed by the anisotropic reaction medium. Adapted with permission from [52]

that the more robust and less dynamic polymer templates might more effectively guide polymerization [53]. However, again phase separation occurred, and the final polymer structures were not directly related to the organization of the lyotropic liquid crystal.

The influence of the lyotropic morphology on polymerization rates was studied for a variety of monomers with different polarity and amphiphilicity [54] in lyotropic phases of dodecyltrimethylammonium bromide in water. The confined geometry of the resulting aggregates has a strong impact on the local monomer concentration and diffusion properties and therefore also on the polymerization kinetics.

A better understanding of such morphologies could arise from fundamental investigations of the ordering effects of phase separation from liquid crystals [55]. For example, colloidal ordering (exactly parallel chains) results when silicon oil droplets separate from a thermotropic liquid crystal [37]. Such ordering is possibly directed by forces between the defects created by the demixing process, yielding strong ordering on the micron scale, rather than on the nanometer scale of the original liquid crystal. The parallels between this result and polymerization within liquid crystalline phases remain to be explored.

4.2 Synergistic Templating Within Lyotropic Mesophases

Synergistic approaches aim at the polymerization of the self-organizing entities themselves (Table 4). The most common approach employs polymerizable surfactants, sometimes together with additional non-amphiphilic comonomers. The aim is chemical fixation of the dynamic self-organized assembly. Although the final materials often show no significant porosity after purification and drying, in many cases the interface becomes accessible by reswelling the gels with solvents. Often increased phase stability against addition of solvents or temperature changes is observed.

Early investigations, e.g., of the polymerization of the system undecenoate/water, showed that phase transitions occur during the course of reaction [56]. Other attempts to polymerize amphiphilic monomers showed no phase transitions and increased phase stability against changes in temperature [57–60]. However, the reactions often proceeded only to low (~40%) conversions. This might be due to the use of amphiphiles with an inactivated terminal olefin as the polymerizable group, such as ω-undecenyltrimethylammonium bromide [59] or 10-undecenoic acid [58, 61]. The reduced extent of polymerization compared to polymerization in solution is generally attributed to the aggregates acting as a "cage" that inhibits polymerization [59]. This effect depends strongly on the position of the polymerizable moiety in the amphiphilic monomer.

High conversions (above 90%) were obtained by polymerization of amphiphilic monomers such as phosphonium dienes, 3,4,5-tris (ω-acryloxyalkoxy)benzoate salts, or styrene ether-modified fatty acids [62–65]. Homopolymerization and copolymerization with DVB as cross-linker in lyotropic states result in freestanding, mechanically stable films. X-ray diffraction and polarized light microscopy show the hexagonal order before and after polymerization. However, there have been no structural investigations (such as BET or electron microscopy) of the isolated polymer matrix after extraction and drying. Imaging methods should provide important information about, for example, the types of disclinations, the degree of order and the formation of any higher order morphologies.

Lee et al. demonstrated the synthesis of nanostructured cubic polymer gels by copolymerization of dienyl substituted lipids [66]. No phase transitions, or changes in dimensions, were observed with temperature changes for the polymerized sample. Furthermore, the polydomain square lattice of the gel was visualized by TEM of ultramicrotomed samples after extraction of the template (Fig. 7). In contrast, copolymerization of monoacylglycerol and 1,2-diacylglycerol in a cubic lyotropic state did not result in a continuous gel structure. Linear polymer chains were obtained instead, and the cubic morphology was destroyed by addition of organic solvent [67]. Similar polymerizations in the inverted

Table 4. Examples of different synergistic approaches towards organic polymer synthesis in lyotropic mesophases

Polymerizable surfactant	Comonomers	Template phase	Phase transitions during polymerization	Remarks	Reference
Sodium 10-undecenoate	–	Hexagonal	Phase transition to a lamellar phase	Complete conversion	[56]
Allyldodecyldimethyl-ammonium bromide	–	Micellar, hexagonal, cubic, lamellar	No phase transitions	Partially polymerized, 11–40% conversion, increased phase stabilities	[60]
Allyldidodecylmethyl-ammonium bromide	–	Lamellar	–	Polymerization did not occur	[60]
Undecenyltrimethyl-ammonium bromide	–	Micellar, hexagonal	No phase transitions	Partially polymerized, 20–47% conversion, no increased phase stability	[59]
Sodium 10-undecenoate	–	Micellar, hexagonal, lamellar	H_I-phase: no transitions, L_α-phase: transition to multiphases	Partially polymerized, ~30% conversion	[58]
Dodecyldimethyl-ammoniumethyl-methacrylate bromide	–	Micellar	Phase separation of polymer from the micellar solution	>95% conversion	[57]
Phosphonium diene	(DVB)	Hexagonal	No phase transition	90% conversion increased phase stability	[62]
Styrene ether modified fatty acids	(DVB)	Hexagonal	No phase transition	Formation of non-dissolvable gels	[65]
Tris(acryloxyalkoxy)-benzoate	–	Hexagonal	No phase transition	Polymerization in the ternary system	[63, 64]
Dienyl substituted lipids	–	Cubic	No phase transition	Cubic polymer morphology (SAXS, TEM)	[66]
Semifluorinated alkyl methacrylic acid	– –	Micellar, lamellar, cubic	Phase transitions at slow polymerization rates	Investigation of polymerization kinetics	[69]

Fig. 7. Transmission electron micrograph of stained ultramicrotomed samples of a polymerized dienyl substituted lipid, cross-linked in a cubic mesophase; scale bar=100 nm. (Adapted with permission from [66])

hexagonal phase [68] yield an increased stability of the lyotropic phase against temperature changes.

The kinetics of polymerization of amphiphilic monomers in lyotropic mesophases has been investigated by Lester and Guymon, who found that the polymerization rate of a semifluorinated alkyl methacrylic acid in a lyotropic liquid crystal increased with the degree of order of the liquid crystalline phase [69].

Recently the successful γ-ray polymerization of polymerizable cationic low molecular weight surfactants (2-methacryloylethyl)dodecyldimethylammonium bromide and (11-methacryloylundecyl)trimethylammonium bromide) in the cubic and the hexagonal phase of binary surfactant/water systems was reported [70]. Nanostructured gels were obtained that retained their morphology during drying and reswelling with hydrophilic or hydrophobic solvents. Cholesteric morphologies of higher molecular weight cellulose derivatives in water were also successfully cross-linked by γ-ray polymerization [71].

Photo-crosslinking of PEO-PDMS-PEO block copolymers in the hexagonal and lamellar lyotropic phase was performed by UV irradiation of films about 10 μm thick [72]. After cross-linking and drying the characteristic textures for each phase were retained. Even though insoluble gels were formed, swelling with organic solvent and drying resulted in isotropic films. This rearrangement could be due to a high degree of local mobility of the polymer chains due to the relatively low cross-linking degree employed.

The cross-linking of poly(oxyethylenes) with polymerizable, hydrophobic endgroups within oriented lamellar phases of the ternary system $C_{12}E_5$/decane/water was reported by Meier [73]. After cross-linking oriented polymer gels were obtained that retained their lamellar morphology after extraction of the template, and which showed one-dimensional swelling by water.

Synergistic templating by cross-linking of PEO-PB block copolymers in lyotropic phases (cubic, lamellar and hexagonal) was studied by Hentze et al. [74]. In this case every repeating unit of the PB block has one polymerizable moiety, thereby allowing a high degree of cross-linking. After cross-linking by γ-irradiation, the order of the lyotropic phase is retained within the mechanically stable polymer gels as shown by SAXS, SANS, and TEM (Fig. 8). No significant porosity could be determined by BET because of the highly compact character of the dried polymer gels. The gel morphology was stable against temperature

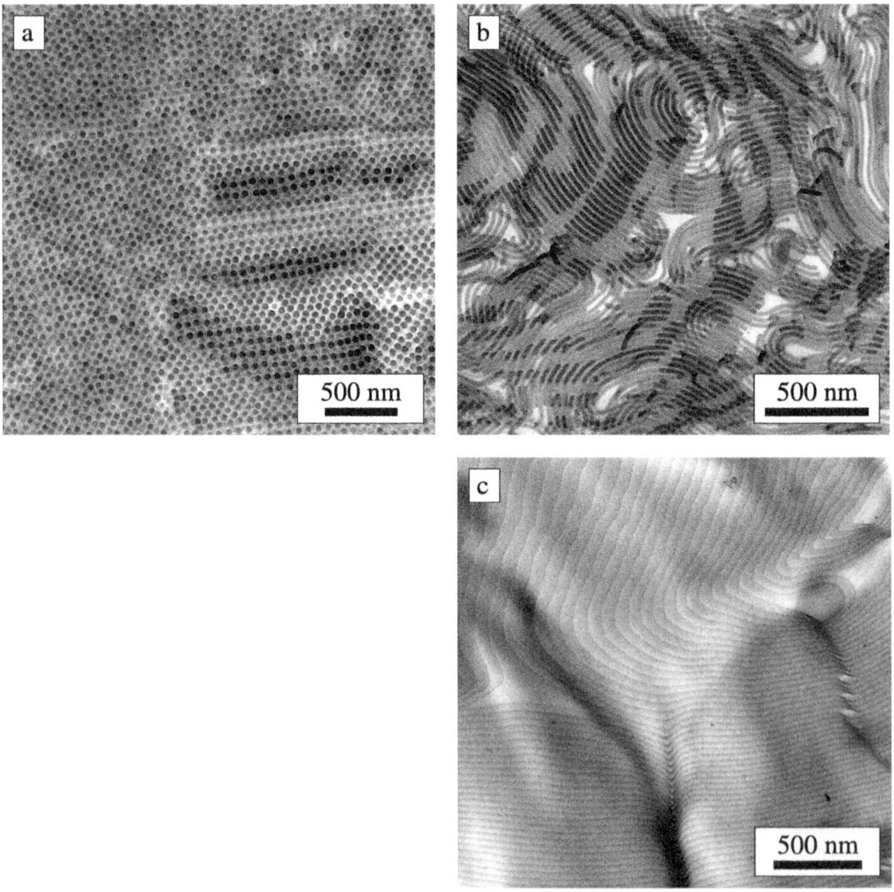

Fig. 8a–c. Micrographs of cross-linked poly(butadiene)-*b*-poly(ethylene oxide) gels, templated in: **a** cubic; **b** hexagonal; **c** lamellar phases. A dense packing of cylinders can be observed by TEM after cross-linking in the hexagonal phase. (Reproduced with permission from [74])

changes, extraction, drying, and reswelling with polar or nonpolar solvents. Electron microscopy of ultramicrotomed samples allowed investigation of lyotropic morphologies, disclinations, domain boundaries, and the topological effects of concentration, block length ratio, and molecular weight of the amphiphilic block copolymers [75].

5 Conclusions

The integration of the natural principles of template synthesis and self-organization into the strategy of chemical synthesis opens up a huge potential for structuring materials over all length scales ranging from molecules to nano- and microstructured materials. The aim is to transfer the superior properties and property combinations of biological materials towards large-scale synthetic materials by mimicking biological principles of supramolecular architecture.

Templated polymer synthesis within vesicular solutions, microemulsions, and lyotropic mesophases offer ways to synthesize hollow nanospheres, nanoparticles, and porous polymers. Besides preservation of the template structure by transcriptive or synergistic synthesis, colloidal ordering can also occur by phase separation from a highly organized initial state (reconstructive synthesis). The final product materials reflect a delicate balance of phase behavior and the reaction and mass transfer parameters controlling structure formation during template synthesis. Key variables are the reaction kinetics, and the roles of monomer partitioning, monomer diffusion, membrane and polymer fluidity, the use of cross-linking monomers and polymerizable monomers, and the compatibility of the polymer and surfactant microstructure all must be understood and controlled. Tuning of these parameters enables structural design of organic polymers with various morphologies on length scales from nano- to micrometer.

Acknowledgements. Support from the Deutscher Akademischer Austauschdienst (DAAD) and the National Science Foundation (CTS-9814399) is gratefully acknowledged. HPH thanks Stephen Mann and Markus Antonietti for helpful discussions.

References

1. Mann S, Burkett SL, Davis SA, Fowler CE, Mendelson NH, Sims SD, Walsh D, Whilton NT (1997) Chem Mater 9:2300
2. Bangham AD, Standish MM, Watkins JC (1965) J Mol Biol 13:238
3. Regen SL (1987) In: Ostro MJ (ed) Liposomes: from biophysics to applications. Marcel Dekker, New York
4. Ringsdorf H, Schlarb B, Venzmer J (1988) Angew Chem Int Ed Engl 27:113
5. Singh A, Schnur JM (1993) In: Cevc G (ed) Phospholipids handbook. Marcel Dekker, New York
6. Paleos CM (1992) In: Paleos CM (ed) Polymerization in organized media. Gordon and Breach Science Publishers, Philadelphia, p 283
7. Murtagh J, Thomas JK (1986) Faraday Discuss Chem Soc 81:127
8. Kurja J, Nolte RJM, Maxwell IA, German AL (1993) Polymer 34:2045

9. Poulain N, Nakache E, Pina A, Levesque G (1996) J Polym Sci Pol Chem 34:729
10. Hotz J, Meier W (1998) Langmuir 14:1031
11. Morgan JD, Johnson CA, Kaler EW (1997) Langmuir 13:6447
12. Hubert DHW (2000) Surfactant vesicles in templating approaches. PhD thesis, Eindhoven University of Technology
13. Hubert DHW, Jung M, Frederik PM, Bomans PHH, Meuldijk J, German AL (2000) Langmuir 16:8973
14. Hubert DHW, Jung M, German AL (2000) Adv Mater 17:1291
15. Jung M (2000) Polymerisation in bilayers. PhD thesis, Eindhoven University of Technology
16. Jung M, den Ouden I, Montoya-Goni A, Hubert DHW, Frederik PM, van Herk AM, German AL (2000) Langmuir 16:4185
17. Jung M, Hubert DHW, Bomans PHH, Frederik P, van Herk AM, German AL (2000) Adv Mater 12:210
18. Jung M, Hubert DHW, Bomans PHH, Frederik PM, Meuldijk J, van Herk AM, Fischer H, German AL (1997) Langmuir 13:6877
19. Jung M, Hubert DHW, van Herk AM, German AL (2000) Macromol Symp 151:393
20. Jung M, Hubert DHW, van Veldhoven E, Frederik P, van Herk AM, German AL (2000) Langmuir 16:3165
21. Jung M, Hubert DHW, van Veldhoven E, Frederik PM, Blandamer MJ, Briggs B, Visser A, van Herk AM, German AL (2000) Langmuir 16:968
22. Jung M, van Casteren I, Monteiro MJ, van Herk AM, German AL (2000) Macromolecules 33:3620
23. Poulain N, Nakache E (1998) J Polym Sci Polym Chem Ed 36:3035
24. McKelvey CA, Kaler EW, Coldren B, Jung H-T, Zasadzinski JA (2000) Langmuir 16:8285
25. McKelvey CA, Kaler EW (2002) J Colloid Interface Sci 245:68
26. McQuarrie DA (1976) Statistical mechanics. Harper Collins, New York
27. Hotz J, Meier W (1998) Adv Mater 10:1387
28. Co CC, Kaler EW (1998) Macromolecules 31:3203
29. de Vries R, Co CC, Kaler EW (2001) Macromolecules 34:3233
30. Co CC, de Vries R, Kaler EW (2001) Macromolecules 34:3224
31. Ming WH, Jones FN, Fu SK (1998) Polym Bull 40:749
32. Xu XJ, Siow KS, Wong MK, Gan LM (2001) Langmuir 17:4519
33. Xu XJ, Chew CH, Siow KS, Wong MK, Gan LM (1999) Langmuir 15:8067
34. Rabelero M, Zacarias M, Mendizabal E, Puig JE, Dominguez JM, Katime I (1997) Polym Bull 38:695
35. Xu W, Siow KS, Gao ZQ, Lee SY, Chow PY, Gan LM (1999) Langmuir 15:4812
36. Liu J, Teo WK, Chew CH, Gan LM (2000) J Appl Polym Sci 77:2785
37. Loudet JC, Barois P, Poulin P (2000) Nature 407:611
38. Chew CH, Li TD, Gan LH, Quek CH, Gan LM (1998) Langmuir 14:6068
39. Chieng TH, Gan LM, Chew CH, Ng SC (1995) Polymer 36:1941
40. Chieng TH, Gan LM, Chew CH, Ng SC, Pey KL (1996) J App Pol Sci 60:1561
41. Gan LM, Chew CH (1997) J Coll Surf A 123/124:681
42. Chow PY, Chew CH, Ong CL, Wang J, Xu G, Gan LM (1999) Langmuir 15:3202
43. Lade O, Beizai K, Sottmann T, Strey R (2000) Langmuir 16:4122
44. Summers M, Eastoe J, Davis S, Du Z, Richardson RM, Heenan RK, Steytler D, Grillo I (2001) Langmuir 17:5388
45. Eastoe J, Summers M, Heenan RK (2000) Chem Mater 12:3533
46. Morkved TL, Stepanek P, Krishnan K, Bates FS, Lodge TP (2001) J Chem Phys 114:7247
47. Miller SA, Ding JH, Gin DL (1999) Curr Opin Colloid Interface Sci 4:338
48. Strom P, Anderson DM (1992) Langmuir 8:691
49. Laversanne R (1992) Macromolecules 25:489
50. Jung M, German AL, Fischer HR (2001) Colloid Polym Sci 279:105
51. Antonietti M, Caruso RA, Goltner CG, Weissenberger MC (1999) Macromolecules 32:1383
52. Antonietti M, Goltner C, Hentze HP (1998) Langmuir 14:2670

53. Hentze HP, Goltner CG, Antonietti M (1997) Ber Bunsenges Phys Chem 101:1699
54. Lester CL, Colson CD, Guymon CA (2001) Macromolecules 34:4430
55. Poulin P (1999) Curr Opin Colloid Interface Sci 4:66
56. Friberg SE, Thundathil R, Stoffer JO (1979) Science 205:607
57. McGrath KM, Drummond CJ (1996) Colloid Polym Sci 274:612
58. McGrath KM (1996) Colloid Polym Sci 274:499
59. McGrath KM (1996) Colloid Polym Sci 274:399
60. McGrath KM, Drummond CJ (1996) Colloid Polym Sci 274:316
61. Shibasaki Y, Fukuda K (1992) Colloids Surf 67:195
62. Pindzola BA, Hoag BP, Gin DL (2001) J Am Chem Soc 123:4617
63. Deng H, Gin DL, Smith RC (1998) J Am Chem Soc 120:3522
64. Resel R, Leising G, Markart P, Kriechbaum M, Smith R, Gin D (2000) Macromol Chem Phys 201:1128
65. Reppy MA, Gray DH, Pindzola BA, Smithers JL, Gin DL (2001) J Am Chem Soc 123:363
66. Lee YS, Yang JZ, Sisson TM, Frankel DA, Gleeson JT, Aksay E, Keller SL, Gruner SM, Obrien DF (1995) J Am Chem Soc 117:5573
67. Srisiri W, Benedicto A, O'Brien DF, Trouard TP, Oradd G, Persson S, Lindblom G (1998) Langmuir 14:1921
68. Srisiri W, Sisson TM, Obrien DF, McGrath KM, Han YQ, Gruner SM (1997) J Am Chem Soc 119:4866
69. Lester CL, Guymon CA (2000) Macromolecules 33:5448
70. Pawlowski D, Haibel A, Tieke B (1998) Ber Bunsenges Phys Chem 102:1865
71. Hohn W, Tieke B (1997) Macromol Chem Phys 198:703
72. Yang JL, Wegner G (1992) Macromolecules 25:1791
73. Meier W (1998) Macromolecules 31:2212
74. Hentze HP, Kramer E, Berton B, Forster S, Antonietti M, Dreja M (1999) Macromolecules 32:5803
75. Forster S, Berton B, Hentze H-P, Kramer E, Antonietti M, Lindner P (2001) Macromolecules 34:4610

Top Curr Chem (2003) 226: 225–246
DOI 10.1007/b10830

Rational Material Design Using Au Core-Shell Nanocrystals

Paul Mulvaney[1], Luis M. Liz-Marzán[2]

[1] Chemistry School, University of Melbourne, Parkville, VIC, 3010, Australia
E-mail: mulvaney@unimelb.edu.au
[2] Departamento de Química Física, Universidad de Vigo, Vigo 36200, Spain

An overview of the construction of complex colloid structures based on coated gold nanocrystals is provided, illustrating how geometry and optical properties are linked. The concept of rational design – building towards a topological structure predicated on desired properties is illustrated.

Keywords. Nanocrystals, Core-shell particles, Topology, Opals, Capsules, Surface plasmon absorption, Bragg diffraction, Arrays, Colloidal crystals

1 Introduction

This chapter provides an overview of our recent research into the creation of designer materials based on the assembly of composite nanocrystals (NCs) with core-shell morphology. In particular, this chapter focuses on gold. The reason for this somewhat parochial choice is the explosion of work on both shape control and nanostructuring of gold particles in the last five years. This system will be a test bed for numerous models of nanomaterial design in the forthcoming decade as increasing sophistication in materials manipulation enables unparalleled development of tailor-made (rational) structures for electronics, optics, catalysis and pseudobiological function. Gold is also a special case for historical reasons. The original work on the optical properties of nanoparticles was inspired by the need to explain the unusual and exotic colours displayed by gold nanocrystals [1–3].

Increasing the topological complexity of nanocrystal based structures opens up the creation of designer materials. A basic outline of the rapidly evolving array of topological objects being devised, designed and synthesized is shown schematically in Fig. 1. By no means comprehensive, it illustrates the four basic concepts for particle modification: size control, shape manipulation, layered structures and nanocrystal ordering. Combinations of these four processes can be readily envisaged. Three of these four pathways from single NCs to more complex materials are well understood mechanistically. Size control is determined by nucleation and growth kinetics, and the efficacy with which capping agents can truncate Ostwald ripening and coalescence. Colloid structuring is largely controlled through manipulation of colloid and surface forces such as double layer repulsion, capillary forces, sedimentation and electro- or magnetophoresis. Finally, layered colloids can be synthesized when there is sufficient control over deposition and nucleation within a nanocrystal system. Shape control is the exception. Shape control requires anisotropic interfacial processes that are not well understood at present and bottom-up synthesis of rods and other non-spherical nanocrystals seems to rely on the vagaries of surfactant adsorption to specific crystal facets in most successful cases.

There are three basic reasons why core-shell structures in particular are attractive for nanostructuring applications. Firstly, the shell layer provides a method for protecting the core particle. The core particles may be chemically reactive and easily oxidized or photocorroded in the presence of solvent molecules. The shell may kinetically inhibit these deleterious processes. Secondly, the shell layer mediates the interaction with the solvent and other particles or surfaces. The particle-particle interactions will be determined by the surface properties of the shell. For example, a silica shell can be made hydrophobic facilitating phase transfer of the core particle. Alternatively, it may be more resistant to coalescence. In this sense, colloids are normally treated as core-shell systems, since there is usually a surfactant or polymer coating to impart colloid stability. Thirdly, the shell can modify coupling between core materials by controlling the distance of closest approach of cores to each other. This is important for metal NCs where dipole coupling is the predominant method of "communication" between particles and is important in designing optical effects. Energy and elec-

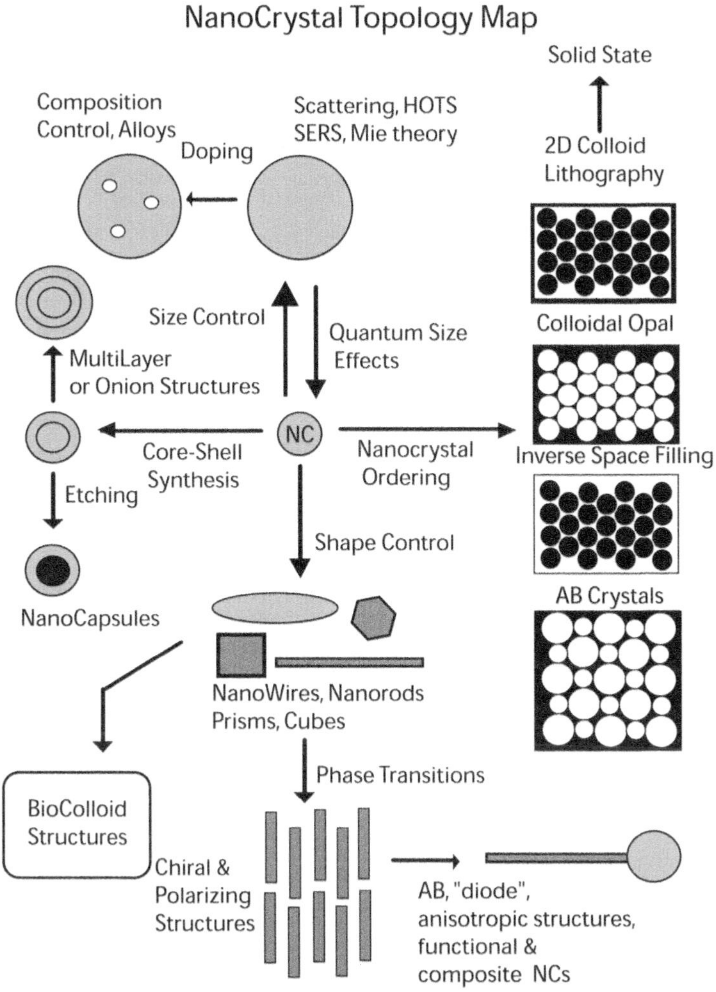

Fig. 1. Scheme showing basic aspects of nanocrystal topology

tron transfer between semiconductor NCs may also be modulated through a shell layer, while control over the band structure can be achieved through core-shell or "quantum well" structures. In this review we restrict ourselves primarily to the manipulation of gold and silica for simplicity.

2
Modelling of Optical Properties

In this section, we describe the basic optical properties of spheres, ellipsoids and coated particles and colloid thin films. Predictions based on these equations will be compared with rationally designed materials in Sects. 3–6.

2.1 Dilute and Concentrated Dispersions

The measured absorbance of light by a sol containing N particles per unit volume with radius R is

$$A = \mathrm{Log}_{10}\, I_o/I_d = \frac{N\, Q_{ext}\, \pi R^2\, d}{\log_{10}\mathrm{e}} \tag{1}$$

with A the absorbance (really extinction) and d the pathlength of the light beam through the solution. Q_{ext} is the normalised extinction cross section of each particle. Mie showed that Q_{ext} is given by [4]

$$Q_{ext} = \frac{2}{R^2 k^2} \sum_{n=1}^{n=\infty} (2n+1)\,(a_n + b_n) \tag{2}$$

where k is the wavenumber. The normalised scattering cross section is

$$Q_{sca} = \frac{2}{R^2 k^2} \sum_{n=1}^{n=\infty} (2n+1)\,(|a_n|^2 + |b_n|^2) \tag{3}$$

Here a_n and b_n are series of Ricatti-Bessel functions [4]. The measured attenuation of light by a colloid (and indeed by molecular solutions) is always due to both scattering and absorption. These effects are additive so

$$Q_{ext} = Q_{abs} + Q_{sca} \tag{4}$$

Since the scattering decreases as R^6, for small enough particles, it can be neglected and

$$Q_{abs} \sim Q_{ext} = \frac{24\pi\, \varepsilon_m^{3/2} R}{\lambda} \frac{\varepsilon''}{(\varepsilon' + 2\varepsilon_m)^2 + \varepsilon''^2} \tag{5}$$

Equations 1 and 5 are fundamental to an understanding of the optical response of nanosized particles, since they directly relate the spectrum of the colloid to the dielectric function of the material. They are valid only for very dilute colloids [5]. For large volume fractions of nanoparticles in glasses, polymers or solutions, dipole coupling becomes important. A more general equation can be derived as follows [6].

Consider a medium with a low volume fraction of nanoparticles ϕ and complex dielectric function $\varepsilon(\omega)$, each with radius R, such that $2\pi R n_m/\lambda \ll 1$. An electric field in the medium polarizes the solvent at the frequency of the field. The key is to determine the average dielectric function of the combined system. The average electric field in the medium is

$$E_{av} = (1 - \phi)\, E_m + \phi E_i \tag{7}$$

where E_m is the field in the solvent, and E_i that within the particles, and assumed to be spherical. Additionally, the field causes a polarization charge density at the surface of the medium

$$P_{av} = \varepsilon_o(\varepsilon_m - 1)\, E_m + \varepsilon_o(\varepsilon - 1)\, E_i = \varepsilon_o(\varepsilon_{av} - 1)\, E_{av} \tag{8}$$

These equations can be solved to find ε_{av} if the relationship between E_i and E_m is known. This can be obtained for a number of simple geometrical shapes. The field differs from the bulk value because the polarized solvent molecules around the sphere create an additional electric field within the particle. For spheres, the result is

$$E_i = \frac{3\varepsilon_m E_m}{\varepsilon + 2\varepsilon_m} \tag{9}$$

Combining Eqs. (7–9) we get:

$$\varepsilon_{av} = \varepsilon_m \frac{\varepsilon(1+2\phi) + 2\varepsilon_m(1-\phi)}{\varepsilon(1-\phi) + \varepsilon_m(2+\phi)} \tag{10}$$

We can now consider the ensemble of particles and medium as a single material characterised by the dielectric function ε_{av}. For $\phi \ll 1$, the imaginary part of ε_{av} is

$$\mathrm{Im}(\varepsilon_{av}) = \varepsilon''_{av} = \frac{9\phi\,\varepsilon_m^2 \varepsilon''}{(\varepsilon' + 2\varepsilon_m)^2 + \varepsilon''^2} \tag{11}$$

The absorption coefficient of a slab (ignoring multiple reflections) can be calculated from

$$I_t = I_o \exp\left[\frac{-\varepsilon''_{av}\,\omega d}{c\, n_{av}}\right] \tag{12}$$

Since the volume fraction of spheres is $\phi = N4\pi R^3/3$, substitution of Eq. (11) into Eq. (12) yields the desired result, Eq. (5). Note this equation is for true absorption, and disappears for any material with a purely real dielectric function, since $\varepsilon''=0$. Equations (11) and (12) validate Beer's law for dilute colloids – the absorbance is proportional to ϕ; however for higher values of ϕ

$$\mathrm{Im}(\varepsilon_{av}) = \varepsilon''_{av} = \frac{9\phi\,\varepsilon_m^2}{(1-\phi)^2} \frac{\varepsilon''}{\left(\varepsilon' + \frac{2+\phi}{1-\phi}\varepsilon_m\right)^2 + \varepsilon''^2} \tag{13}$$

This equation will be used to model thin films composed of $Au@SiO_2$ particles in Sect. 3.

2.2 Core-Shell Nanocrystals

For core-shell particles, Aden derived the equation [7]:

$$Q_{ext} = 4k\,\mathrm{Im}\left\{\frac{(\varepsilon_{shell} - \varepsilon_m)(\varepsilon_{core} - 2\varepsilon_{shell}) + (1-g)(\varepsilon_{core} - \varepsilon_{shell})(\varepsilon_m + 2\varepsilon_{shell})}{(\varepsilon_{shell} + 2\varepsilon_m)(\varepsilon_{core} + 2\varepsilon_{shell}) + (1-g)(\varepsilon_{shell} - 2\varepsilon_m)(\varepsilon_{core} - \varepsilon_{shell})}\right\} \tag{14}$$

where ε_{core} is the complex dielectric function of the core material, ε_{shell} is that of the shell, g is the volume fraction of the shell layer and R is the radius of the coated particle. As expected, when $g=0$, Eq. (14) reduces to Eq. (5) for an uncoated sphere, and for g=1, Eq. (14) yields the extinction cross section for a sphere of the shell material.

Equation (14) can be used to model hollow particles, as well as core-shell particles. It describes the basic optical effects associated with changes in the medium refractive index and the effects of different core-shell thicknesses (Sect. 4.1). Note that these equations are limited by the fact they are predicated upon dipole models and work accurately only for particles such that $kR \ll 1$, which in practice restricts them to particles less than 30–40 nm in total size.

The above equations can be used to calculate the optical properties of various structures; however, a knowledge of the dielectric properties of the material are needed.

2.3 Ellipsoids

The optical properties of shapes other than spheres are generally more complex. If the metal nanocrystal is in the form of an ellipsoid or rod, then its surface plasmon frequency shifts drastically because the restoring force on the conduction electrons is extraordinarily sensitive to particle curvature [2]. For the general case of an ellipsoid, the extinction cross-section of a rod-like particle is given by

$$Q_{\text{ext}} = k\,\text{Im}\{\alpha_{x,y,z}\} = k\,\text{Im}\left\{\frac{4\pi\, abc(\varepsilon_{\text{Au}} - \varepsilon_{\text{m}})}{3\varepsilon_{\text{m}} + 2L_{x,y,z}(\varepsilon_{\text{Au}} - \varepsilon_{\text{m}})}\right\} \tag{15}$$

Here a, b and c refer to the length of the rod along the x, y and z axes (a>b=c), ε_{Au} is the dielectric function of Au and $L_{x,y,z}$ is the depolarisation factor for the respective axis, given by

$$L_x = \frac{1-e^2}{e^2}\left(-1 + \frac{1}{2e}\ln\frac{1+e}{1-e}\right); \quad L_{y,z} = (1 - L_x)/2 \tag{16}$$

where e is the rod ellipticity given by

$$e = 1 - (b/a)^2 \tag{17}$$

For a sphere e=0 and L=1/3.

2.4 Thin Films

A key issue in nanostructured materials is the dipole coupling between nanocrystals which will cause the optical properties of a nanocrystal ensemble to become like those of the bulk material. There has been extensive investigation of the interactions between particles embedded within media for a range of boundary conditions. We have found that the effective dielectric function given by Eq. (10), based on the Maxwell-Garnett model [1] is very accurate for quite dense nanocrystal arrays. In practice, one measures the transmittance of a thin film of the dense nanoparticle based film. Conventional solutions are simply opaque at moderate volume fractions! The transmittance of radiation with a frequency ω through a thin film is

$$T_{\text{film}} = \frac{(1 - R)^2 + 4R \sin^2 \psi}{R^2 \exp(-\alpha h) + \exp(\alpha h) - 2R \cos(\zeta + 2\psi)} \tag{18}$$

where h is the film thickness, R is the reflectance at normal incidence,

$$R = \frac{(n_{\text{av}} - 1)^2 + k_{\text{av}}^2}{(n_{\text{av}} - 1)^2 + k_{\text{av}}^2} \tag{19}$$

and $\alpha = \omega \text{Im}(\varepsilon_{\text{av}})/cn_{\text{av}}$ is the absorption coefficient [calculated from Eq. (10)]. Finally, we define the parameters

$$\zeta = 4\pi\, n_{\text{av}} h/\lambda, \tag{20}$$

and

$$\psi = \tan^{-1}(2k_{\text{av}}/(n_{\text{av}}^2 + k_{\text{av}}^2 - 1)) \tag{21}$$

Note that the expression for the transmittance takes into account not only the absorption by the metal centres, but also the reflection losses. Therefore, it can be considered as the extinction coefficient of the film.

Mean free path effects affect the band width primarily. For a sphere, the mean free path is equal to the radius of the sphere [8]. Granqvist et al. proposed that the mean free path in the shell layer of a layered particle should be given by [9]

$$\text{MFP} = \frac{(d_{\text{shell}} - d_{\text{core}})\,(d_{\text{shell}}^2 - d_{\text{core}}^2)^{1/3}}{2} \tag{22}$$

where d_{shell} is the diameter of the coated particle and d_{core} the diameter of the core. In the following calculated spectra, the effect of a different free path on the actual peak position is quite small; however the peaks will be generally broader than calculated. Equation (22) was shown to work well for Pd@Ag core-shell particles [10]. It slightly underestimated the mean free path in the shell layer.

3
Disordered Nanocrystal thin Films

Although numerous kinds of core-shell nanoparticles with various properties have been assembled into thin films using the LbL method (see [11] and [12] and references therein), we only discuss here thin films made of monodisperse gold nanoparticles coated with uniform, thin shells of amorphous silica. The interest of the assembly of silica-coated gold nanoparticles relies on the possibility of controlling the particle volume fraction by means of the variation of the silica shell thickness, provided that the assembled films are close-packed. In close-packing conditions, the separation between metal cores is just twice the thickness of the coating shell, which can be controlled during the synthesis of the colloids.

In Fig. 2, the experimental and calculated absorbance spectra are shown for thin films with different gold volume fractions. It is observed that, as the gold

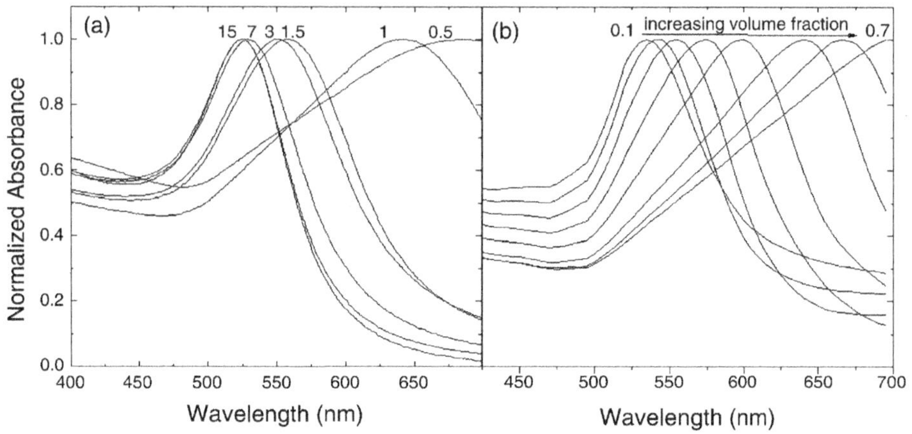

Fig. 2. **a** Experimental spectra as a function of the particle spacing (nm), as indicated. **b** Calculated and normalized extinction spectra of the 13.2 nm particles as a function of the metal volume fraction using Eqs. (10), (18) and (19). Film thickness used was 100 nm. Refractive index of silica used was 1.46. The dielectric response of particles was corrected for surface scattering. Adapted with permission from Ref. 26. Copyright 2002, Elsevier

volume fraction increases (the separation between nanoparticles decreases), there is a red-shift of the plasmon resonance, as well as a broadening of the band. This effect originates in the dipole-dipole interactions between neighbouring nanoparticles, and we can see here that a separation of just 15 nm is sufficient to screen such interactions, so that the thin films display basically the same properties as a dilute dispersion of the same nanoparticles in water.

The agreement between the experimental results and the calculated spectra demonstrates that Maxwell-Garnett effective medium theory effectively accounts for such dipole-dipole interactions. Similarly, the reflectance properties of the films are also affected by the interparticle spacing. This is supported by the data shown in Fig. 3, where the specular reflectance for films with different interparticle separation are compared to the spectra calculated by means of Eqs. (11), (18) and (19) [13]. Again, as the particles approach each other, the reflectance red-shifts and broadens. The effect is not as dramatic as seen for absorption, but the film changes from a green reflected hue to a metallic gold as ϕ increases.

Thus, both the transmission and reflection properties of thin gold films can be easily controlled by means of an adjustment of the thickness of the silica shell surrounding each NC, so that dipole-dipole interparticle interactions are effectively screened. Detailed studies using different metal core sizes have not been carried out yet, but it is expected that the distance at which interactions are effectively screened will scale up with particle size, being of the order of one particle diameter.

It should be mentioned here that a similar effect has been observed for the magnetic properties of thin films of silica-coated magnetite nanoparticles [14].

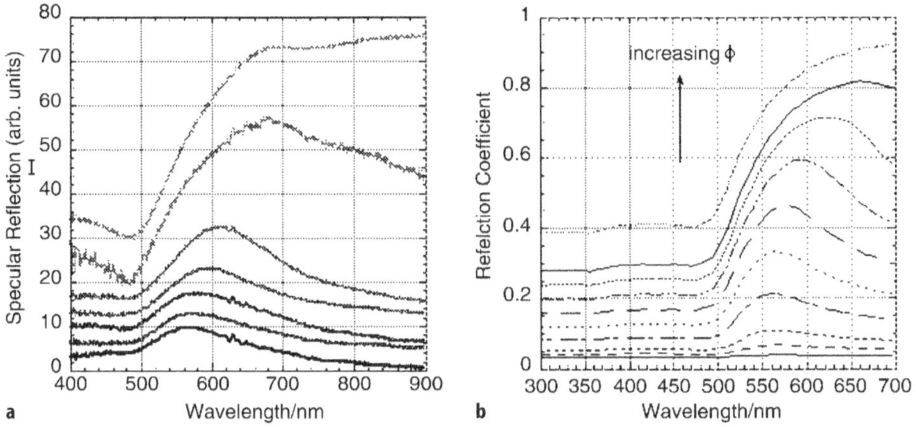

Fig. 3. **a** Experimentally measured specular reflectance spectra for a series of Au@SiO_2 films as a function of the metal volume fraction. **b** Calculated reflectance spectra for gold colloid films of different volume fractions, using Eqs. (11) and (19). Reproduced with permission from Ref. [13]. Copyright 2002, Elsevier

Screening of interparticle interactions using LbL assembly has also been demonstrated by depositing inert monolayers between each two consecutive metal [11] or magnetic [15] nanoparticle layers. In such methods, only interlayer interactions are screened, so that spatially modulated coupling can be achieved and polarization effects on the reflection coefficient can be obtained.

4 Nanoparticle Assembly on Colloid Spheres

4.1 Silica Coated Gold

It was suggested by Birnholm, Haus and others in 1988 [16], that the surface plasmon resonance would be drastically tunable if the metal was the shell in a core-shell particle, rather than the core. It was also predicted that enhanced non-linear optical properties would ensue. We now compare the predictions for the surface plasmon mode for these two cases. Firstly, we look at silica coated gold nanocrystals.

The resonance condition for surface plasmon excitation is found by setting the denominator in Eq. (14) to zero.

$$\varepsilon_{\mathrm{core}} = -2\varepsilon_{\mathrm{shell}} \frac{\varepsilon_{\mathrm{shell}} g + \varepsilon_{\mathrm{m}}(3-g)}{\varepsilon_{\mathrm{shell}}(3-2g) + 2\varepsilon_{\mathrm{m}} g} \tag{23}$$

Around monolayer coverage, $g \ll 1$ and we can simplify this to

$$\varepsilon_{\mathrm{core}} = -2\varepsilon_{\mathrm{m}} - \frac{2g(\varepsilon_{\mathrm{Shell}} - \varepsilon_{\mathrm{m}})}{3} \tag{24}$$

Inserting a Drude dielectric function, we get

$$\lambda_{\mathrm{peak}}^2 = \lambda_{\mathrm{p}}^2 \left(\varepsilon^{\infty} + 2\varepsilon_{\mathrm{m}} + \frac{2g(\varepsilon_{\mathrm{SiO_2}} - \varepsilon_{\mathrm{m}})}{3} \right) \tag{25}$$

which shows that the peak position may red-shift or blue-shift depending on whether the shell has a higher or lower refractive index than the medium. The shift is proportional to the volume fraction of the polarizing shell, and for $g=0$, yields the usual resonance of an uncoated material. This limiting form is valid only for very thin shells. The shifts though very easily discerned by eye are not overly dramatic. The shell material does not change the polarizability significantly. More drastic effects can be obtained by depositing a layer of a metal, but since these have high absorption coefficients, drastic shifts in the plasmon band are also accompanied by damping of the band.

4.2 Gold Coated Silica

Now we consider the inverse structure, that of a metal coated insulator. For this case, we find that

$$\lambda_{\mathrm{peak}}^2 = \lambda_{\mathrm{p}}^2 \left(\varepsilon^{\infty} + \varepsilon_{\mathrm{m}} + \frac{3\varepsilon_{\mathrm{c}} + 2\varepsilon_{\mathrm{m}}}{2g} \right) \tag{26}$$

The properties of this equation are quite different. For very small values of g, the plasmon mode is shifted drastically to longer wavelengths. and converges to the resonance condition for a gold shell immersed in the medium as g increases. (i.e. the core plays no role as the shell thickness increases). Thus provided the shell volume fraction is very small, indeed only a few monolayers thick, incredible tunability is feasible for these core-shell structures.

In Fig. 4, the plasmon resonance for gold coated silica is plotted as a function of the shell thickness (for fixed core thickness of 10 nm). In Fig. 6 we see the effects of changing either the core or medium refractive index.

An experimental difficulty is coating a colloid particle with a metal homogenously. Electrochemical deposition onto an insulating surface is difficult. In some cases, adsorption of catalytic ions such as Pd(II), Ag(I) or Pt(II) can assist. An alternative is to controllably deposit small gold nanoparticles onto a larger silica or latex substrate. This can be done using LbL processes as outlined in the introduction.

In Fig. 6, we illustrate some different ways that the core-shell topology could be varied for silica and gold. So far we have considered the two normal core-shell structures. We now focus on the third example: the assembly of Au@SiO_2 nanoparticles onto spherical polystyrene latex colloids. The resulting spheres are also essentially different to continuous metal shells grown on colloid templates, which have been reported by Halas and colleagues [17] and by van Blaaderen and coworkers [18]. Such continuous shells display optical properties associated with resonances of the whole shell, and are therefore extremely sensitive to both core size and shell thickness, while in the system presented here

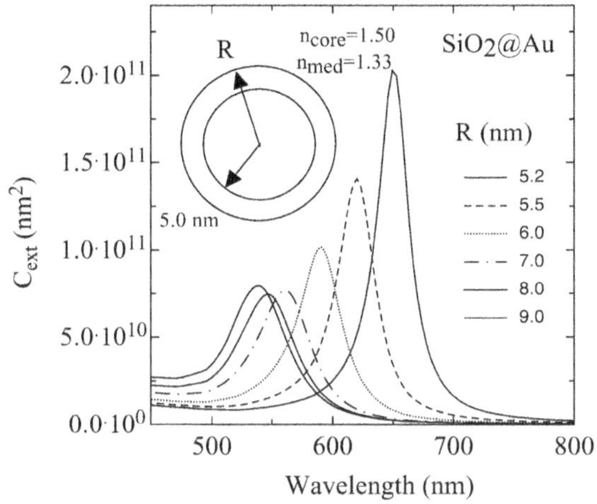

Fig. 4. Calculated spectra using Eq. (15) for the extinction cross section of 5 nm silica particles coated with shells of 0.2 to 4.0 nm of gold in water

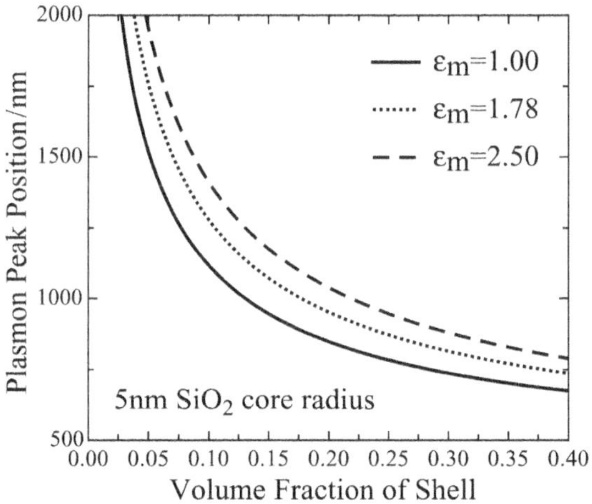

Fig. 5. Predicted position of the Au surface plasmon resonance for 5 nm SiO_2@Au particles in vacuum, water and a high refractive index liquid such as CS_2 as a function of the gold shell volume fraction. Equation (15) strictly only applies for $\phi \ll 1$

the optical properties only depend on the nature and dimensions of the constituting units.

The assembly was initially performed for Au@SiO_2 with very thin shells (ca. 2 nm) on 640 nm polystyrene spheres [19] and later extended for latex cores of various sizes and gold nanoparticles coated with thicker shells. Examples of nanostructured colloids formed by assembly of Au@SiO_2 with identical cores but silica shells with various thickness are shown in Fig. 7. It is clear that, although the surface of the particles is rough, the resultant layers are very compact

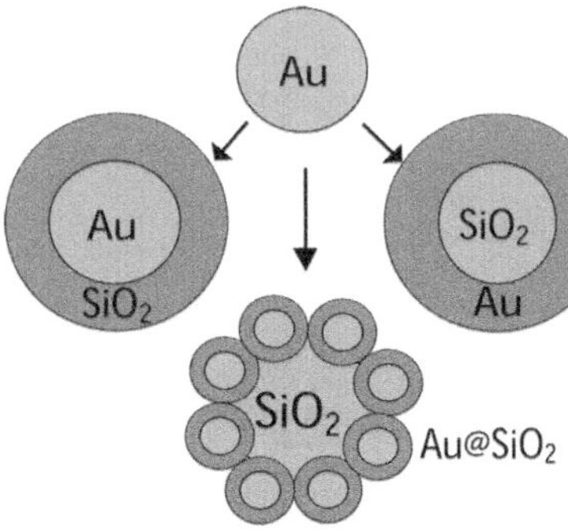

Fig. 6. Three ways to construct silica-gold core shell particles of different topology

and uniform, so that, again, the interparticle separation will be determined by the corresponding thickness of the silica shell.

An ansatz for treating the resultant optical properties is to treat the shell using MG theory to yield an approximate average dielectric function for the shell, and then to use this function in the core-shell geometry. The predictions of this model are shown in Fig. 8. A fixed latex or silica core of 50 nm is shown surrounded by a 5 nm shell of Au@SiO_2 where the Au volume fraction is varied. The medium is taken to be water for simplicity.

The experimental UV-visible spectra of 640 nm polystyrene colloids coated with five monolayers of Au@SiO_2 nanoparticles of various shell thickness are shown in Fig. 9. Again, as the separation between gold nanoparticles increases, the plasmon resonance blue-shifts towards that characteristic of isolated Au nanoparticles. As opposed to assemblies of the same coated nanoparticles on flat substrates, it has also been observed in the present case that the short-wavelength tail of the spectra dramatically rises as the size of the deposited nanoparticles increases, which is due to a larger effect on light scattering by the larger, composite spheres. These scattering effects also slightly affect the actual position of the plasmon band, as demonstrated by a larger shift as the size of the polystyrene cores is increased.

The versatility of this system is clearly greater than that of thin films on flat substrates, since the composite colloid spheres can still be further modified for the synthesis of complex nanostructures. An example of further modification can be found in Fig. 10, showing transversal sections of calcined coated spheres. Calcination leads to sintering of the silica shells, so that they become more stable, but also leads to the combustion of the polystyrene spheres, so that hollow Au-SiO_2 shells are finally obtained. Such hollow spheres are extremely interesting for the construction of photonic materials with a large dielectric contrast. Additionally, it is also possible to remove the insulating silica shells by gentle dissolution with hydrofluoric acid, so that shells made of closely packed gold nanoparticles are obtained.

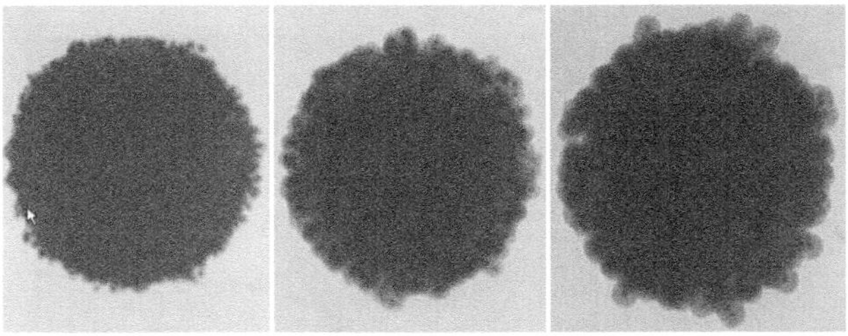

Fig. 7. Typical transmission electron micrographs of 640 nm polystyrene spheres on which one monolayer of $Au@SiO_2$ nanoparticles has been assembled. The size of the Au cores is 15 nm in all cases. From left to right, the silica shell thicknesses are 8, 18, and 28 nm

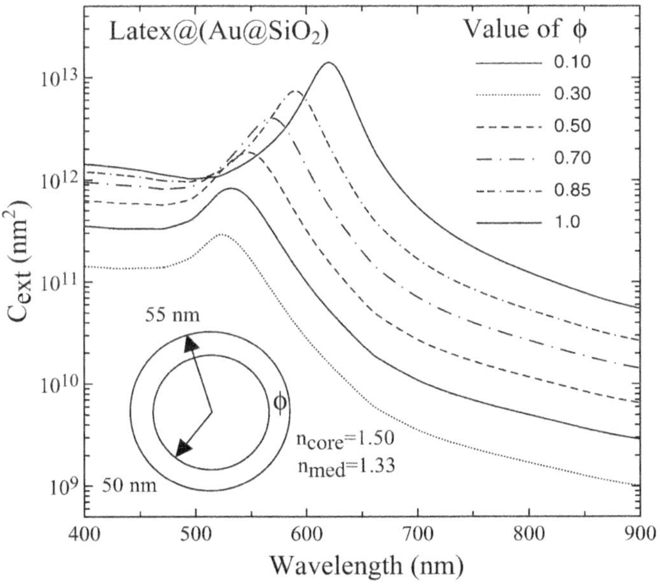

Fig. 8. Calculated extinction spectra of latex@($Au@SiO_2$) particles in water. The latex or silica core has a radius of 50 nm. The shell is fixed to be just 5 nm thick, which is impractical, but useful for gauging the sensitivity of the effect. The volume fraction of gold in the silica coated particles adsorbed onto the silica core is shown in the figure

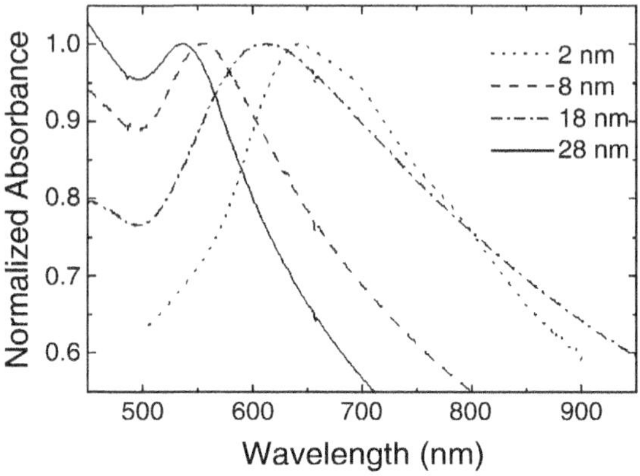

Fig. 9. Normalized UV-visible spectra of dilute dispersions of 640 nm latex spheres coated with five monolayers of Au@SiO_2 nanoparticles. The thickness of the corresponding silica shells is indicated. The trends are consistent with the predictions of Eq. (15) but quantitative agreement is not possible due to the higher volume fraction of the shells in experiments

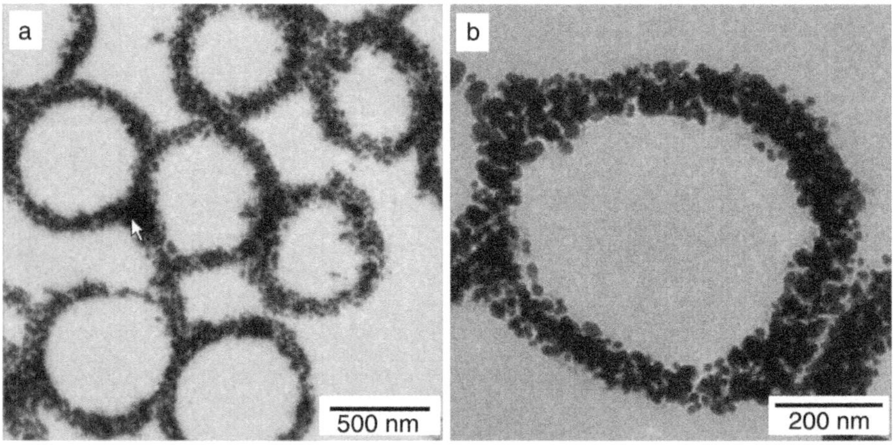

Fig. 10a, b. Transmission electron micrographs of cross-sections of hollow Au-SiO_2 spheres, prepared by calcination of 640 nm latex coated with five monolayers of Au@SiO_2 nanoparticles (core size 15 nm, shell thickness 2 nm). Reproduced with permission from Ref. [19]. Copyright 2001, Wiley-VCH

5 Doped Glass from Coated Au Nanocrystals

The use of glass doped with metal nanoparticles for the construction of materials with non-linear optical properties has long been implemented. Most of the methods used for the doping of glasses are comprised of the reduction of infiltrated metal salts within the wet or dry gel. Although successful in most cases, such methods fail to control the particle size and shape adequately and, of course are unable to allow for the modification of nanoparticle composition. These difficulties from the in-situ preparation methods can be overcome if premade nanoparticles are used for incorporation into the glass matrix. It was recently demonstrated that silica-coated nanoparticles are ideal candidates for the fabrication of silica gels and glasses homogeneously doped with different kinds of nanoparticles [20]. The reason for this is that the silica shell provides the core-shell nanoparticles with precisely the same surface properties as those of the silica units that are assembled during the sol-gel process, so that they do not have any preferred tendency to mutual coagulation and thus will be randomly distributed within the final nanostructure.

Since silica coating has been demonstrated for a number of nanoparticles with different composition, size and shape, the use of silica-coated nanoparticles offers an almost unlimited range of possibilities for the synthesis of doped gels and glasses with various properties. Examples are shown in Fig. 11 of 15 nm Au nanoparticles coated with 10 nm thick silica shells, which are homogeneously distributed within silica gels prepared by citrate acid catalysed hydrolysis of tetramethoxysilane in water/methanol mixtures, in the presence of the Au@SiO_2 nanoparticles. The concentration of the nanoparticles within the gels was varied by simple adjustment of the concentration of the colloid prior to the sol-gel process. The advantage of such a homogeneous distribution, apart from a high uniformity in particle size (which can be controlled prior to the sol-gel processing) is once again that the (linear) optical properties of the material can be predicted before the material is prepared, since there is no coupling between the plasmon resonances of the gold nanoparticles (see above). This is exemplified in

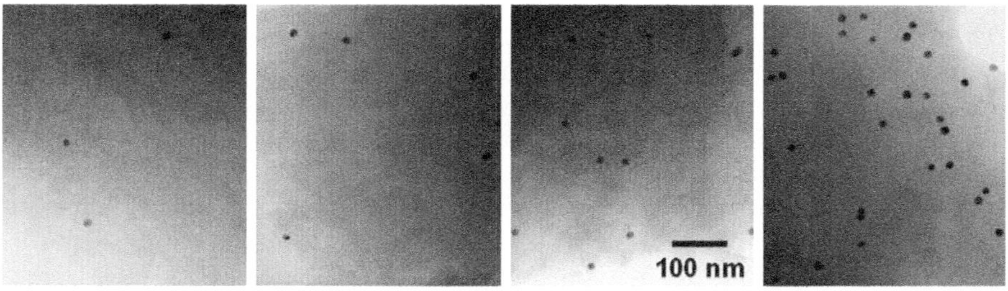

Fig. 11. Transmission electron micrographs of silica gels doped with Au@SiO_2 nanoparticles. From left to right, the weight percentages of Au in the gels are 0.06, 0.14, 0.29 and 0.61. Reproduced with permission from Ref. [20]. Copyright 2001, American Chemical Society

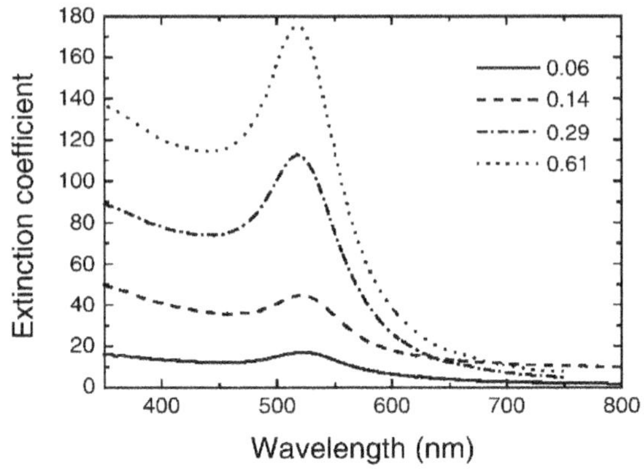

Fig. 12. UV-visible spectra of xerogels doped with Au@SiO_2 nanoparticles (15 nm core, 10 nm shell). The Au content is indicated in the labels as w/w%

Fig. 12, in which the spectra of thin xerogels corresponding to the micrographs in Fig. 11 have been plotted. It is remarkable that the plasmon band is centred precisely at the same position, regardless of concentration. It has also been shown that for larger gold cores the optical properties are equally preserved during the sol-gel transition.

The third-order non-linear optical properties of such gels have been measured by degenerate four-wave mixing [21]. For a thin xerogel containing 0.14 wt% Au the measured value was of 2.3×10^{-10} esu, while after sintering at 600 °C to convert it into a glass, the $|\chi^{(3)}|$ value was 4.6×10^{-10} esu. These values are over 10 times larger than those of Au particles in glasses prepared by the conventional melt-quenching method.

6 Ordered Nanocrystals

6.1 Wet and Dry Au@SiO_2 Opals

The study of colloidal crystals was initiated as part of research into the determination of phase diagrams for colloids, which itself was perceived as a means to model phase behaviour in molecular systems [22]. Extensive literature is available on the dynamics of colloidal crystal formation, as a function of several parameters, such as the nature of the solvent, surface charge, particle size and concentration. The results described here refer to the formation of colloidal crystals from dispersions of silica-coated gold nanoparticles in ethanol, after silica surface functionalization with 3-(trimethoxysilyl)propyl methacrylate (TPM). Earlier studies by Philipse and Vrij [23] showed that TPM adsorption leads to a reduction in surface charge, so that the particles are stable in organic solvents with low polarity, such as ethanol, toluene or DMF. This means that the particle be-

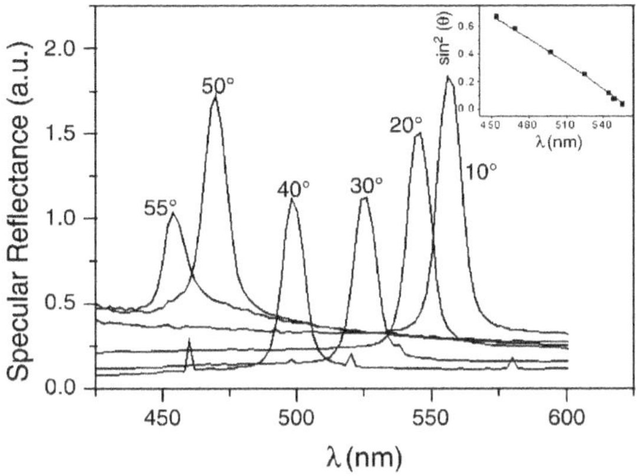

Fig. 13. Specular reflectance spectra at different incident angles for a colloidal crystal made of TPM-modified $Au@SiO_2$ particles with 15 nm core diameter and 150 nm total diameter. The particle volume fraction in this sample was 15.8 vol%, corresponding to 0.31 wt% Au content. The *inset* shows the fit of the experimental Bragg peak positions with Eq. (28)

haviour is intermediate between that of hard and soft spheres and thus the crystallization concentration is also intermediate. Dhont et al. [24] showed that for silica spheres coated with TPM, crystallization takes place around 20 vol%. Since the colloid chemistry of our $Au@SiO_2$ particles is governed by the silica surface (the Au volume fraction in each particle is small enough to consider that it hardly affects the Hamaker constant), we expect crystallization to occur at similar concentration values.

Results are shown in Fig. 13 for 150 nm coated particles, with a particle volume fraction close to 16%. This choice was motivated by the similarity in the Au content (0.31 w%) with respect to that of the gels described above. The optical characterization was performed through specular reflectance measurements, so that the spectra in Fig. 9 for different incident angles show the crystallinity of the sample, through intense Bragg reflection peaks. The peak positions for different incident angles have been fitted using the following equation for Bragg's law:

$$\lambda = 2d_{111} \left(\langle n^2 \rangle \sin^2\theta\right)^{1/2} \tag{27}$$

where $\langle n^2 \rangle$ is the average dielectric constant of the sample estimated using

$$\langle n^2 \rangle = n_1^2 \phi + n_2^2 (1 - \phi) \tag{28}$$

where n_1 and n_2 are the refractive indices of silica and ethanol, respectively. d_{111} is the lattice parameter for an fcc structure, calculated as $0.739\phi^{1/3}d$, d being the particle diameter. The Au cores have not been included in the calculation, since they only occupy about 0.1% of the volume of each particle, and therefore their contribution to diffraction should be negligible. However, there is still a contribution from absorption, as will be shown below. It is interesting to note here that

Asher and co-workers have recently reported [25] the formation of colloidal crystals made from silica particles doped with a random distribution of Ag nanoparticles. These authors found that the plasmon band of the Ag nanoparticles varied during the transition from a disordered to an ordered state, which suggests the existence of some sort of coupling between the two optical responses of the system.

Despite the simplicity of the formation of colloidal crystals in solution, which is a spontaneous process at the right concentration and temperature conditions, such systems suffer from a number of problems for their use as photonic crystals. One of the major drawbacks of colloidal crystals is their large polycrystallinity, i.e., small crystallites form in solution, rather than a large, well-ordered, single crystal, which strongly limits their efficiency for the final goal of obtaining a full photonic band-gap. One simple route to opal formation is slow sedimentation of the colloid particles in a suitable solvent, so that the interparticle interactions during deposition result in ordered structures with a face-centred-cubic (fcc) crystal lattice [26]. The same method can be used for the formation of opals using Au@SiO_2 particles as the constituent units, since the nature of the surface is identical. As opposed to Asher's report, the optical characterization of such opals suggests that the plasmon resonance and Bragg diffraction are altered through independent mechanisms. Clear evidence for this is shown in Fig. 14, where optical extinction spectra have been plotted for a single Au@SiO_2 opal, but with different materials filling the interstices between the spheres [27]. While the position of the Bragg peak is red-shifted and its intensity reduced as the refractive index of the medium is increased, the plasmon band position remains basically unchanged, since the silica shell around the Au cores is thick enough that they do not "see" the outer medium any longer.

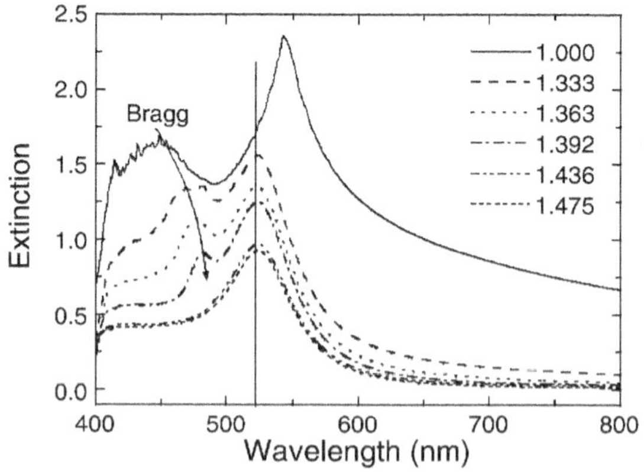

Fig. 14. Extinction spectra in different media (air, water and water-glycerol mixtures) for sintered opals made of Au@SiO_2 particles with 15 nm core diameter and 225 nm total diameter. The corresponding refractive indices are indicated. The extinction spectrum consists of a surface plasmon mode from embedded gold particles and a weaker Bragg peak. The position of both modes is dependent upon solvent refractive index and (core and shell) particle volume fractions

The major requirement for long range ordering is that the particles experience repulsive interactions or attractive interactions of ~kT at most. However van der Waals forces between nanocrystals are already ≫kT, so that this process is possible only for nanocrystals stabilized by repulsive interactions, which can offset the attractive interactions which would otherwise produce disordered or fractal like structures. NCs experiencing repulsive interaction forces will spontaneously order when subjected to a cohesive pressure. The difficulty arises that once the pressure is released, the system disorders. For example, centrifugation can lead to ordering of quite small colloid particles; however thermal diffusion rapidly occurs once the field is switched off. What is needed is a method to apply a cohesive force after ordering. One approach is to induce ordering then use sol-gel processing to lock the structure into place. In Fig. 15, we show SEMs of Au@SiO_2 particles that have been sedimented in ethanol using TPM coating. Then ammonia and TMOS have been injected and the centrifugation continued [28]. The TMOS hydrolysis occurs over an hour or so, and the oligomeric silanes bridge the colloidal opal. Solvent is displaced gradually though not completely. The gelation can be accelerated using an alkylamine. The resultant opal is a deep transparent red colour. The method yields single crystals several hundred microns in size as determined by SAXS.

6.2 Inverse Opals from Au@SiO_2 Particles

One final example of nanostructures created from the assembly of silica-coated nanoparticles is the formation of inverse opals. These structures are similar to opals, in the sense that they comprise solids with an ordered structure whose lattice constant lies in the optical wavelength range. The only difference lies in the nature of the crystal centres and the embedding medium. While "direct" opals are composed of solid (or liquid) particles typically embedded in air, "inverse" opals comprise air bubbles embedded in a solid material. One of the most

Fig. 15. Large-scale fcc Au@SiO_2 lattice created by simultaneous sol-gel processing and centrifugation using TMOS, octylamine and ethanol [27]. The sample has been fractured for SEM imaging. On the right, smaller blobs of condensed silane are seen which hold the opal together

promising pathways toward the fabrication of inverse opals is the infiltration of the interstices of direct opals with the desired material, followed by removal of the colloid templates [29].

Theoretical modelling shows that this configuration is more favourable for the achievement of a full photonic band-gap, provided that there is a dielectric contrast of at least 2.8 [30]. This requirement is hard to achieve, and it has only been reported for silicon [31] and germanium [32] inverse opals. An alternative can be the use of metals with a Drude-like behavior of the dielectric function. When $\omega \approx \omega_p$, the dielectric contrast of the photonic crystal becomes extremely large, which favours the opening of a complete photonic bandgap, even in a simple face-centred cubic (fcc) structure [33]. It is therefore desirable to use metals with a well-defined plasmon frequency, which is the case for metal nanoparticles. Thus, one suitable approach is the infiltration of preformed opals with the previously described Au@SiO_2 nanoparticles, so that the optical features of single nanoparticles are preserved in the final, inverse opal. Upon removal of the template, rigid and mechanically stable nanostructures can be easily obtained through sintering of the silica shells, while retaining the optical properties of the individual gold nanoparticles. An example of Au/SiO_2 based inverse opals prepared by infiltration of polystyrene latex opals and template removal by calcination, is shown in Fig. 16. The figure shows low and high resolution scanning electron micrographs of an inverse opal formed by Au@SiO_2 nanoparticles with 15 nm Au cores surrounded by 28 nm thick silica shells [34]. The spherical particles can still be recognized within the nanostructured system. The optical characterisation of these inverse opals shows again that the position of both the plasmon band and the Bragg peak can be modulated independently, so that a gate is open to search for the coupling needed to achieve a full photonic bandgap.

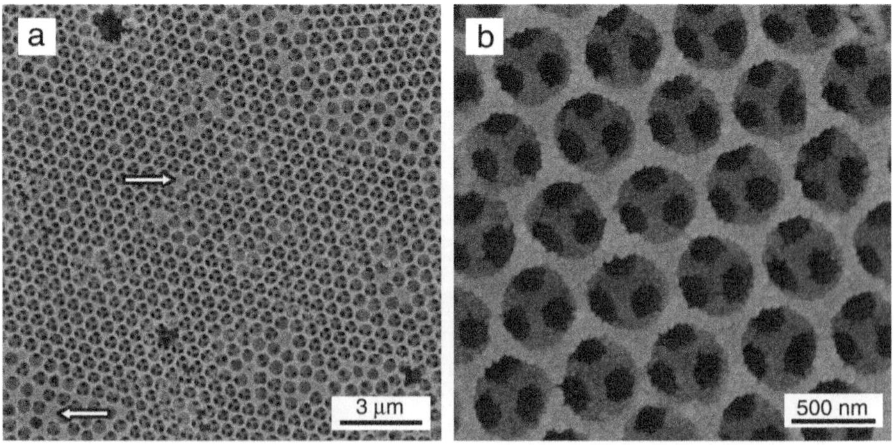

Fig. 16a,b. Scanning electron micrographs of gold-silica inverse opals formed by infiltrating colloidal crystals made of polystyrene spheres (diameter 640 nm) with Au@SiO_2 nanoparticles (core 15 nm; shell 28 nm). *Arrows* show point defects within the crystalline structure

7 Conclusions

The main thrust of this chapter has been quite simple. Taking as starting materials transparent glass and gold metal, new structures have been designed and constructed through a rational approach based on colloid interaction forces and the predictive power of basic light scattering theory. An enormous range of colours and topologies emerge from this synthesis of spectroscopy and nanoscale construction. With the advent of increasing shape control, another avenue to designer optical materials is also opening up. The ideas here can readily be extended to semiconductors, magnetic and chiral materials. Necessarily, reliable synthetic routes underpin any more complex optical structures. Unless monodisperse starting materials can be synthesized, the creation of many more ambitious structures will remain elusive. Perhaps most exciting, given the predictions of Moroz [32], would be the discovery of templating methods for crystallising colloids in zincblende and diamond lattice structures. Such a breakthrough would have implications not just for photonic materials but for biomineralization and biomimetic tissue engineering.

Acknowledgement. PM wishes to thank the Humboldt Foundation for support. Financial support to LMLM from the Spanish Ministerio de Ciencia y Tecnología and Xunta de Galicia, through several research grants is also greatly appreciated.

References

1. Maxwell-Garnett JC (1904) Phil Trans Roy Soc 203:385
2. Gans R (1911) Ann Phys 37:881
3. Gans R (1915) Ann Phys 47:270
4. Kerker M (1969) The Scattering of Light and Other Electromagnetic Radiation. Academic Press, New York
5. Creighton JA, Eadon DG (1991) J Chem Soc Faraday Trans 87:3881
6. Woggon U (1997) Optical Properties of Semiconductor Quantum Dots. Springer, Berlin Heidelberg New York
7. Aden AL, Kerker M (1951) J Appl Phys 22:1242
8. Euler J (1954) Z Phys 137:318
9. Granqvist CG, Hunderi O (1978) Z Phys B30:47
10. Michaelis M, Henglein A, Mulvaney P (1994) J Phys Chem 98:6212
11. Malikova N, Pastoriza-Santos I, Schierhorn M, Kotov NA, Liz-Marzán LM (2002) Langmuir 18:3694
12. Liz-Marzán LM, Correa-Duarte MA, Pastoriza-Santos I, Mulvaney P, Ung T, Giersig M, Kotov NA (2001) In: Nalwa HN (ed) Handbook of Surfaces and Interfaces of Materials, vol 3. Academic Press, New York, p 189
13. Ung T, Liz-Marzan LM, Mulvaney P (2002) Colloids & Surfaces A Physicochemical & Engineering Aspects 202:119
14. Aliev F, Correa-Duarte MA, Mamedov A, Ostrander J, Giersig M, Liz-Marzán LM, Kotov NA (1999) Adv Mat 11:1006
15. Mamedov A, Ostrander JW, Aliev F, Kotov NA (2000) Langmuir 16:3941
16. Special Edition on NLO Properties of Particles (1988) J Opt Soc Am B 7
17. Oldenburg SJ, Averitt RD, Westcott SL, Halas NJ (1998) Chem Phys Lett 288:243
18. Graf C, Van Blaaderen A (2002) Langmuir 18:524

19. Caruso F, Spasova M, Salgueiriño-Maceira V, Liz-Marzán LM (2001) Adv Mat 13:1090
20. Kobayashi Y, Correa-Duarte MA, Liz-Marzán LM (2001) Langmuir 17:6375
21. Tamil Selvan S, Hayakawa T, Nogami M, Kobayashi Y, Liz-Marzán LM, Hamanaka Y, Nakamura A (2002) J Phys Chem B (submitted)
22. Arora AK, Tata BVRE (1996) Ordering and Phase Transitions in Charged Colloids. VCH-Wiley, New York
23. Philipse AP, Vrij A (1989) J Chem Phys 88:6459
24. Dhont JKG, Smits C, Lekkerkerker HNW (1991) J Colloid Interface Sci 152:386
25. Wang W, Asher SA (2001) J Am Chem Soc 123:12528
26. Míguez H, Meseguer F, López C, Mifsud A, Moya JS, Vázquez L (1997) Langmuir 13:6009
27. García-Santamaria F, Salgueiriño-Maceira V, López C, Liz-Marzán LM (2002) Langmuir 18:4519
28 Nakamura H, Mulvaney P, Hyde ST (2002) Langmuir (submitted)
29. Special Edition on Photonic Crystals (2001) Adv Mater 13(6)
30. Norris DJ, Vlasov YA (2001) Adv Mater 13:371
31. Blanco A, Chomski E, Grabtchak S, Ibisate M, John S, Leonard SW, Lopez C, Meseguer F, Miguez H, Mondia JP, Ozin GA, Toader O, van Driel HM (2000) Nature 405:437
32. Míguez H, Chomski E, García-Santamaría F, Ibisate M, John S, Lopez C, Meseguer F, Mondia JP, Ozin GA, Toader O, van Driel HM (2001) Adv Mater 13:1634
33. Moroz A (2000) Europhys Lett 50:466
34. Wang D, Salgueiriño-Maceira V, Liz-Marzán LM, Caruso F (2002) Adv. Mat. 13:908

Author Index Volume 201–226

Author Index Vols. 26–50 see Vol. 50
Author Index Vols. 51–100 see Vol. 100
Author Index Vols. 101–150 see Vol. 150
Author Index Vols. 151–200 see Vol. 200

The volume numbers are printed in italics

Subject Index

GPSR Compliance

The European Union's (EU) General Product Safety Regulation (GPSR) is a set of rules that requires consumer products to be safe and our obligations to ensure this.

If you have any concerns about our products, you can contact us on ProductSafety@springernature.com

In case Publisher is established outside the EU, the EU authorized representative is:

Springer Nature Customer Service Center GmbH
Europaplatz 3
69115 Heidelberg, Germany

Batch number: 10755442

Printed by Printforce, the Netherlands